AF868517

How to Fulfil the UN Sustainability Goals

Felix Fuders

How to Fulfil the UN Sustainability Goals

Rethinking the Role and Concept of Money in the Light of Sustainability

Springer

Felix Fuders
Instituto de Economía
Universidad Austral de Chile
Valdivia, Chile

ISBN 978-3-031-37767-9 ISBN 978-3-031-37768-6 (eBook)
https://doi.org/10.1007/978-3-031-37768-6

© The Editor(s) (if applicable) and The Author(s), under exclusive license to Springer Nature Switzerland AG 2023
This work is subject to copyright. All rights are solely and exclusively licensed by the Publisher, whether the whole or part of the material is concerned, specifically the rights of translation, reprinting, reuse of illustrations, recitation, broadcasting, reproduction on microfilms or in any other physical way, and transmission or information storage and retrieval, electronic adaptation, computer software, or by similar or dissimilar methodology now known or hereafter developed.
The use of general descriptive names, registered names, trademarks, service marks, etc. in this publication does not imply, even in the absence of a specific statement, that such names are exempt from the relevant protective laws and regulations and therefore free for general use.
The publisher, the authors, and the editors are safe to assume that the advice and information in this book are believed to be true and accurate at the date of publication. Neither the publisher nor the authors or the editors give a warranty, expressed or implied, with respect to the material contained herein or for any errors or omissions that may have been made. The publisher remains neutral with regard to jurisdictional claims in published maps and institutional affiliations.

This Springer imprint is published by the registered company Springer Nature Switzerland AG
The registered company address is: Gewerbestrasse 11, 6330 Cham, Switzerland

Paper in this product is recyclable.

Dedicated to
Magdalena, Maximiliam, and Vinzent

Foreword

As the author indicates, this book is about the truth. The truth has become not only more necessary to our survival but also more dangerous. This is because we are an experimental species like any other living species of biodiversity. But an experiment conducted on the basis of development models that are absolutely divorced from nature and its rules of coexistence. That divorce continues and is successfully maintained globally, in all countries without exception, thanks to the persistence of grossly wrong recipes, tools, and visions. The worst of these mistakes is to believe in continuous growth as the objective of all societies and all economies.

The human experiment began 150,000 years ago in Africa. Homo sapiens combines a nervous system that allows large amounts of information, which is continuously stored and related, with a series of less exclusive physical characteristics. For example, walking upright (bipedalism), opposable thumbs, stereoscopic vision, gregariousness, and permanent sexual response after puberty. These characteristics, combined with a powerful brain, have allowed us to have expansive and amazing success as species.

However, there is a problem being partly ignored, minimized, and even hidden. Almost all non-human species, a biodiversity that is being rapidly destroyed by human expansion, have predictable ecological roles and behaviors. Each generation of those species inherits behavior patterns from the previous generation with little change. Evolution and natural selection never stop. Terrestrial and aquatic natural ecosystems have thousands of species, from viruses, bacteria, and fungi to large vertebrates, green plants, and invertebrates. Being predictable, all of them have been able to adapt to coexist with one another.

Each human generation transmits to the next the accumulated cultural information, and as a result of this transmission, facilitated by languages of all kinds, each generation has on average more information than the previous one. People in a society, and societies in different countries, have different information contents, different behaviors, and therefore, unpredictable individuals and societies. Education, laws, religions, public policies, and the customs of each place cannot reduce such growing differences. Thus, a serious collision arises: how to coexist as a species being unpredictable, and driver of continuous growth, with ecosystems and biodiversity based on relatively predictable behaviors, and temporary and adaptive growth?

The tragic answer is that this strategy, which has generated cultures of astonishingly increasing complexity and innovation, has little chance of long-term survival. As part of the experiment, groups of humans are in bloody wars with other humans, and rich humans continue to exploit poorer humans. Homo sapiens has become responsible for the sixth global spasm of extinctions that have impacted the biosphere. In the past, such spasms were produced by meteorite crash and volcanic activity. Since the successive cultural revolutions that began 6000–10,000 years ago with the invention of agriculture in seven different regions, today we are the main agent of destruction.

Global climate change and the collapse of vital ecosystem services—such as the provision of water by water basins, or the regeneration of soils by forests and natural grasslands—are examples of the crises that we ourselves cause. With total irresponsibility, we continue to naturalize poverty and wealth, and destroy ecosystems with bulldozers, hunting, large-scale fishing, fire, millions of new chemicals, non-ionizing radiation, and pesticides.

Currency and monetarism are part of a maladaptive strategy, increasingly divorced from the rules and cycles of natural ecosystems, and by extension, from the biosphere. That is why Felix Fuder's expression, that the monetary system is unnatural, is so accurate.

The vectors of our development, most of them pushed by wrong guidelines, are so colossally strong that we cannot replace their exponential growth curves by sigmoid curves that respect the carrying capacities of the biosphere, and of the society ("K"). But we can decrease the slope of those exponential curves. Every small mitigation and change, socially, economically, and ecologically positive, proportionately reduce human suffering. The challenge posed by the author of the book is indispensable and courageous, since it deals with necessary truths. And he proposes revisiting Gesell's solutions, which helps us to achieve, with new approaches and old truths, an adaptive coexistence with nature and with ourselves.

Campus Cordoba of the Right Livelihood College
National University of Cordoba
Cordoba, Argentina
April 2023

Raul Montenegro

Preface

Sir Josiah Stamp, a former director of the Bank of England, is quoted on the first page of Irving Fisher's (1932) book "Booms and Depressions" as saying, "Money, as a physical medium of exchange, made a diversified civilization possible … And yet it is money, in its mechanical more than in its spiritual effects, which may well, having brought us to the present level, actually destroy society." This sentence, written almost 100 years ago, has lost none of its relevance. Our money, as it is designed today, is unnatural, and so is the phenomenon of interest to which it gives rise. This insight goes as far back as Aristotle, who characterized money as a problematic man-made construct which could be subject to change. Like everything that deviates from the laws of nature, money causes many diseases that plague our world.

One might ask why I characterize our monetary system as unnatural. Well, while everything in nature is subject to the rhythmic cycle of growth and decay, only money seems to escape the transience of all earthly things, as Silvio Gesell (1949) wrote it in his masterpiece "The Natural Economic Order." The unnatural storability of money gives rise to the economics of accumulation that Aristotle referred to as "chrematistics": earning money as an end in itself. Since profits can be stored easily in the form of monetary units, this creates an incentive to produce more than is actually needed to satisfy fundamental human needs. The storability of money is also what gives rise to the phenomenon of interest, which Keynes described as a "reward for parting with liquidity." Interest not only incentivizes the production of more than is needed but essentially requires us to do so. This is the root cause of many of the main challenges the world faces today, including unsustainability and income inequality. Unfortunately, very few scholars understand the linkage between these problems and the design of our monetary system. This book connects these dots and shows how these problems are rooted in our current monetary system. It also presents a solution.

One reason why these connections are not generally understood is because the conventional view of monetary theory is inherently flawed. In fact, the prevailing view of the relationship between interest rates and money supply (and economic growth) is exactly opposite to reality. The only way to rid ourselves of the so-called growth imperative is to replace our existing form of money with a truly neutral (and natural) monetary medium that does not automatically give rise to monetary growth. Incidentally, this also sheds light on the phenomena of inflation and financial crises. If our monetary system reflected a more rational understanding of the effects of interest rates on the money supply, inflation as well as financial crises could be averted. Without a proper understanding of money, we cannot possibly achieve what my dear colleague Manfred Max-Neef, laureate of the Right Livelihood Award, once called "development at a human scale". Toward that end, this book offers some completely new insights:

- A comprehensive examination of the different views of money creation. One might assume that this topic is well understood, but it is not. In fact, there are two competing theories explaining money creation. This book explains where these different viewpoints come from.
- The standard view of the relationship between interest rates and the money supply will be challenged, and it will be shown that the truth is exactly the opposite of what is usually taught – i.e., contrary to conventional wisdom, strong arguments exist that high interest rates actually promote the growth of the money supply, aggregate demand, inflation, and GDP (and *vice versa*). Central banks should be thought of as trend followers rather than trend setters.
- Regarding the cause of inflation, we will survey the various explanatory theories and show how they all boil down to one primary phenomenon viewed from different perspectives. In the end, there is only one true explanation for inflation – i.e., an increase in the supply of money relative to the quantity of goods and services in the economy. None of the existing inflation theories explain why money supply constantly grows. This book will demonstrate that the primary responsibility for this fact does not lie with central banks.
- It will be explained that the so-called "Keynesian multiplier" does not actually exist but is rather conflated with the "money multiplier." This is to say that the stimulative effect on demand (and GDP and inflation) does not come from spending alone but is rather caused by the expansion of the money supply associated with increasing government indebtedness. This is an important distinction because it shows that "expansionary" fiscal policy that is not based on deficit spending has no expansionary effect.
- The aforementioned points help us to understand why the constant growth of money supply gives rise to the so-called economic growth imperative. Perpetual economic growth ultimately leads to the over exploitation of natural resources and can be considered the most important obstacle to achieving the UN Sustainable Development Goals.

The solution presented in this book, based on Silvio Gesell's book, "The Natural Economic Order," would not only mitigate but *fight at their roots* the great challenges the world faces. It would pave the way toward development at human scale; development that allows everyone to satisfy their fundamental needs in the best way possible. It would create an economy that serves the people rather than the other way around. I invite the reader to join me on this journey to rethink the role and concept of money, particularly through the lens of sustainability. This book is not about advocating for a political orientation, either of the left or of the right. It is about the truth. The truth is neither right nor left.

Finally, I would like to make some acknowledgments. First of and above all, I give special thanks to the Almighty God for blessing me with good health, perseverance, and the time necessary to make this work possible. Then I owe deepest thanks and appreciation to Josh Sidman who did the tedious work of proofreading. Josh not only corrected spelling and grammar but considerably improved the phrasing. No one could have done this task better. Thank you so much Josh! Thanks also to Rafael Sanhueza, Manuel Carrasco, Claes Wickzen, and Daphne Gho-illanes for preparing some of the figures. Rafael and Manuel also helped me with literature research. Martina Fraißler-Eller was so kind as to provide her highly valuable opinion on biblical matters. Rafael Galaz contributed the illustrations. Thanks to all of you!

Valdivia, Chile
February 2023

Felix Fuders

Contents

Part I
Money and Its Role in the Economy

Chapter 1
Introduction

If we take a look at any newspaper in any country, two major problems frequently are addressed: **inequality** and the **increasing destruction of the natural environment**, that is unsustainability in the *stricto sensu*. In 2015, more than 190 world leaders recognized that the world is on a "collision course" (Max-Neef, 2010) and committed to 17 Sustainable Development Goals (SDGs). Here, too, all but one goals are either linked to the unsustainability of our current lifestyle (goals 9, 11–15) or to inequality (goals 1–8, 10, 16).

Many conferences and high-level meetings have been held since then, and one of the major topics is how to finance these goals. However, in this book we will argue that what is really needed is not more money, but a **different kind of money**, a more natural kind of money. In fact, both the overexploitation of the natural environment and income inequality are directly linked to the unnatural design of our money and reflect a misunderstanding of what money is and what it should be (Fuders et al., 2013; Fuders, 2016, 2017; Fuders & Max-Neef, 2014; Lietaer et al., 2013). Furthermore, many other chronic economic problems, such as inflation and deflation, unemployment, speculation, are directly linked to the unnatural design of money. Since all but one of the goals are either linked to the unsustainability of the *status quo* or to inequality, a substantial part of the book will deal with how our monetary system gives rise to the economic growth imperative and income inequality. In addition, there will be a separate chapter on the relevance of the findings to each of the SDGs.

We will examine what money currently is, what it should be and why the big concerns the world faces today are directly linked to it. We will see how to solve these problems and how by doing so we can build an economy that truly serves people and not *vice versa*, an economy in which everyone earns according to their efforts and in which economic growth is possible (e.g., to improve the satisfaction of fundamental human needs), but not mandatory. This would be an economy that approaches what Manfred Max-Neef et al. (1991) referred to as *development at human scale*, development where everyone can best satisfy their fundamental

© The Author(s), under exclusive license to Springer Nature Switzerland AG 2023
F. Fuders, *How to Fulfil the UN Sustainability Goals*,
https://doi.org/10.1007/978-3-031-37768-6_1

human needs. It would also be an economy that generates a long-term equilibrium of input and output. Under our current monetary system neither development at human scale nor true sustainability are possible.

To understand why these problems are rooted in our monetary system we need to first understand what money is and how it is created (Chap. 2). We will then discuss what money should be (Chap. 3), the problems which arise from the current design of money (Chaps. 4, 5, 6, 7, 8, and 9) and how we can reform it (Chaps. 10, 11, 12, 13, and 14). Chapters 15 and 16 provide an outlook for the future and Chap. 17 contains some final considerations.

References

Fuders, F. (2016). Smarter money for smarter cities: How regional currencies can help to promote a decentralised and sustainable regional development. In *Decentralisation and regional development – Experiences and lessons from four continents over three decades* (pp. 155–185). Springer.

Fuders, F. (2017). Neues Geld für eine neue Ökonomie – Die Reform des Geldwesens als Voraussetzung für eine Marktwirtschaft, die den Menschen dient. In *Finanzwirtschaft in ethischer Verantwortung – Erfolgskonzepte für Social Banking und Social Finance* (pp. 121–183). Springer-Gabler.

Fuders, F., & Max-Neef, M. (2014). Local money as solution to a capitalistic global financial crisis. In *From capitalistic to humanistic business*. Palgrave Macmillan.

Fuders, F., Mondaca, C., & Azungah Haruna, M. (2013). The central bank's dilemma, the inflation-deflation paradox and a new interpretation of the Kondratieff waves. *Economía (U. de los Andes), 36*, 33–66.

Lietaer, B., Arnsperger, C., Goerner, S., & Brunnhuber, S. (2013). *Geld und Nachhaltigkeit – Von einem überholten Finanzsystem zu einem monetären Ökosystem. Ein Bericht des Club of Rome/EU-Chapter*. Europa Verlag.

Max-Neef, M. (2010). The world on a collision course and the need for a new economy. *Ambio, 39*, 200–210.

Max-Neef, M., Elizalde, A., & Hopenhayn, M. (1991). *Human scale development – Conception, application and further reflections*. Apex.

Chapter 2
What Is Money?

2.1 The Functions of Money

Money is known to have three functions (see any textbook, e.g., Mankiw & Taylor, 2014; Krugman et al., 2008; see also IMF, 2000; Bundesbank, 2019):

- Facilitate the exchange of goods and services (payment)
- Store of value
- Measure of value and unit of account

Money thus serves primarily to simplify the exchange of goods (see already Aristotle, [384 BC] 1995; Steiner, 1979). In nature, when people exchanged goods for goods, trade was rather complicated. With money, it is easier to trade. So, money as a medium of exchange makes perfect sense. However, the fact that money is also used as a store of value **contradicts the first function** of money as a medium of exchange. Because if everyone hoards their money under the mattress, it cannot circulate at the same time in the economy and thus function as a medium of exchange. This contradiction is, as we will see, the cause of regularly occurring financial and debt crises (including the one that is almost upon us, but for which many in the aftermath will blame the Covid 19 pandemic or the war in Ukraine), for inflation, for the **growth imperative of the economy** and for the fact that the **income inequality everywhere** in the world is ever expanding.

Money as unit of account helps to serve its first and primordial function: to facilitate the exchange of goods and services. Assigning a value in numbers to goods and services helps us to identify how many units of one good should be exchanged for another. For example, if we were to visit a so-called "barter faire", where goods and services (or information) are exchanged directly without using money as intermediary, we would probably catch ourselves assessing the value of the goods & services

© The Author(s), under exclusive license to Springer Nature Switzerland AG 2023
F. Fuders, *How to Fulfil the UN Sustainability Goals*,
https://doi.org/10.1007/978-3-031-37768-6_2

offered at the faire by mentally comparing them to well-known monetary units. Compared with other units of measure that we use in other fields, for example engineering, money performs its function rather poorly, since its exchange value changes constantly; in particular it loses value due to inflation. Just imagine if a standard yard stick were to constantly change its length. Would it be possible to specify standard, agreed-upon lengths? Engineering would not be possible, bridges could end up too short or too long, etc. Nevertheless, in economics everything is priced in monetary units that are subject to variation.

In the following pages, we will analyze why money is such a poor measure of value, why money actually has no value at all (even though it may be difficult or costly to obtain), and how it is created. In this context, the origin of the German word for money, "Geld", is quite interesting. "Geld" derives from "vergelten", that is, settling a penance or debt (Pfeiffer, 1997). And that leads us to the next point.

2.2 Money Creation: Some Preliminary Thoughts

Most people believe that money is created by the central bank. However, this is only true for a very small fraction of the total money supply. What Milton Friedman (1971) once observed is still true today: it is not well understood even amongst experts how money is created. One could expand upon by observing that the relationship between interest rates, money supply, demand and economic growth are not generally understood (see Fuders et al., 2013). Although most economic models express prices in monetary units, there is much ignorance and confusion about what money actually is and how it is created. This ignorance is probably a major reason for the misunderstanding and misinterpretations about the causes of (i) regularly recurring financial crises, (ii) the existence of inflation and deflation, (iii) the widening gap between the rich and the poor and (iv) the growth imperative and the unsustainable nature of our existing economic growth model.

Among those scholars who have conducted in-depth studies on money creation, two major currents can be discerned. Both agree that the central banks produce only a small fraction of the total money supply (original central bank money or monetary base), while the largest part of what we call money is, in fact, created by lending. However, how this takes place is a point of doctrinal dispute. While one view assumes that the money supply multiplies **in the banking sector as a whole** by the granting of loans and the subsequent deposit of these funds in other banks (*money multiplier*—see, e.g., Phillips, 1931; Stiglitz, 1998; Larroulet & Mochón, 2003; Samuelson & Nordhaus, 2009; Mankiw & Taylor, 2014; Krugman & Wells, 2018), the other view assumes that **banks themselves create money** *ex nihilo*—out of nothing (see e.g., Huber, 1998; Huber & Robertson, 2008; Werner, 2014; Jakab & Kumhof, 2015; Gross & Siebenbrunner, 2019) by a simple bookkeeping entry, limited only by the reserve requirement imposed by the central bank. Both views have existed alongside each other ever since books on banking have been written.[1]

Even though the money multiplier model is still predominantly used in textbooks, nowadays the idea of money creation *ex nihilo* finds again—it was already *in vogue* between 1900 and 1930 (Werner, 2014)—ever more followers. Even some central banks as, for example, the Bank of England (McLeay et al., 2014) or the German Bundesbank (2017, 2019) today seem to adhere to this concept, while in former times they followed the money multiplier model. Sometimes the statements on their websites and official publications are formulated in an ambiguous way so that it is not clear as to which theory they adhere to (Werner, 2014). This chapter discusses both concepts in order to determine where the two views actually came from. It will be concluded that the *money multiplier* theory is to be considered the true theory, while the idea of money creation *ex nihilo* is based on a **misinterpretation of an accounting record**. In this context it will be explained why the "empirical test" undertaken by Werner (2014) that allegedly "proves" the theory of money creation *ex nihilo* is invalid since it is based on the same misinterpretation.

2.3 Money Creation: The Money Multiplier Theory

2.3.1 Banks Lend Out Customer's Money

The classic business model of any bank is to lend out its customers' money (see e.g., Mises, 1912; Keynes, 1930, 1936; Tobin, 1963, 1969; Gurley & Shaw, 1976; Bernanke, 1993; Stiglitz, 1998; Samuelson & Nordhaus, 2009; Mankiw & Taylor, 2014; Krugman et al., 2015; Krugman & Wells, 2018; see also references in Werner, 2014). In fact, this is not just a business model for banks but a **crucial service that is necessary for the economy to function**, since otherwise people would hoard surpluses at home "under the mattress". In consequence money would not circulate well, people would not (or much less) be able to exchange goods and services, thus strongly affecting allocative efficiency, i.e., causing net losses in societal welfare; in other words, the economy would plunge into a permanent depression (Fuders, 2021). What we generally are not aware of is that through the granting of loans the money supply grows—at least if money is understood, as central banks do and as

[1] Werner (2014) and Gross and Siebenbrunner (2019) identified a third view. Authors who follow this third view believe that money supply would not increase at all by credit expansion. The common denominator of authors who follow this idea is that they reduce the definition of money supply merely to the so-called monetary base (bills and coins and reserves held by commercial banks at the central bank), without taking into account deposits. The monetary base (also called base money) is exogenously produced by the central bank and does, of course, not differ by credit expansion. How to define money supply and what can be regarded as money has also been a point of doctrinal dispute, but to analyze this in detail goes beyond the scope of this book. Scholars throughout the last century have traditionally defined money as cash plus deposits, including time deposits (on this discussion and for further references see Demeulemeester, 2018).

monetary theory today usually does (for references see authors cited in Demeulemeester, 2018), as the sum of cash in circulation plus deposits (the most used money supply definition "M3" also includes time deposits).

2.3.2 *Lending Customer's Money Does Not Increase the Balance Sheet Total*

If someone owns, let's say, 100 € originally created by the central bank and deposits this sum in a bank (Bank 1 in Fig. 2.1), the bank will then lend this money, minus a reserve, to one customer. Even though the biggest share of the funds deposited in bank accounts will usually not be withdrawn by the customers, the bank will not lend out all its liabilities but will hold a reserve, because some costumers will wish to withdraw some or all of their funds. Let's assume that the bank holds a cash reserve of 10%. It will then lend 90 €, while the sum of 100 € apparently remains available for use via the customer's deposit account at Bank 1 (Fig. 2.1). I use the word "apparently" since, in fact, 90% is loaned out. But as long as the total of all withdrawals is less than the bank's cash reserve (or deposits due on demand that the bank holds at other financial institutions or the central bank), no one will take notice.

In fact, any customer can still withdraw the total amount of their own deposits as long as not all costumers do so at the same time, i.e., provided that the total of all withdrawals does not exceed the reserves of the bank. In the event that such a thing did occur, perhaps due to a loss of confidence in the solvency of the bank, the bank would then have to declare itself bankrupt. Note that by granting and paying out the loan on the asset side of the bank's balance sheet the cash desk will record 90 € less while the receivables from costumers increase respectively (transaction 2 in Fig. 2.1). In accounting we call this transaction an 'asset swap'. **The balance sheet does not change** by granting the loan. The latter is an important point to highlight, since those who believe in money creation *ex nihilo* mistakenly believe that the balance sheet increases by granting a loan.

2.3.3 *Lending Increases the Money Supply Throughout the Banking Sector*

Let us now look at the customer who takes the loan and who now holds the 90 € and who will spend the money somewhere. Maybe he will buy a new cell phone. The cell phone vendor now deposits the 90 € at his bank, let us call it "Bank 2"

Fig. 2.1 Money multiplier (bank accounts). (Source: Own work)

The Money Multiplier

Transaction 1: Customer deposits money at Bank 1

Bank 1

Assets	**Liabilities**
Cash account 100	Liabilities to customers 100

Posting record: Cash account 100 against liabilities to customers 100

Transaction 2: Bank 1 grants loan from customer deposit:

Bank 1

Assets	**Liabilities**
Cash account 10	Liabilities to customers 100
Receivables from custom. 90	

Posting record: Receivable from customers against Cash account 90

Transaction 3: Vendor of the cellphone deposits money at Bank 2

Bank 2

Assets	**Liabilities**
Cash account 90	Liabilities to customers 90

Posting record: Cash account against Liabilities to customers 90

Transaction 4: Bank 2 grants loan to another customer:

Bank 2

Assets	**Liabilities**
Cash account 9	Liabilities to customers 90
Receivables from custom. 81	

Posting record: Receivables from custom against cash account 81

(transaction 3 in Fig. 2.1).[2] Until now, the amount of original 100 € central bank money has already multiplied to 190 € on the two bank accounts of both banks taken together. Bank 2 will most likely do as Bank 1 did. It will lend out the deposit of 90 € while retaining a reserve, which will cause the money supply to continue to grow. If all banks hold a liquidity reserve of 10% of all deposits, and all monetary units loaned out will again end up in another bank account, the original 100 € put into circulation by the central bank can multiply to as much as 1000 €. With a reserve of

[2] He might also spend the money, but at some point this money is likely to reappear as a deposit in some other bank account.

1% (which is the current reserve requirement in the eurozone), 100 € can multiply in the banking sector up to 10,000 €. This mechanism of money creation is usually referred to as the *money multiplier* (see any current textbook, e.g., Stiglitz, 1998; Larroulet & Mochón, 2003; Mankiw & Taylor, 2014; Krugman & Wells, 2018). Many countries currently have no reserve requirement at all.

The money multiplier can also be understood in the following way: the money supply grows when loans are granted because the original deposit **remains in circulation via the loan**, while the original depositor still holds a demand deposit, savings deposit or time deposit. Bank deposits are not money in the strict sense but a **money-equivalent claim** since deposits can be used as money. Most people would not even differentiate between a bank deposit and cash (see Fig. 2.2).

2.3.4 *Every Loan Expands the Money Supply*

The fact that banks lend out funds deposited on demand (also called sight deposits, current accounts or checkbook accounts) is the reason why it is not possible for everyone to withdraw their money at the same time. If bank customers want to withdraw more of their money than the bank has liquidity reserves, the bank must close and a feared "bank run" can occur, in which the loss of confidence leads to panic-like mass withdrawals from other banks, which then also have to close. The last time this happened on a large scale in the industrialized world was in 1929 in the USA, which triggered (but didn't cause—Fuders et al., 2013; Fuders & Max-Neef, 2014a, b)

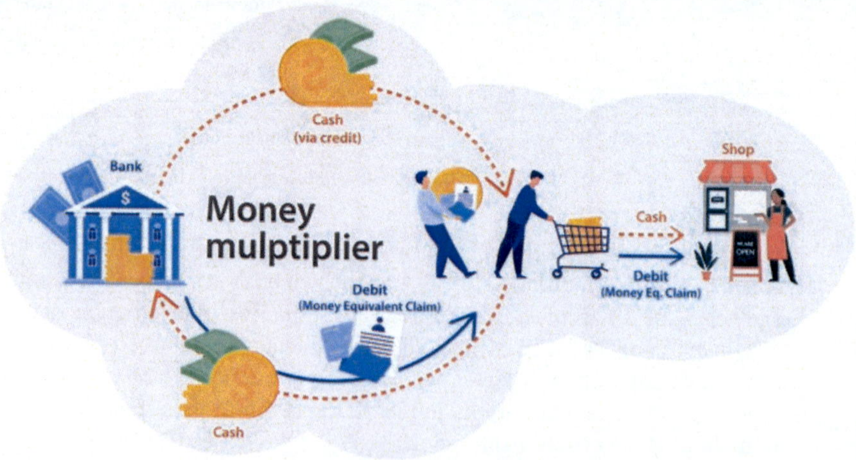

Fig. 2.2 The money multiplier (scheme). (Source: Own work)

the Great Depression. It almost happened in Cyprus (euro zone) in 2013, although this was not widely reported in the media.[3]

Banks being allowed to lend on demand deposits was not always the case (Fisher, 1936a). However, the central bank can impose a minimum reserve requirement that commercial banks must retain. The reserve requirement (10% in Fig. 2.1) is the cash reserve that each bank must hold in their vaults or at their central bank current account relative to certain types of deposits (usually demand deposits, i.e., subject to check, fulfilling a payment function). This requirement is intended to keep the financial system liquid. A monetary system operating according to these rules is therefore referred to as a "fractional reserve system" or "fractional reserve banking". The higher the fractional reserve requirement, the less likely it is that a bank run will occur, and the less money is generated throughout the banking sector by lending out demand deposits. The reserve requirement therefore limits the money multiplier effect, but **only in relation to loans funded by demand deposits**.

Unfortunately, the explanation of money creation via the multiplier in textbooks is usually limited to money creation via loans that are granted out of those deposits that underlie the reserve requirement, or textbooks they do not differentiate between loans funded by demand deposits versus time deposits.[4] This may contribute to misunderstandings. The expansion of the money supply takes place through **every loan**, i.e., also including loans funded by short-term or long-term time deposits (maturity of 2 years or more) or from inflows generated by issuing bearer bonds or any other debt securities. These latter financial assets can also be regarded as money, although less liquid than cash or demand deposits. After all, one could exchange time deposits, or in case of bearer bonds use them directly as payment for goods or services (in the current system possibly having to admit some discount). Obligations, debt securities, bonds, time deposits, etc. can therefore be used as money as well and are consequently included in the broader definitions of money supply (M3—see Fig. 2.3). The money supply definition M3 is the one most frequently used by central banks worldwide. In addition to cash and demand deposits, it also includes savings deposits and other short- and medium-term debt securities (time deposits, bonds, obligations).[5] Interestingly, the German word for time deposit is "Festgeld", which contains the expression "Geld" which means money.

[3] A panic-like run on banks was averted by a government-imposed moratorium. Banks closed for a few days and did not reopen until the panic had subsided.

[4] A good explanation can be found in Stiglitz (1998).

[5] It should be noted that this definition has also always been a point of dispute. While according to some authors only cash and demand deposits count as money, the vast majority usually include also time deposits, obligations and debt securities or an even broader definitions of money: everything that can be used to exchange goods and services are money (Mises, 1912; on this discussion see, e.g., Demeulemeester, 2018). There is no uniform definition that is used worldwide to specify the money supply, M3.

Fig. 2.3 Money supply M3. (Source: Own work)

2.3.5 Summary

To sum up, the money supply grows when loans are granted because the money originally deposited in a bank account remains in circulation as a loan, while the bank customers who brought the money to the bank can either dispose of their deposits (if loans were granted from demand deposits) or hold a **money-equivalent claim** against the bank (time deposit, bearer bond, obligation, etc.). The latter, although less liquid than cash, can still be understood as money and are therefore also included in the most commonly used money definition M3 (similarly also Keynes, 1936; Anderegg, 2007). That is to say, if we only understand cash (bills and coins) as money, then money supply would not grow via lending. But the moment that deposits (even if only demand deposits) are included in the definition of money—which is usually the case and makes sense—, then money supply grows via lending in between the banking sector. The money multiplier increases money supply **within the entire banking sector**, in contrast to the idea of money creation *ex nihilo* as explained in the next section, where the same multiplication of monetary units (from 100 to 1000 € in case of 10% reserve requirement) is mistakenly believed to happen **within each bank**.

2.4 Money Creation "Ex Nihilo" (Out of Nothing)?

2.4.1 Supposed Balance Sheet Expansion

A growing number of authors and scholars in financial economics are convinced that the money supply does not grow in the banking sector as a whole by the money multiplier effect described above, but rather believe that when banks extend loans they are simply creating money "out of thin air" (Werner, 2014: 16) applying a kind of **magic (but legal) trick**. On a lending bank's balance sheet, the loan appears on the assets side as "receivables from customers". If the loan amount is paid into the bank client's current account, the same amount appears on the liabilities side as a "liability to customers" (see Fig. 2.4). Because when viewed this way the balance sheet total increases by exactly the amount of the loan, this would represent money creation "*ex nihilo*", i.e., out of nothing (Huber, 1998: 13; Huber & Robertson, 2008). This is sometimes also referred to as *book money creation*. This view was and still is represented by some economists (Fisher, 1936a; Huber, 1998; Huber & Robertson, 2008; Werner, 2014; McLeay et al., 2014; Bundesbank, 2017; Gross & Siebenbrunner, 2019), notably by the famous Irving Fisher, who described the act of money creation by banks in his book "100% Money" (Fisher, 1936a: 7) as follows:

> When a bank grants me a $1,000 loan, and so adds $1,000 to my checking deposit, that $1,000 of "money I have in the bank" is new. It was freshly manufactured by the bank out of my loan and written by pen and ink on the stub of my check book and on the books of the bank.

2.4.2 Magic Money Creation?

According to this theory, banks are no financial intermediaries but literally "create" the money they lend, i.e., loans create deposits and not the other way around (Fig. 2.5). Money is created by a sort of "air entry"[6] and this form of money creation

Money creation '*ex nihilo*'

Assets	Liabilities
Receivables from customers 1000	Liabilities to customers 1000

Fig. 2.4 Supposed money creation '*ex nihilo*'. (Source: Own work)

[6] In accounting, an "air entry" describes an entry without an actual corresponding business activity. The entry is therefore only made for appearance, and often with fraudulent intent.

Fig. 2.5 Supposed magic money creation by individual banks

would only be limited by the reserve requirement of the central bank. That is, even according to this theory, money can be generated by banks only if they have deposits from customers. However, they do not need the deposits in order to lend them out, but only to meet the reserve requirement. According to this idea, if the reserve requirement is 1%, a bank can create loans of up to 10,000 € from a deposit of 100 €. Accordingly, a **single bank** can generate as much book money as can be maximally generated by the *money multiplier* in the entire banking sector (Stiglitz, 1998). For deposits that are not subject to any reserve requirement, banks could create money without any limit, as long as customers demand loans. That the money supply can only grow as long as customers demand loans is also true for the money creation via *money multiplier*. The difference is that in the *money multiplier* theory this happens within the entire banking sector, not inside of one bank.

2.4.3 Contradictory Views in Literature

It should be noted that even 1 year after the publication of the book "100% Money", Irving Fisher published an essay in which he still advocated—as he did in the book—for a minimum reserve requirement of 100%, however he no longer mentioned the concept of creating money out of nothing and even expressly stated that banks lend money of their customers (Fisher, 1936b). Similarly, John M. Keynes originally advocated the theory of money creation *ex nihilo* (Keynes, 1924) but later disavowed it (Keynes, 1930, 1936). Countless books on banking and monetary theory published during the last century leave no doubt that money is created via lending but also make statements that could be viewed as consistent with either theory, thus creating a great deal of confusion.[7]

Some central banks such as the German Bundesbank (part of the Eurosystem) seem to use both concepts. Contradictory statements can be found on the Bundesbank website as well as other official publications. The same can be said for publications of the Bank of England (Werner, 2014). Indeed, until some years ago the Bundesbank website described the process of money creation via the *money multiplier* explained above. In newer publications the Bundesbank seems to adhere to the concept of money creation out of nothing (Bundesbank, 2017, 2019). However, some affirmations or conclusions they draw even within these newer publications can hardly be seen as coherent with this view. For example, in a publication from 2017 in which the Bundesbank discusses the idea of a 100% reserve requirement it seems to be clear that the Bundesbank subscribes to the idea of money creation *ex nihilo*, stating that:

> (…) a bank can grant loans without any prior inflows of customer deposits. In fact, book money is created as a result of an accounting entry: when a bank grants a loan, it posts the associated credit entry for the customer as a sight deposit by the latter and therefore as a liability on the liability side of its own balance sheet. This refutes a popular misconception that banks act simply as intermediaries at the time of lending – i.e. that banks can only grant loans using funds placed with them previously as deposits by other customers (Bundesbank, 2017: 17).

Nevertheless, in the same text it is argued that in case of a 100% reserve requirement the so-called "maturity transformation service"[8] that usually is applied to loans granted **out of** demand deposits would not be possible anymore, which can be seen as a contradiction to the *ex-nihilo* concept, since a maturity transformation function would not exist if money could be created out of nothing.

2.5 How Do the Different Views Come About?

2.5.1 Where Is the 99% Interest Margin?

That banks cannot create money out of nothing can be understood from a simple fact. If banks really could create money by a simple entry in their books, their interest margins would be exorbitantly high; after all, they could collect interest practically without having to pay any themselves. Since the minimum reserve is only 1% today in the eurozone, banks would only have to hold back 1% of the credit volume

[7] To just give one example, Wicksell, 1907: 214 writes: "In a pure system of credit, where all payments were made by transference in the bankbooks, the banks would be able to grant at any moment any amount of loans at any, however diminutive, rate of interest". It is not clear which theory Wicksell follows.

[8] Maturity transformation is the practice by financial institutions of borrowing money on shorter timeframes than they lend money out.

as customer deposits. In some countries as, e.g., the USA, currently there is no reserve requirement at all. The interest margin would have to exceed 99%.[9] In fact, the interest margin is "only" 25% (Creutz, 2013). Also, if that was possible, **monetary policy of central banks would be quite meaningless**, since commercial banks would be in no (or very little) need of loans from the central bank, they could simply air entry the credit in their books, especially nowadays since many countries do not have any reserve requirement at all.

2.5.2 *No Balance Sheet Expansion Through Loans*

The confusion about how money is generated by loans may have been promoted by the fact that in accounting textbooks as well as in business encyclopedias, the granting of loans is often erroneously presented as an example of a so-called balance sheet expansion (an increase of the balance sheet total). For example, in an economic encyclopedia (Wirtschaftslexikon24, 2021), as a definition for "**balance sheet expansion**", we find:

> Increase in the sum of assets and liabilities by the same amount. Example: a bank grants a loan and credits the borrower this amount at his disposal on his personal account (Original text in German, author's translation).

Indeed, if the balance sheet of a bank increases without the balance sheet of another bank decreasing, the money supply in the economy has grown. It is, at first sight, understandable to assume that money has been generated simply by a bookkeeping entry. However, and this following point is crucial, it is **NOT the granting of the loan** itself that results in an expansion of the balance sheet. The balance sheet is only expanded when a bank grants a loan AND then credits the money loaned to the cash account of the customer **at the same bank**, which is usually the case. In the balance sheet depicted in Fig. 2.4 (which is typically used to argue in favor of the theory of money creation *ex nihilo*) we do not see this, because it is a simplified balance sheet where the cash account is left out.

The balance sheet expansion **does not come about through the granting of the loan** but, instead, through the crediting of the money to the borrower's account at the same bank. Any deposit into any bank account always represents an expansion of the balance sheet. It is not the loan that generates the expansion of the balance sheet but the crediting of the money to the customer's account held **at the same bank**. If the customer did not wish to receive the loan in the form of a credit to his

[9] Accordingly, if a bank has € 100 in customer deposits, it could air book € 100/0.01 = € 10,000 in loans. At an annual interest rate of 5%, this would therefore be € 500 in interest income compared with less than € 5 that the bank would have to pay in interest on the € 100 deposit. Even if the interest to be paid by the bank on credit balances were as high as the interest on loans, the interest margin would therefore be at least € 495. That is 99 times as much as the bank had to pay on the deposit. For every euro that a bank has to pay in interest on credit balances, it would gain € 99 according to the theory of money creation *ex nihilo* (Creutz, 2012, 2013).

account but, instead, took the loan directly in cash, the expansion of the balance sheet would not occur. It is not that the accounting record is wrong, but the interpretation that the act of granting a loan represents money creation out of nothing is incorrect.

2.5.3 Granting of Loans Provokes an Asset Swap

Banks can only lend out the money they receive as deposits from customers or that they borrow from other institutions (Keynes, 1930; Tobin, 1963; Stiglitz, 1998; Krugman & Wells, 2018; Mankiw & Taylor, 2014; see authors cited in Werner, 2014). To understand this, let us look again at transaction 1 in Fig. 2.1. When a customer deposits money in a bank, the bank's cash holdings increase on the assets side of the balance sheet. On the liabilities side, the bank posts a corresponding increase in the form of liabilities to customers, in our example 100 € (posting record: cash desk against liabilities to customers). If the bank now decides to grant a loan of 90 € from this deposit (transaction 2 in Fig. 2.1), it takes some of the money from the cash desk and grants the loan from it. This means that the cash balance in the cash desk decreases by the amount of the loan. The corresponding offsetting entry is the increase in the item "receivables from customers" on the **asset side**.

As already discussed, this is an asset swap, **the balance sheet has not been expanded** (posting record: receivables from customers against cash desk 90 €). However, borrowers usually do not choose to receive their loans in the form of cash (bills and coins) but rather ask that it be credited to their accounts. Then of course the cash desk of the bank **is not reduced, despite the granting of the loan** (meanwhile the customer does not dispose of the money).[10] On the liabilities side of the balance sheet, the liabilities to customers then increase accordingly: there is still the account of the person who originally deposited money (liabilities to customer 1), 100 €, **plus** the account of borrower (liabilities to customer 2 in Fig. 2.6), 90 €. The balance sheet total expanded by the amount of the loan.

The balance sheet expansion therefore does not come about through the granting of the loan, but through the fact that the customer keeps the money at the bank, i.e., the bank immediately redeposits it after the loan has been issued. This deposit of the borrowed money into an account at the same bank is a **second transaction** (transaction 2.2 in Fig. 2.6) that has to be considered separately form the granting of the loan. This second transaction, not the granting of the loan itself, then generates, **as any cash deposit** does, an expansion of the balance sheet. Again, if the borrower chose to receive the loan in the form of cash, there would be no expansion of the balance sheet. Conversely, any expansion of the balance sheet by crediting of the

[10] This is important to understand since some mistakenly argue that granting a loan without the cash desk (or reserve balance) being reduced nullifies the idea of money creation via money multiplier (Werner, 2014).

Money creation *ex nihilo* does not exist.

Transaction 2.2: Bank 1 deposits the sum of the loan granted on current account of borrower:

Bank 1

Assets		Liabilities	
Cash desk	100	Liabilities to customer 1	100
Receivable from custom.	90	Liabilities to customer 2	90

Posting record: Cash desk against liabilities to customers 90

Fig. 2.6 Money creation *ex nihilo* does not exist. (Source: Own work)

loan to an account held at the same bank will be reduced the very moment the customer disposes of the money credited (strangely this is even admitted by Bundesbank (2017) who nowadays seems to adhere to the concept of money creation out of nothing). Of course, this money will end up in some other bank account and expand the balance sheet of the other bank, and the economy's money supply will increase. But this is what was above described as the *money multiplier* and not money creation out of nothing.

2.5.4 *The Origin of the Confusion: Banks Are Not Only Financial Intermediaries*

What has been said so far could also be visualized the following way. Any bank deposit can be seen as a credit to the bank, which is why deposits appear on the liability side of the balance sheet. As long as borrowers do not spend the proceeds of their loans and keep them on deposit at the same bank, their relation to the bank is twofold: they are **not only borrowers but at the same time creditors**, for their deposit represents a credit to the bank. This dual role of the bank's customer as both borrower and creditor, is what produces the confusion.

Here we can also see the reason why some adherents of the idea of money creation *ex nihilo* are convinced that only banks can create money, while other financial institutions which do not offer demand deposits do not (Tobin, 1963; for further references see, e.g., Gurley & Shaw, 1976). Well, the difference between banks and other institutions that are mere brokers of credits is that banks offer **both functions**: checking accounts (utility banking function of settling trades) and also intermediate between savers and borrowers. If banks would only serve as financial intermediaries and didn't offer checking accounts, it would not be possible to credit the proceeds of loans to accounts at the same institution and thus no increase in the balance sheet total would occur.

2.5.5 Werner's "Empirical Test"

Richard Werner, an outstanding economist to whose academic contributions I usually fully subscribe, here unfortunately falls prey to the same misinterpretation (Werner, 2014). He purports to have proven the theory of money creation *ex nihilo* (he calls it *credit creation theory*) by exercising an interesting empirical test. Basically, what he did is the following. He considered a € 200,000 loan issued by a local bank in Germany and then compared the bank's balance sheet from the day the loan was granted with the balance sheet from the day before. Since the balance sheet total expanded, his interpretation was that this would only be consistent with the theory of money creation *ex nihilo*, not taking into account that, if the loan had been paid out in cash (or if it had been transferred immediately to an account at another bank) this expansion of the balance sheet would not have occurred.

Werner further mistakenly argues that the *money multiplier* theory (he calls it *fractional reserve hypothesis*) requires an immediate "increase in deposits" (Werner, 2014: 15) to precede the bank loan, as a precondition for being able to grant the bank loan. If such an increase did not occur, this would prove that the bank created the deposit out of nothing. However, loans can of course also be granted out of "older" deposits that the bank invested elsewhere. In fact, no bank holds huge cash reserves but rather invests money deposited by customers in order to be able to pay the interest due on those deposits, thus increasing their claims on financial institutions. Here, the local bank had reduced its "claims on financial institutions" (see table 6 in Werner, 2014: 15) in order to be able to grant the loan. That is to say, the bank granted the loan by mobilizing its demand deposits held with another banks, most probably the central bank.[11]

2.5.6 A Clear Definition of What We Call Money Is Crucial

A clear definition of what we call money is crucial to understand the *money multiplier* concept. Here we have defined money broadly as cash plus demand deposits plus savings accounts plus time deposits plus obligations and bearer bonds.

[11] The underlying transactions were first to mobilize deposits held at other financial institutions and book this on the cash account (posting record: cash against claims on financial institutions 200,000). Then the loan was granted as shown in Fig. 2.1 (transaction 2) increasing the receivables from customers (in Werner's text called claims on customers) and decreasing cash respectively (posting record: claims on customers against cash 200,000). But since Werner did not immediately withdraw the money in cash, the third transaction is (as shown in Fig. 2.6) to deposit the sum lent out to the cash account increasing the cash balance as well as the liabilities against customers (posting record: cash against liabilities to customers 200,000). That the difference in the accounts in the balance sheet is not exactly 200,000 is because other transactions have occurred on the same day, i.e., even though the loan granted to Werner was the most significant transaction that day, it was not the only one.

Certainly, if we only count bills and coins or the reserves that commercial banks hold at the central bank (so-called "M0" or "base money", sometimes also referred to as "central bank money" or "lawful money") as money, then lending does not influence money supply or has very little effect. If deposits are not counted as money, then money supply will of course not grow since the base money is exogenously produced by the central bank.

This explains why there is a **third view** alongside money creation via *money multiplier* and the money creation *ex nihilo* (for references see authors cited in Werner, 2014 and Gross & Siebenbrunner, 2019). According to this third view banks do not engage in money creation at all, either individually or in the industry as a whole. James Tobin (1963) is one of the most prominent advocates of this view. According to this view, banks function as mere brokers of credits and in doing so no money is created. In so far as money is **narrowly defined as only base money** (excluding deposits, obligations and other debt securities), this view can be regarded as correct. However, this narrow definition of money is not meaningful when it comes to describing the **problems stemming from the steady growth of deposits**. The rest of this book will deal with those problems.

In addition, this third view is in fact only partly correct since a strong credit expansion might indirectly put the central bank in a situation of being forced to increase the monetary base. That can happen if customers of commercial banks want to withdraw more money than the bank holds as cash reserves. In this case the commercial bank is likely to ask for a credit form the central bank and consequently the monetary base will increase. That is to say, credit expansion can also here lead to an increase of the monetary base.

2.5.7 *Summary*

In accounting, the corresponding counterpart of a bank's claim against its customers resulting from a loan ("receivables from customers") on the asset side of the balance sheet is not an increase in liabilities to customers, as those who believe in money creation *ex nihilo* mistakenly argue, but the reduction in the bank's cash desk. Even though the borrowing customers do not notice this, the granting of the loan and the subsequent deposit of this sum to their account, which makes the cash desk increase again, are theoretically **two separate transactions**. The latter is the one that—like any deposit to a bank account—leads to the balance sheet expansion, **not the granting of the loan itself**.

2.6 Discussion

2.6.1 The Effect of a 100% Reserve Requirement (Full Reserve Banking)

A 100% reserve requirement on **demand deposits**, as proposed in the 1930s by Irving Fisher and others (the so-called Chicago Plan, see, e.g., Phillips, 1992; Demeulemeester, 2018) and later also by many renowned authors such as Milton Friedman (1960) James Tobin (1985) and Hyman Minsky (1994)—most of them actually not adhering to the theory of money creation out of nothing—could prevent a bank run, since banks would not be allowed to lend out the money on demand deposits (checking account), i.e., money due on demand (Fisher, 1936a; Huber, 1998; Huber & Robertson, 2008; Bundesbank, 2017). Any person would then always be able to withdraw money on their checking account, and banks would never lack liquidity in regard to deposits due on demand. This is a powerful argument in favor of a 100% reserve requirement.[12] However, as already mentioned above, a 100% reserve requirement on demand deposits does not impede the *money multiplier* effect regarding loans funded by time deposits or the issuance of bonds, obligations or other debt instruments, a fact that was highlighted even by Irving Fisher (1936a) (see also Demeulemeester, 2018).

A 100% reserve requirement would therefore **not automatically produce a stable money supply**, nor would it avoid inflationary and deflationary episodes. However, money supply growth through the *money multiplier* effect would be substantially restricted, compared with today's situation since it is precisely the demand deposits that banks are keen to lend out because they pay customers little or no interest and therefore gain a particularly high interest margin on those loans. In fact, about half of all loans in the Eurozone are funded by demand deposits (Bundesbank, 2017, see Fig. 2.7). From this point of view a system with a 100% reserve requirement on demand deposits would be an improvement compared to today's situation, even though it would not solve the most important problems of our financial system.

It is worth noting that a 100% reserve requirement on demand deposits would not prevent the money supply from growing **even** if one adopted the (wrong) interpretation of money creation out of nothing. Commercial banks would (if this theory was true) still be able to create book money *ex nihilo* on savings and time deposits. This is important to note because a 100% reserve requirement nowadays is strongly advocated, especially by those scholars who believe in money creation *ex nihilo* and who view this as a solution to uncontrolled increase in the money supply as well as a way to change the fact that under the current system money is counterpart of debt

[12] Although some argue that this positive effect of a 100% reserve requirement for demand deposits would be offset by the fact that the maturity transformation function for loans granted out of demand deposits then no longer exist (Bundesbank, Monthly Report April 2017, 2017).

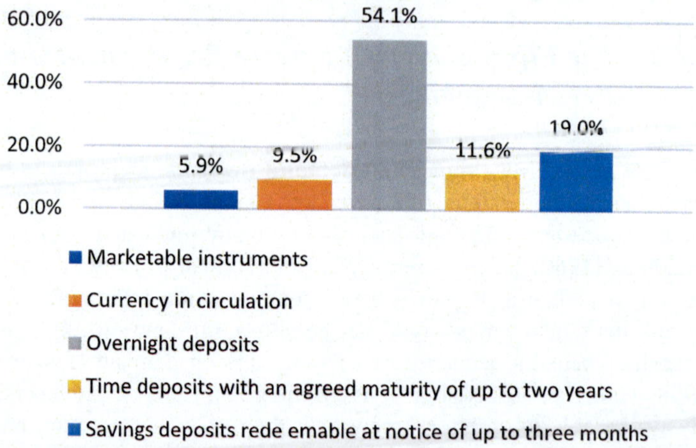

Fig. 2.7 Components of euro-area M3. (Source: own work, based on data provided by Bundesbank (2017))

(see e.g., Demeulemeester, 2018). This leads us to the next point that should be discussed.[13]

2.6.2 *Money Is Always the Counterpart of Debt*

What has been said so far gives an excellent opportunity to eradicate some important misunderstandings regarding money and debt. Except for the money originally created by the central bank (base money) **held physically** in cash (bills and coins), money is always the counterpart of debt. That money is the counterpart of debt is **not possible to avoid** as some think (Fisher, 1936a, Huber, 1998, see also so-called "positive money"[14] or "sovereign money"[15] movements). For example, if customers were to deposit money into a bank account, then this represents an asset from the customer's point-of-view (money), while for the bank it is a liability, a debt (see Fig. 2.1). If someone buys a government bond, then this is an asset for the buyer, while for the government it is a debt. An old proverb expresses this beautifully: "one man's financial assets are another man's debts". Also, the

[13] Not all who follow the idea of money creation *ex nihilo* advocate for a 100% reserve requirement on demand deposits. For example, the German central bank (Bundesbank, 2017)—even though having recently adopted the misinterpretation of money creation *ex nihilo*—argues that a 100% reserve requirement would mean that some essential functions such as the maturity transformation would no longer exist.

[14] https://positivemoney.org/

[15] https://sovereignmoney.site/what-is-sovereign-money

Fig. 2.8 Left upper corner of a US dollar bill. (Photo: Felix Fuders)

English expression "credit" contains this ambiguity, since it is a synonym for a loan while at the same time it can also express a positive value, for example in a bank account. The German word for money ("Geld") contains a similar ambiguity, mentioned previously. This term derives from "vergelten", which means repaying a debt (Pfeiffer, 1997). Interestingly, that money is the counterpart of debt is clearly stated on the upper left corner of any US dollar bill (Fig. 2.8): "This note is legal tender for all debts public and private". According to this, a dollar is not for purchasing goods or services, but rather for the repayment of a debts or obligations (noting that the latter might, of course, result from a purchase agreement).

There is nothing that can change this ambiguity. Even in a system with a 100% reserve requirement on demand deposits, any deposit would represent money for the holder of the deposit while it would be a debt for the bank. And of course, also in a banking system with a 100% reserve requirement on demand deposits people could still deposit money as time deposits or invest in bonds and obligations. Here again, while for the holder of the time deposit or bond it represents an asset, for the bank or the issuer of the bond it is a liability. Again, except for physical bills and coins which have not yet been deposited into a bank account, so-called **"positive" money does not exist.**

And even bills and coins and any other monetary instruments directly created by the central bank (base money) can theoretically be regarded as counterparts of debt, since on the central bank's balance sheet, cash as well as reserves held by commercial banks represent liabilities. That is to say, from the point of view of the central bank, cash circulating in the economy represents debt. Even so-called "time banks" where people exchange services without using conventional money crediting hours worked to be used later as payment for services from other individuals, are still not positive money as some might believe, because even in such cases the people who provide services extend credit (from Latin "credere" = to believe, trust) to the time bank community, i.e., they trust that they will receive services from other people with the same (time) value as the services they previously provided. To put it in the words of Frederick Soddy (1934: 25) "Money is a credit-debt relation from which none can effectually escape".

2.6.3 *Money Being the Counterpart of Debt Is Not a Problem*

It is worth noting that neither the fact that the money supply increases within the banking sector by the granting of loans, nor the fact that money is the counterpart of debt is a problem. In fact, both could be beneficial since the money supply would probably **automatically adjust to the business cycle**. For example, in economic boom times, there would be more money circulating due to an increased demand for loans, while an economic downturn would be accompanied by shrinking demand for loans and thus less money circulating. We could call this in accordance with Fullarton (1845) the *law of reflux*. Probably we would neither see inflation nor deflation. In fact, central banks might not even be necessary anymore.

The reason for why this mechanism does not work—and many other problems that concern society today are connected herewith—is that **banks are obliged to grant an ever-larger volume of loans**, which is why both money supply and debt grow incessantly and usually faster than real economic production. The productive economy therefore has to keep pace with this steady growth in money supply and debt (Fuders et al., 2013; Fuders & Max-Neef, 2014a, b; Fuders, 2016, 2017, 2020, 2021). This latter aspect will be discussed in the following chapters.

Many of the problems addressed by the "positive money" or "full sovereign money" movements, e.g., price bubbles and growing income inequality, are indeed rooted in our monetary system, and these movements are correct in arguing that the understanding of what money is and how money is created is crucial in order to solve our problems. Nevertheless, both movements mistakenly believe in money creation out of nothing and that with a reserve requirement of 100% on demand deposits these problems could be solved.

2.7 Conclusions

1. Banks, as well as other financial institutions, when acting as financial intermediaries lend out customers' deposits. Performing this crucial function increases the money supply, defined as cash plus demand and time deposits, obligations, bearer bonds, etc.
2. The growth of the money supply does not come about by some sort of magic accounting trick or "air entry" in the bank's books, but by the fact that deposits remain in circulation because of loans, while customers depositing money hold a **money-equivalent debt instrument**. This can be a demand deposit (current or checking account), a savings account, a time deposit or even a bearer bond, obligation, or any other debt instrument. Thus, money is created by banks acting collectively, i.e., by the **baking sector as a whole**.
3. The idea of money creation out of nothing or *ex nihilo* stems from a **misinterpretation of accounting records** and how the granting of loans represents an expansion of banks' balance sheets. However, the expansion of balance sheets

associated with the issuance of a loan **does not result from the loan itself**, but from the fact that the amount loaned out is credited to the current or savings account of the borrower **held at the same bank**. It is this second transaction that results in the increase of the balance sheet of the bank. If the borrower withdraws the proceeds of the loan, the balance sheet expansion doesn't occur. Of course, then the money is likely to be deposited into another bank, thus leading to an expansion of that bank's balance sheet and an increase in the money supply **throughout the banking sector**. However, this is not money creation *ex nihilo*, but the effect of the correct interpretation of money creation via the *money multiplier*.

4. The distinction sometimes made between the so-called "currency theory" and "banking theory" becomes irrelevant, since this distinction is based on the different views on how to deal with money creation *ex nihilo*, which does not exist.
5. Except for the money originally issued by the central bank and held as cash, all money is **always the counterpart of debt**. This is true even in a system with a 100% reserve requirement on demand deposits, which is essentially the same thing as saying that banks cannot issue loans funded by demand deposits.
6. A 100% reserve requirement on demand deposits is to be advocated for since (i) the risk of a so-called bank run can be avoided, and the stability of the banking system would be improved and (ii) the *money multiplier* effect would be reduced significantly because the granting of loans out of demand deposits would not be possible. However, a 100% reserve requirement on demand deposits would not completely eliminate the money multiplier effect, since banks could still lend out money from time deposits or inflows from issuing bonds, obligations, or other debt securities.
7. Independently of how money creation is understood, and even if the theory of money creation out of nothing was deemed to be true, the problems which will be the subject of the rest of this book **do not depend on how money is created,** but rather on the unnatural design of money that tries to serve two functions at the same time: to act as a medium of exchange and also a store of value.

References

Anderegg, R. (2007). *Grundzüge der Geldtheorie und Geldpolitik*. Oldenbourg.

Aristotle. (1995). Politik 1st Book. In F. Meiner (Ed.), *Aristoteles Philosophische Schriften in sechs Bänden. Band 4*. Felix Meiner Verlag.

Bernanke, B. S. (1993). Credit in the macroeconomy. *Federal Reserve Bank of St. Louis, Quarterly Review, 18*, 50–70.

Bundesbank. (2017). *Monthly report April 2017*. Bundesbank.

Bundesbank. (2019). *Geld und Geldpolitik*. Bundebank.

Creutz, H. (2012). Vollgeld – Ein Beitrag zur Lösung unserer Geldprobleme oder zur weiteren Verwirrung? – Versuch einer Klärung. *Humane Wirtschaft, 6*, 28–35.

Creutz, H. (2013). Die Fehlvorstellungen der Vollgeld-Initiative. *Sozialimpulse, 1*, 26–32.

Demeulemeester, S. (2018). The 100% money proposal and its implications for banking: the Currie–Fisher approach versus the Chicago Plan approach. *European Journal of the History of Economic Thought, 25*(2), 357–387.

Fisher, I. (1936a). *100% money* (2nd ed.). Adelphi.

Fisher, I. (1936b, Spring). 100% money and the public debt. *Economic Forum*, pp. 406–420.

Friedman, M. (1960). *A program for monetary stability*. Fordham University Press.

Friedman, M. (1971). The euro-dollar market – Some first principles. *Review, 53*, 16–24.

Fuders, F. (2016). Smarter money for smarter cities: How regional currencies can help to promote a decentralised and sustainable regional development. In E. Dick, K. Gaesing, D. Inkoom, & T. Kausel (Eds.), *Decentralisation and regional development – Experiences and lessons from four continents over three decades* (pp. 155–185). Springer.

Fuders, F. (2017). Neues Geld für eine neue Ökonomie – Die Reform des Geldwesens als Voraussetzung für eine Marktwirtschaft, die den Menschen dient. In G. Krämer (Ed.), *Finanzwirtschaft in ethischer Verantwortung – Erfolgskonzepte für Social Banking und Social Finance* (pp. 121–183). Springer-Gabler.

Fuders, F. (2020). *The 'tragedy of the commons' and the role of the money interest rate*. Springer.

Fuders, F. (2021). The role of money for a healthy economy. In *Urban health and wellbeing programme – Policy briefs: Volume 2* (pp. 17–22). Springer Nature.

Fuders, F., & Max-Neef, M. (2014a). Local money as solution to a capitalistic global financial crisis. In M. Pirson, U. Steinvorth, C. Largacha-Martinez, & C. Dierksmeier (Eds.), *From capitalistic to humanistic business*. Palgrave Macmillan.

Fuders, F., & Max-Neef, M. (2014b). Dinero, deuda y crisis financieras. Propuestas teórico-prácticas en pos de la sostenibilidad del sistema financiero internacional. In J. Fernández (Ed.), *Economía Internacional. Claves teórica-prácticas sobre la inserción de Latinoamérica en el mundo*. LATIn.

Fuders, F., Mondaca, C., & Azungah Haruna, M. (2013). The central bank's dilemma, the inflation-deflation paradox and a new interpretation of the Kondratieff waves. *Economía (U. de los Andes), 36*, 33–66.

Fullarton, J. (1845). *On the regulation of currencies* (2nd ed.). John Murray.

Gross, M., & Siebenbrunner, C. (2019). *Money creation in fiat and digital currency systems* (Working Paper No. 285). IMF.

Gurley, J. G., & Shaw, E. S. (1976). *Money in a theory of finance* (8th ed.). Brookings Inst.

Huber, J. (1998). *Vollgeld – Beschäftigung, Grundsicherung und weniger Staatsquote durch eine modernisierte Geldordnung*. Duncker & Humblot.

Huber, J., & Robertson, J. (2008). *Joseph Huber, James Robertson: Geldschöpfung in öffentlicher Hand – Weg zu einer gerechten Geldordnung im Informationszeitalter*. Gauke Verlag für Sozialökonomie.

International Monetary Fund (IMF). (2000). *Manual de estadísticas monetarias y financieras*. IMF.

Jakab, Z., & Kumhof, M. (2015). *Banks are not intermediaries of loanable funds—And why this matters*. Bank of England.

Keynes, J. M. (1924). *A tract on monetary reform*. Macmillan.

Keynes, J. M. (1930). *A treatise on money – Vol. 1. The pure theory of money*. Cambridge University Press.

Keynes, J. (1936). *General theory of employment, interest and money*. Hartcourt.

Krugman, P., & Wells, R. (2018). *Macroeconomics* (5th ed.). Macmillan Learning.

Krugman, P., Olney, M., & Wells, R. (2008). *Fundamentos de economía*. Reverté.

Krugman, P. R., Obstfeld, M., & Melitz, M. J. (2015). *International economics – Theory and policy* (10th ed.). Pearson.

Larroulet, C., & Mochón, F. (2003). *Economía*. McGraw-Hill.

Mankiw, G., & Taylor, M. (2014). *Economics*. Cengage Learning.

McLeay, M., Amar, R., & Ryland, T. (2014). Money creation in the modern economy. *Quarterly Bulletin, Q1*, 14–27.

Minsky, H. P. (1994). *Financial instability and the decline (?) of banking: Future policy implications* (Working Paper No. 127). The Jerome Levy Research Institute of Bard College.
Mises, L. (1912). *Theorie des Geldes und der Umlaufmittel*. Duncker & Humblot.
Pfeiffer, W. (1997). *Etymologisches Wörterbuch des Deutschen*. Dtv.
Phillips, C. A. (1931). *Bank credit: A study of the principles and factors underlying advances made by banks to borrowers*. Macmillan.
Phillips, R. J. (1992). *The 'Chicago plan' and New deal banking reform* (Working Paper No. 76). The Jerome Levy Economics Institute of Bard College.
Samuelson, P., & Nordhaus, W. (2009). *Economics* (19th ed.). McGraw-Hill.
Soddy, F. (1934). *The role of money – What it should be contrasted with what it has become*. Routledge.
Steiner, R. (1979). *Die soziale Grundforderung unserer Zeit – In geänderter Zeitlage, Zwölf Vorträge, gehalten in Dornach und Bern v. 29. 11. bis 21.12.1918*. Rudolf-Steiner-Verlag.
Stiglitz, J. E. (1998). *Macroeconomía* (2nd ed.). Ariel.
Tobin, J. (1963). Commercial banks as creators of "money". *Cowles Foundation for Reserch in Economics, 159*, 1–16.
Tobin, J. (1969). A general equilibrium approach to monetary theory. *Journal of Money, Credit and Banking, 1*(1), 15–29.
Tobin, J. (1985). Financial innovation and deregulation in perspective. In Bank of Japan (Ed.), *Bank of Japan monetary and economic studies, III* (pp. 19–29). Tokyo.
Werner, R. A. (2014). Can banks individually create money out of nothing?—The theories and the empirical evidence. *International Review of Financial Analysis, 36*, 1–19.
Wicksell, K. (1907). *The influence of the rate of interest on prices* (Vol. 17, pp. 213–220). Economic Journal.
Wirtschaftslexikon24. (2021). *Wirtschaftslexikon24*. Retrieved 2 February 2021, from http://www.wirtschaftslexikon24.com/d/bilanzverl%C3%A4ngerung/bilanzverl%C3%A4ngerung.htm

Chapter 3
Money Is Like the 'Blood' of the Economy

3.1 Introduction

Neither the fact that money is the counterpart of debt, nor the fact that the money supply increases via the granting of loans, is a problem in itself. The problem is the unnatural design of our money. Unlike real goods, money is easy to store and therefore does not circulate easily throughout the economy. However, circulating is money's most important function. Money can be described as the 'blood' of our economy. Blood is the medium for delivering nutrients and oxygen to the cells. It is the medium of exchange of our body. Blood needs to circulate, otherwise the body gets sick. Like the circulation of blood in the human body the economy also gets sick if money does not circulate well. Money is supposed to be a medium to facilitate the exchange of goods and services (already Aristotle, [384 BC] 1995; Steiner, 1979; Mankiw & Taylor, 2014). But, because people tend to hold what Keynes (1936: 165 ff., 194 ff.) referred to as the "*liquidity-preference*" we like to accumulate money rather than allowing it to circulate.

However, in nature it is only possible to hoard goods in a very limited way since real goods perish. Any excessive hoarding would, in time, result in the loss of the hoarded goods. But our money, as it is designed today, makes it possible to hoard almost without limit. In fact, the Keynesian notion of the *liquidity-preference* can be regarded as **just a symptom** of the fact that money is easily storable. The function ascribed to money as a store of value is due to the unnatural design of our money whereby unlimited hoarding is made possible. This in turn provides a strong incentive to **produce more than is actually needed**.

Before we examine further the devastating effects of this contradiction between the essential function of money as the blood of the economy and the function of store of value, it is of the utmost importance to understand that money does not have any intrinsic value. This understanding is also crucial for an analysis of the moral-ethical problems inherent in this system addressed later in Chap. 9.

© The Author(s), under exclusive license to Springer Nature Switzerland AG 2023
F. Fuders, *How to Fulfil the UN Sustainability Goals*,
https://doi.org/10.1007/978-3-031-37768-6_3

3.2 Virtual Wealth: Why Money Has No Intrinsic Value

During the last two centuries, much thought has been given to the value of money and how it can be measured. Many theories have been offered on the subject. It would require a book of its own to explore this issue in detail, if one were to cite all that has been written on this matter, and in the end, we would probably be none the wiser for it. A good starting point for exploring the question of whether or not money has value is to ask whether a check has value. Because that's what money originally was: a check from a bank promising to deliver a certain amount of gold or silver coins. The gold might have intrinsic value[1] but the check does not. It is only a promise to hand over a certain amount of gold silver.

Let us explore this question a little bit more and take a quick look at the history of money. Initially banks were institutions where gold was kept. An example of such early banks were goldsmiths who used their secure storage facilities to provide custodial services for valuable items such as gold (Samuelson & Nordhaus, 2010; Bundesbank, 2019). Over time, some goldsmiths came to specialize in this custodial service; thus "banks" were born. Later, instead of people having to withdraw gold every time they needed it to settle a transaction, the practice developed of paying via a check drawn on the bank. This was certainly less cumbersome and above all safer than transacting in physical gold. The next step in the evolution of money was that people started to endorse the checks received instead of redeeming them for physical gold and depositing it in their own bank. Later, to make transactions even easier and to eliminate the need for endorsement, checks payable to the bearer were invented, and people began to trade and store the 'checks' that today are called banknotes.[2] A banknote did not have any value in itself but was only a promise to deliver an item (usually gold or silver) with actual value.

Today, the gold standard no longer exists—i.e., no central bank in the world promises to deliver gold against banknotes. So then, what promise is inherent in a modern banknote? The answer is: **none whatsoever**. In other words, even if we still

---

[1] In fact, one might even ask if it is accurate to state that gold has intrinsic value, since what is true for paper money is also true for gold: we cannot eat it, live in it, nor does it serve any other use, except maybe for certain limited industrial uses or for the manufacture of jewelry. The value we ascribe to gold is mostly psychological and depends on the amount of fiat (paper) money circulating, i.e., the reason why gold is so expensive is not because of its intrinsic value but because there are ever more units of fiat money in circulation, against which the value of gold is measured. That gold has no intrinsic value was discussed by Aristotle who used the fable of King Midas—a monarch with a special gift which caused everything he touched to turn to gold, but then ended up starving because his food also turned to gold—to illustrate this point.

[2] Later this practice of issuance bank notes by private banks was prohibited and this power was granted exclusively to the central bank (cf. US Bank Charter Act of 1844 forbade banks to "make any engagement for the payment of money payable to bearer on demand"). Note that a 'check' nowadays is something different. It is not a claim against gold or silver but against banknotes. And the latter (the banknote) today is not a claim against gold or silver since a gold standard no longer exists.

used the gold standard, a banknote would still have no intrinsic value, but would only **represent a value**.

3.3 Money Is the Measure of Value

3.3.1 The Exchange Value and the Monopolistic Character of Money

The question arises, why, if money has no intrinsic value, is it considered a means of storing value? Alongside facilitating the exchange of goods and services, this is considered money's main function. The answer is, it is due to the "monopolistic" character (Gesell, 1949: 237) that the law assigns to money as the **only means of payment**. If we want to acquire a certain good or service or to pay taxes, we need the official currency. However, neither the law nor any central bank states how much of a given good or service we should receive for one monetary unit. Why, for example, can we presently buy one liter of fresh milk for one euro, while some time ago we were able to buy two or three liters for the same amount of money? Who determines this?

The answer is that the **exchange value** of a banknote depends simply on the amount of real goods or services that we, the consumers, choose to accept in exchange for a banknote. Any banknote has the (exchange) value that consumers attach to it. Here we find the reason why "in God we trust" is inscribed on every US Dollar bill. The exchange value corresponding to a monetary unit depends on the quantity of products and services that can be acquired in exchange for that money, which obviously bears a relationship to the number of monetary units in circulation. In an economic void where no products or services exist, all money would be worthless. Money only represents the value of real goods and services for which it can be exchanged. **It is the real goods and services that give money its "value"**. This is a very important point to understand especially when considering the moral-ethical dimension, see Chap. 9.

Therefore, it could be said that each monetary unit represents a theoretical (but not exactly specified) fraction of the Gross National Product (GDP), i.e., of all goods and services produced in an economy. In other words: money does not have value but is the **measure of value.** Products and services have (real intrinsic) value, that we measure in monetary units. To put it in the words of Keynes:

> Money is the measure of value, but to regard it as having value itself is a relic of the view that the value of money is regulated by the value of the substance of which it is made, and is like confusing a theatre ticket with the performance. (Keynes, 1914: 418)

As said above, one of the three functions of money is that of an accounting unit. Money as an accounting unit helps us to express (to measure) the value of real goods and services. This capability is necessary in order for money to perform its main function, i.e., facilitating the exchange of goods and services. To exchange goods and services directly (barter) without using money as measure of value would be quite cumbersome.

3.3.2 The Role of Markets

We just established that it is we, the consumers, who determine the value of real goods and services in terms of monetary units. In other words, we define the exchange value of money. But basic economics classes teach that markets determine prices. Any economics textbook explains that the interaction of supply and demand determines the price of any good or service (Fig. 3.1). It is true that markets determine prices for goods and services. But this is essentially equivalent to saying that it is consumers and producers who determine prices, since a market is the location where consumers and producers meet to negotiate price. The demand curve depends not only on the consumer's personal preferences, as many microeconomics textbooks would suggest, but above all on how much money they possess. After all we pay with monetary units and not with, e.g., eggs. An increase in the consumer's rent, for example, will shift the consumer's demand curve to the right. An increase in the economy's total money supply will shift the economy's aggregate demand curve to the right and the general price level will rise. We call this phenomenon inflation. To say that "markets determine price" is essentially equivalent to saying that it is we, the people (the consumers in negotiations with the vendors), who determine price.

Fig. 3.1 Supply and demand. (Source: Own work)

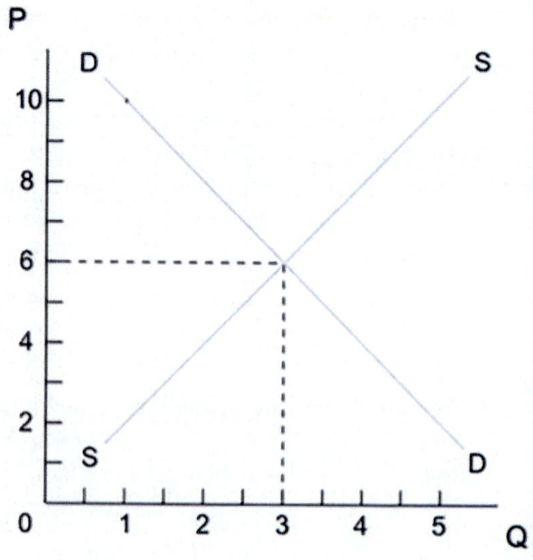

3.3.3 Money Does Not Have Value but Represents Value

Unfortunately, even Silvio Gesell (1949), whose proposal will be presented as the solution to the problems inherent in our monetary system, tried to define the value of money. However, all discussion of the "value" of money are nonsense. Money does not have value but represents value. This would be true even if we used money made of a material with intrinsic value, e.g., gold. Such gold-standard money would nevertheless be only a representation of the value of the goods and services that are measured by it. So how could we possibly measure the "value" of this money? Today the value of gold is measured in terms of fiat currencies, e.g., US dollars, euros, Chilean pesos, etc. But if gold is the unit we use to express prices (to measure the value of goods), how can we express the value of the gold contained in gold currency? I will come back to this tricky case of circular reasoning at the end of this book, where I will explain why gold money is no cure for the problems inherent in our monetary system.

3.4 The *Money Illusion* and the Importance of Perception

As we just discussed, it is the people (consumers in negotiations with vendors) who define how much milk we can buy for one monetary unit. We, the people assign exchange value to each monetary unit. Of course, indirectly this exchange value depends on how many units of a particular good exist in relation to all monetary units. But do consumers know how many monetary units exit? The answer is: they do not. Therefore, the price of a good (depending on the exchange value we assign to each monetary unit) depends on **our perception** of the relation of real values (real goods and services) and monetary units in circulation. If we lose confidence in the monetary system, money becomes worthless instantly.

We can consider ourselves lucky that most people are not aware of the fact that the money supply grows at a much faster pace than the production of real goods and services (I will come back this point later). This is what Irving Fisher (1928) once referred to as the "*money illusion*". People are not aware of how many monetary units exist. If they were, they would probably lose confidence in the monetary system. Unfortunately, although Irving Fisher introduced this interesting concept to the world of economics, he was not able to convincingly explain why the money supply constantly increases at a faster pace than the production of goods and services and where the *money illusion* really originates from. Chapter 4 will offer an explanation. We will see that this phenomenon is related to the fact that our economy is obliged to constantly grow.

Since, as will be shown in Chap. 4 the dynamics of compound interest require the money supply to grow steadily and, above all, exponentially, there must come a point in time when the money supply grows faster than the productivity of the economy. Money supply growth **decouples** from the growth of the real (productive) economy. Deposits grow faster and faster without these deposits being backed by a corresponding growth in economic output. However, people are still used to the idea that one unit of money can buy a certain quantity of products. Therefore, they continue to believe the money in their savings accounts still has the same purchasing power. However, this is becoming more and more "*virtual wealth*" (Soddy, 1933) for in fact, only what is produced can be bought. All monetary assets taken together can only be worth as much as the products that can be bought with it. But if the money supply grows faster than the quantity of goods, each monetary unit must lose purchasing power. Consumers usually are not immediately aware of this. This is also due to the fact that the growing cash balances are generally not spent for consumption, but are largely invested, e.g., in the stock market or in real estate, which is why high inflation rates (as measured by price changes in consumer goods markets) do not occur immediately. I will come back to this *inflation-deflation paradox* in Sect. 4.4.5. If people were to use their growing credit balances exclusively to purchase consumer goods and food, the nominal increase in demand for consumer goods and food caused by the steadily growing money supply would lead to substantial price increases in these markets. Households would then perceive high inflation rates.

3.5 The Problem Is the Unnatural Design of Our Money

The fact that money has no intrinsic value is not a problem in itself, in the same way as it is no problem, as already outlined above, that money is the counterpart of debt, and that money supply grows through the granting of loans. We could use paper money that perfectly serves as medium of exchange without that it would bring along any side effects whatsoever. The problem is the unnatural design of our money, especially its unnatural storability. While everything in nature is subject to the rhythmic change of growth and decay, only money seems to escape the transience of all earthly things (Gesell, 1949). Money makes it possible to store surplus wealth almost without limitation. Saving for the future is a natural behavior. Animals display this same behavior (e.g., hamsters store food in their cheeks, bears in their belly). However, in nature it is only possible to hoard goods to a very limited degree since real goods perish. Food spoils rapidly, technology loses value due to obsolescence, and services are not possible to hoard at all. Any excessive hoarding would, in time, result in the loss of the hoarded goods. This means that **nature keeps our natural instinct to hoard in check**. But our money, as it is designed today, makes it possible to hoard any surplus without limit. Our instinctual impulse to hoard is perverted by the fact that our money does not perish and therefore has a special, monopolistic-like position compared to real perishable goods (Gesell, 1949), which creates a strong incentive to **produce more than is actually needed**.

References

Aristotle. (1995). Politik 1st Book. In: *Aristoteles Philosophische Schriften in sechs Bänden. Band 4.* Felix Meiner Verlag.

Bundesbank. (2019). *Geld und Geldpolitik.* Bundebank.

Fisher, I. (1928). *The money illusion.* Adelphi.

Gesell, S. (1949). *Die natürliche Wirtschaftsordnung durch Freiland und Freigeld.* Rudolf Zitzmann.

Keynes, J. M. (1914). Review of Theorie des Geldes und der Umlaufsmittel' by Ludwig von Mises and 'geld und Kapital' by Friedrich Bendixen. *Economic Journal, 24*(95), 417–419.

Keynes, J. M. (1936). *General theory of employment, interest and money.* Hartcourt.

Mankiw, G., & Taylor, M. (2014). *Economics.* Cengage Learning.

Samuelson, P., & Nordhaus, W. (2010). *Economía con aplicaciones a Latinoamérica.* 19 Hrsg. Mc Graw-Hill.

Soddy, F. (1933, 1961). *Wealth, virtual wealth and debt—The solution of the economic paradox.* Briton.

Steiner, R. (1979). *Die soziale Grundforderung unserer Zeit—In geänderter Zeitlage, Zwölf Vorträge, gehalten in Dornach und Bern v. 29. 11. bis 21.12.1918.* Dornach: Rudolf-Steiner-Verlag.

Part II
Money and Unsustainability in *Stricto Sensu*

Chapter 4
The Growth Imperative Inherent in Our Financial System

Money, as a physical medium of exchange, made a diversified civilization possible. . .
And yet it is money, in its mechanical more than in its spiritual effects,
which may well, having brought us to the present level, actually destroy society.

Sir Josiah Stamp (quoted from Fisher, 1932: VI).

4.1 The Unnatural Storability of Money as Origin of Interest

The unnatural design of our money does not only give a powerful incentive to produce more than is needed for immediate consumption; it even **obliges us to do so** and here, again, the root cause is the unnatural storability of money (for similar argumentation see Kennedy, 1990, 2011; Creutz, 1993; Costanza, 2012; Löhr, 2012; Lietaer et al., 2013; Farley, 2013; Fuders, 2009, 2016a, b, 2020a, b, 2021). Money stored under the mattress does not circulate and, therefore, cannot function as a medium to facilitate the exchange of goods and services. If everyone hoards money at home, demand for goods and services will decrease, prices will decline creating an incentive to postpone consumption since tomorrow products and services will be even cheaper. This downward spiral is known as deflation, and it is the worst possible economic scenario. Even healthy companies are forced into bankruptcy or to lay off employees and engage in other cost cutting measures, causing increased anxiety amongst households and business owners, increasing the Keynesian "*liquidity-preference*" (Keynes, 1936: 168), i.e., making people hoard even more, driving the economy into an even deeper depression.

The reason why we do not hoard our money under the mattress but bring it to a bank which then lends it out and causes it to circulate in the economy is the interest rate. Interest is a "*reward for parting with liquidity*" (Keynes, 1936: 167), creating an incentive to lend money out (or to bring it to the bank which then lends it out for us).

© The Author(s), under exclusive license to Springer Nature Switzerland AG 2023
F. Fuders, *How to Fulfil the UN Sustainability Goals*,
https://doi.org/10.1007/978-3-031-37768-6_4

To express it in the more extreme language of Silvio Gesell, it is the unnatural storability of money that puts the owner of money in a monopolistic-like position in which he can either choke off economic activity or extort those who are in need of money and "press" interest as a form of ransom (Gesell, 1949: 205, 344). If money could not be stored easily and (almost) without risk, holders of money would be unable to extract interest from borrowers. Seen in this light, interest results from a **contradiction** between money's two main functions: it is supposed to facilitate the exchange of goods and services while and at the same time serving as a medium to store value.

To return to the previous metaphor, if money is the 'blood of the economy', then interest is the **drug that makes the blood circulate**. The drug is necessary since money has a design flaw and does not circulate well. And as with any drug that is used for a prolonged period, interest causes significant negative side effects for the health of our economies. Especially the obligation to grow economically and a constantly growing income inequality are caused by it. But before we can understand this dynamic, we need to explore the origin and the role of interest.

4.2 Demystifying the Role of Interest

4.2.1 Interest Is Not Caused Be the Keynesian Liquidity-Preference

The origin of interest is not the preference for liquidity as Keynes believed (Keynes, 1936), but the fact that money—compared to real goods and services—is easy to store. If money was as perishable as real goods are, we would not prefer to hold money over goods. We would then not have a psychological penchant to save as much of our surpluses as we can. In other words, the Keynesian *liquidity-preference* can be understood as just a **symptom of the unnatural storability of money**. This point is crucial to understand, since the solution that will be presented in Chap. 10 will be precisely based on the introduction of money that is not indefinitely storable.

4.2.2 Interest Is Not Natural

Contrary to a widespread belief interest is not natural. In fact, going back as far as Aristotle ([384 BC] 1995) it was recognized that interest is a completely unnatural human construct that contradicts the laws of nature, especially physics (Soddy, 1934). As discussed above, interest can be regarded as symptom or side effect of the unnatural storability of money. It is the unnatural storability of money that makes us use money for purposes of hoarding and that causes us to demand more money than we need for immediate consumption. If money was not unnaturally easy to hoard, we would not have a preference for liquidity. Instead, we would spend it as fast as we could. No one would be willing to pay interest as the price of borrowing money if money didn't have this special unnatural "non-rusting" (Gesell, 1949: 238) character compared to real goods. In other words, interest as a price for borrowing money only exists because money is unnaturally storable, i.e., interest is the 'natural' consequence of money that is unnaturally storable.

4.2.3 Interest Is Not the Price of Money

In economics, interest is often referred to as the price of money (e.g., Wienert, 2001; Alegre Saz et al., 2010). Since, as has been pointed out, money is not a good but a voucher representing goods, it makes little sense to speak of interest as the "price" of money, as Keynes (1936: 167) correctly understood. Goods have prices which are expressed in monetary units. Money as a unit of account in which we express prices **cannot itself also have a price**. Interest therefore should be understood as the price of **renting** money. But does it make sense to pay rent on something that has no intrinsic value? Steiner called this a "complete nonsense" (Steiner, 1979: 50). The reason why we are willing to rent money is because we need money while at the same time it is easily storable. We need money because of the "monopolistic" position (Gesell, 1949: 237) that is bestowed on money by law when it is designated as the only legal medium to facilitate the exchange of goods and to pay taxes. Both the monopolistic position of money (compared to real goods) together with its unnatural storability put the money holder in a position to be able to "press" (Gesell, 1949: 205, 344) interest, or to "exploit the scarcity-value of capital" to put it in the words of Keynes (1936: 376). This understanding is crucial for the discussion of the moral-ethical aspects of our monetary system in Chap. 9.

4.2.4 Interest Is Not a Premium for Abstaining from Consumption

Renting real assets such as cars or machines can be considered a **service**. The foregone use of the equipment and its maintenance can justify the rental income. A similar argument is often made when money is lent. The lender forgoes consumption and thus interest is a compensation for abstaining from consumption (e.g., already Carver, 1883; Süchting, 1995; Sell et al., 2020). However, the comparison with real goods does not hold up because money is not a real good or service with intrinsic value but rather a voucher that by itself does not provide any direct benefit. A voucher cannot generate any income or benefit; therefore, the owner does not forego anything by refraining from using it. Only if the voucher is first exchanged for a real goods, e.g., a car, and then those goods are lent out rather than being used by their owner, it is accurate to speak of abstaining from consumption. Persons who lend their money **neither abstain from, nor renounce, nor sacrifice consumption** when they lend it, because they as of yet do not hold anything with real intrinsic value whose use they can abstain from, renounce or sacrifice (Fuders, 2009, 2010b, 2017). Instead of a premium for abstaining from consumption, interest is a liquidity or non-hoarding premium as Keynes (1936: 167) correctly understood. Recognizing that interest is neither a premium for abstaining from consumption, nor a risk premium (see next point) is crucial to understanding the moral-ethical problems that results from interest.

One might object that money is not perfectly storable because of inflation, and therefore we need interest to compensate for inflation. However, as will be shown in Sect. 4.4, this reasoning is backwards since inflation is only a side effect of interest.

4.2.5 *Interest Is in Most Cases Not a Risk Premium*

In monetary theory we find another justification for the existence of interest. Interest is considered a reward for assuming risk, a risk premium (e.g., Bundesbank, 2019). However, usually it is the borrower who bears the risk inherent in a loan, while the lender secures loans with some form of collateral. The bank (or any other creditor) does not care if the project financed with the money does not perform as expected. The difference between a loan and an equity investment lies precisely in the fact that the loan requires the payment of interest **irrespective of the success of the business** financed by it. Accordingly, the difference between the interest rate and the profit that an equity investment yields is that the lender unlike the equity investor receives interest regardless of the success of the business that is financed by the loan.

To define interest as a risk premium is accurate in only two cases: (i) government bonds and (ii) venture or risk "capital". Governments can usually borrow without offering any security whatsoever, and in case of default the creditors lose their investment. Risk capital refers to those loans which cannot be secured (or not completely) via mortgages or other guarantees. Usually, creditors require high rates of interest to compensate for the risk in such situations. But in such a situation the interest rate is **only partly** to be considered a risk premium, namely in insofar it exceeds the "reward for parting with liquidity" (Keynes, 1936: 167) stemming from the possibility to "exploit the scarcity value" of money (Keynes, 1936: 376). We could therefore define the interest rate of venture capital r_v as composed of the interest rate as reward for parting with liquidity r_l plus interest rate as risk premium r_r (see Eq. 4.1).

$$r_v = r_l + r_r \tag{4.1}$$

4.2.6 *The Interest Margins of Banks Are Not the Main Problem*

Some readers might think that the difference between deposit rates and lending rates are the biggest problem of our monetary system. But this is not correct. The difference between deposit and lending rates may be quite high (especially in developing countries due to poorly developed banking sectors) thus allowing banks to make huge profits. But this is only a secondary consequence of the lack of competition in the banking sector. Even if banks were purely altruistic and the deposit rate were equal to the lending rate, the problems that result from the monetary system discussed in this book would still arise.

4.3 Interest Makes the Money Supply and Debt Grow Exponentially

Interest makes deposits and thus the money supply (cash plus deposits) grow. But the money supply and debt do not grow in a linear fashion, but rather according to an exponentially function. This is because any amount deposited in an interest-bearing bank account will double after a certain period. Any amount that doubles at regular intervals grows according to an exponential function: slowly at first, but then faster and faster (see Fig. 4.1). Note that even at low interest rates, compound interest grows exponentially, because eventually the original amount will have doubled. In the case of a low interest rate, the flat part of the exponential curve (black curve in Fig. 4.1) is longer, but at some point, the curve will steepen and eventually become almost vertical.

In fact, banks use this feature of interest as a selling point to attract customers. For example, the European Investment Fund Deka uses a metaphor to explain exponential growth by comparing the growth of compound interest with chickens laying eggs (Deka, 2014). Yet, such a metaphor is inappropriate, since nothing in nature

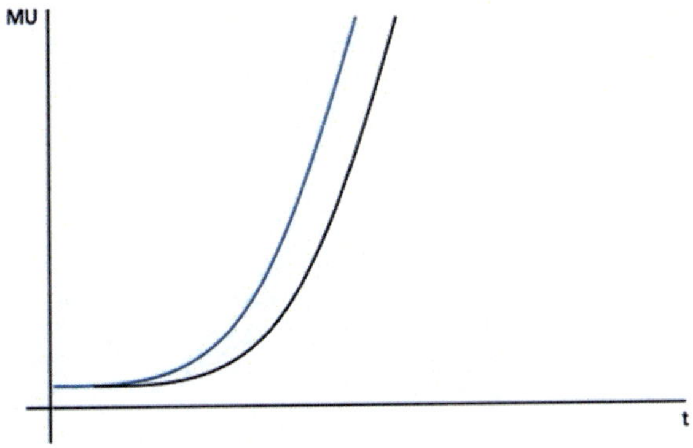

Fig. 4.1 Exponential development of monetary units (MU) on bank accounts. Blue curve: higher interest, black curve: lower interest. (Source: own work)

grows exponentially forever. And because nothing in nature grows endlessly—and certainly not exponentially—the lack of constraints on the growth of financial assets is fundamentally different from anything that exists in nature. A good way to visualize this effect is the so-called "Joseph's penny". If Joseph had bequeathed just one cent to Jesus, and if it had been invested at 5% interest, in the year 2000 the investment would have grown to a ball of gold 589 billion times larger than earth (Kennedy, 2006).[1] Even the most powerful computers in the world would at some point be unable to calculate the amount of money that is accumulated in this way, due to a lack of memory. This is the logic of an exponential function (Kennedy, 1990, 2011; Creutz, 1993; Costanza, 2012; Lietaer et al., 2013). Yet this is the logic that underlies our current unnatural (Soddy, 1934) monetary system, a fact that was remarked upon as far back as Aristotle ([384 BC] 1995).

Some readers might make the objection that exponential growth of deposits only occurs if interest is reinvested, and that compounding and exponential growth do not take effect if the proceeds are spent instead. However, even if an individual depositor chooses to withdraw the proceeds of interest rather than allowing them to compound, the money will in all likelihood end up in someone else's bank account and continue earning interest there. This means, even though this exponential function might not take effect when viewed from the perspective of one specific bank account, all bank accounts in an economy taken together will show exponential growth over time. Hence, for the **economy as a whole** the money supply (defined as cash plus bank deposits) will always grow exponentially as long as interest rates are positive (Kennedy, 1990, 2006, 2011; Farley, 2013; Costanza, 2012).

Since as outlined above (Chap. 2) money is the counterpart of debt, the **money supply and total debt** of an economy must grow in this manner. That deposits and debt grow in tandem can also be understood by keeping in mind the simple fact that **there is no interest without debt**. To recall once again the words of Frederick Soddy (1934: 25) "Money is a credit-debt relation from which none can effectually escape". Figure 4.2 depicts the money supply and total public debt for the US economy. We can recognize that not only do the money supply and debt grow in tandem but also that they grow according to an exponential function. The same exponential growth of money supply and debt can be observed in every country on earth over a long period of time.[2]

[1] The calculation presumes a fixed gold price. It is worth mentioning that Jesus most probably would not have deposited money in a bank account that yields interest since the Bible, as well as the Tanakh and the Koran prohibit charging interest (Tanakh: Deuteronomy 23:19; 23:20–21; 24:10; Exodus 22:25; Leviticus 25:35–37; Ezekiel 18:8; 18:13; 22:12; Bible: Deuteronomy 23:19; 23:20–21; 24:10; Exodus 22:25; Leviticus 25:35–37; Ezekiel 18:8; 18:13; 22:12; Psalm 15:5; 112:5; Proverbs 28:8; Luke 6:30; 6:35; 1 Timothy 6:9. Koran: 2.275).

[2] Even a Wikipedia search for the term "Money Supply" turns up charts of the exponential growth of the money supply in numerous countries (Wikipedia, 2022).

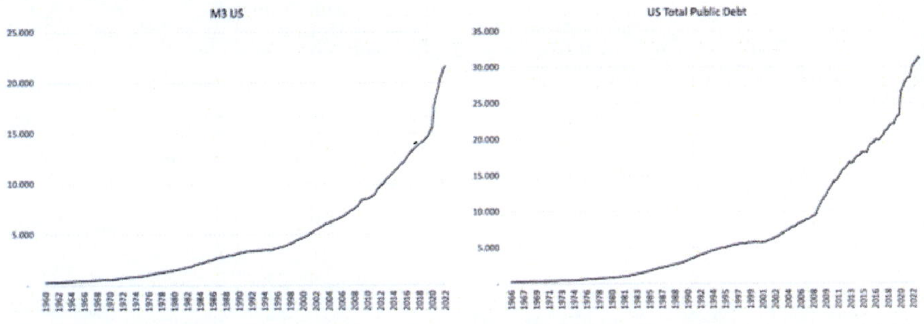

Fig. 4.2 US money supply and public debt. Left: US money supply M3 (in billions of USD). Right: US total public debt (in billions of USD). (Source: own work. Data provided by OECD (2023), and US Department of Treasury). Retrieved from Federal Reserve Bank of St. Louis (https://fred.stlouisfed.org)

4.4 Inflation, Speculation, Crisis and Deflation

Before discussing the growth imperative and the problems this creates in terms of sustainability, it is useful to examine some related phenomena: inflation, deflation, speculation and financial crisis. We will see that all of these problems can be traced back to one common cause, i.e., the non-neutrality of money and its consequence, interest. The reason why this problem remains unresolved after centuries of economic history is because the nature of the problem is still not well understood among professional economists. I am convinced that these problems can be solved, or at least significantly mitigated, by replacing our existing form of money with one that is more in keeping with the laws of nature.

4.4.1 Inflation

Inflation along with unemployment, is one of the primary concerns of politicians all around the world, and economics textbooks devote significant discussion to monetary policy measures designed to avoid both scenarios. Inflation is a general and sustained increase in prices of goods and services throughout an economy (Mankiw & Taylor, 2014; Cuadrado, et al., 2015). One can intuitively grasp that an exponentially growing money supply must generate inflation, since total production of an economy cannot, in the long run, keep pace with the exponential growth of the money supply. Even as far back as Aristotle it was understood that infinite growth of production is not possible and that if the money supply grows faster than the productive output of the economy, the exchange value of each monetary unit must decrease. We can only buy what is produced. If the amount of money in circulation increases but the quantity of goods and services remains unchanged (or does not grow at the same pace), each monetary unit must decline in terms of purchasing power. This dynamic was described as far back as Jean Bodin and David Hume and was formalized into a more modern theory by Irving Fisher (1912). Next to Irving Fisher's *quantity theory*—what Milton Friedman referred to as *monetarism*—there are two other major theories of inflation: demand inflation and cost inflation. Even though the respective authors of these theories started off by criticizing Fisher's quantity theory, claiming they could explain inflation independently of the money supply, both theories essentially say the same thing as Fisher in different words.

4.4.1.1 Quantity Theory

Irving Fisher (2012: 24 ff.) gave us his famous quantity Eq. (4.2), where M represents the money supply, V the velocity of money circulation, i.e., how fast one monetary unit passes from one person to another, and T the volume of economic transactions representing the productive output of an economy and P the price level.

$$MV = PT \tag{4.2}$$

Nowadays T in Fisher's equation often is replaced by the Y, the letter that is typically used to represent gross domestic product (GDP). We can rewrite the equation as:

$$P = \frac{MV}{Y} \tag{4.3}$$

Fisher's equation explains what we all know intuitively and what has been said above, if in relation to the quantity of products and services (Y) that exist there are more monetary units circulating, and/or the velocity of circulation increases, each monetary unit is worth less. This is the reason why the growth of the money supply should not decouple from the growth of the real productive economy. As long as the money supply and total production move in the same direction and in proportion to one another and the velocity stays the same (people do not hoard more or less money under the mattress) neither inflation nor deflation will result. However, as has been explained in the previous chapter, the money supply in all modern economies is subject to constant growth caused by interest. For example, in the last 40 years, the supply of US dollars grew 130 times faster than the productive output of the USA. Similar divergence between money supply growth and GDP growth can be seen in any economy. The reason is because our monetary system generates a constant increase in money supply.[3]

[3] Some might argue that interest has been very low in many industrialized countries for the last 15 years. As will be explained in Sect. 5.5 the interest rate actually shows a trend in the very long run to decrease. However, if interest rates are low, there is little incentive to lend money, or to deposit it in the bank which then lends it out. This is what is happening in Europe right now (March 2022). People hoard money literally under the mattress. Cash hoarding has increased exponentially and increased from 0.66 trillion in 2008 to 1.4 trillion in 2021 (Statista, 2021). Money does not circulate well and does not serve the economy in the way it is supposed to. In a desperate attempt to avoid deflation, central banks in industrialized countries started to drastically increase the money supply, i.e., to print money, so-called "Quantitative Easing". Hence, the money supply keeps growing, in this case, not via interest but due to central banks money printing.

4.4.1.2 Demand Inflation Theory

The demand inflation theory is based on the early writings of Keynes (1930). According to this theory inflation is caused by an excess of demand for goods and services and the result is rising prices. The idea is simple: The demand curve shifts to the right (see again Fig. 3.1). If the supply of goods does not increase in the same proportion, i.e., if the supply curve does not also shift to the right, prices will rise according to the law of supply and demand. In other words, inflation is caused by the demand for goods and services growing faster than the supply of goods and services.

But does this theory really give us an explanation that is independent of money supply? Is it possible that aggregate demand (the demand for all goods and services taken together) grows without corresponding growth of the money supply? The answer is: no, it is not possible. The reason is that we do not create economic demand simply by desiring a product or service. In order to create economic demand, we must be willing and able to offer money in exchange for the goods & services we desire. Meanwhile the number of monetary units circulating in an economy does not grow, it is not possible for demand to increase so as to produce a generalized rise in prices. Of course, the demand for individual goods & services may increase without an increase in the total money supply. But in such a scenario the demand for other goods must decrease because any individual unit of money can be spent only once. Inflation, however, means that the aggregate supply of consumer goods becomes more expensive. This is **only possible if the money supply grows** and accordingly people have more money in their wallets. For the same reason population growth or immigration do not necessarily contribute to inflation, since an increasing number of people in an economy does not automatically mean that there are more monetary units.

In fact, Keynes (1930) himself seems to have recognized this and developed the following equation that shows some interesting similarities to Fisher's equation. In this Eq. (4.4), P is the price level, E is the remuneration of production factors (salaries and business earnings), S is savings, I is investment and O is global production.

$$P = \frac{E - S + I}{O} \tag{4.4}$$

According to this equation, the surplus demand is conditioned by a stronger increase in the remuneration of production factors E and/or an increase in investments **relative to global production**. Again, we can ask if it is possible to increase the numerator of the equation without increasing the money supply? And the answer is again: no, it is not possible. A divergence between investment and savings (I – S in Eq. 4.4) cannot exist while the banking sector is working properly since money deposited in a bank will be kept in circulation by the bank (Sect. 7.3 will discuss this in more detail). The numerator can only increase if the money supply increases. If the money supply remains unchanged, and employees gain, then the remuneration of other production factors (and thus the profit of businesses) must decline by a corresponding amount, and *vice versa*. This means, also according to Keynes' formula the money supply must increase, or productivity must decrease (meaning less goods and services are produced) in order for inflation to occur.

4.4.1.3 Cost Inflation Theory

Similar reasoning applies to the so-called cost inflation theory. Cost inflation occurs when production costs increase (e.g., salaries, or energy prices) and in turn cause the prices of goods and services to increase. However, the passing on of increased production costs to consumers is only possible if consumers (or firms in the case of intermediate products) have more monetary units at their disposal, i.e., if there is an increase in nominal demand, which in turn requires an increase in the money supply, as discussed above (similar argumentation in McCulloch, 1975). Therefore, it is correct to state that cost inflation and demand inflation are strictly correlated (Fernández Díaz et al., 2011). Usually, in textbooks we find an Eq. (4.5) to explain inflation via cost inflation theory (Fernández Díaz et al., 2011):

$$\mathrm{P} = \frac{\mathrm{w} + (\pi + \mathrm{r}) q\alpha}{\beta} \tag{4.5}$$

In this equation P is the price level, w is the nominal average salary, π the average profit rate, r the market interest rate, q measures α, the relation between real capital and labor (human workforce) and β is labor's productivity. Very similar to the equation explaining demand inflation, cost-induced inflation is caused by a larger increase in the remuneration of the factors of production relative to labor productivity. As explained above, it is not possible for the average remuneration of production factors to increase without an increase in the money supply. Hence, also according to this theory, a rising price level requires an increase in the money supply relative to the quantity of products on offer.

Note that this equation contains the interest rate r in the numerator, which is correct since interest is a cost. Higher interest rates cause higher inflation. However, according to conventional wisdom the interest rate has the exact opposite relation to inflation: high interest rates would restrain inflation while lowering rates would increase inflation. This is probably why textbooks explain that r in this equation is considered as a constant (Fernández Díaz et al., 2011), otherwise it would contradict the conventional wisdom regarding the effect that the interest rate is supposed to have on the money supply and inflation. Chapter 5 will examine in greater depth the effect that the interest rate has on the money supply. We will see that some strong arguments exist to support the notion that the effect is exactly the opposite of what is conventionally understood and, therefore, there is no contradiction in the cost inflation theory.

4.4.2 In the Current System Stable Money Does Not Exist

So, the three most common inflation theories basically say the same thing, although they appear to be different at first glance. In fact, they describe the cause of inflation from different perspectives. If you look at the same phenomenon from different angles, it will appear different. But ultimately, all three theories state that inflation results from the money supply growing faster than the quantity of goods and services available. However, **none of the theories explains how or why this happens**. Usually, the money supply is regarded as an exogenous variable controlled by the central bank, and if inflation occurs it is because the central bank has performed its job poorly. However, it is in many cases not the fault of the central bank that inflation occurs. The reason why the money supply usually grows faster than the productive output of the economy is because the money supply grows exponentially according to the dynamics of interest. Even though we try to keep pace, the interest rate sets the rhythm, and the growth of the money supply always seems to grow a bit faster than the productive economy. As long as there is a positive interest rate stable money cannot exist (Fuders et al., 2013). We either see inflation or—if the system seizes up and banks go bankrupt on a large scale and cannot perform the essential function of keeping money in circulation—deflation. There is nothing in between. In the current system stable money does not exist.

Only if we introduce money that circulates without interest as a "*reward for parting with liquidity*" (Keynes, 1936: 167) can we achieve truly stable money. For example, in economic boom times, there would be more money circulating due to an increased demand for loans and consequently more money circulating, while an economic downturn would be accompanied by a shrinking demand for loans and less money circulating. We could call this in accordance with Fullarton (1845: 67 ff.) *law of "reflux"*. Probably we would neither see inflation nor deflation, central banks might not even be necessary anymore, at least not to combat inflation. We would finally achieve **monetary neutrality**. However, we might witness deflationary tendency already before the system has broken up. I call this effect the *inflation-deflation paradox* (Fuders, 2011; Fuders et al., 2013; Fuders & Max-Neef, 2014a). Section 4.4.5 will discuss this point in more detail.

4.4.3 Purchasing Power Depends on the Productivity of the National Economy

We have now established that the exchange value (purchasing power) of a currency depends on the relationship between the number of units of the currency in circulation (money supply) and the productive output of the national economy. It should be noted that "national economy" refers to the economy to which the currency relates, e.g., the euro area in case of the euro currency. This clarification is important because it is often argued in discussions on this topic that there are still enough goods to buy abroad, which is why there would be no immediate threat of a loss of purchasing power if the money supply was to grow faster than the productivity of the economy.

However, this objection overlooks the fact that the value of a currency depends exclusively on the productivity of the economy that it belongs to. The productivity of foreign countries is irrelevant because foreign manufacturers certainly would not want to exchange their products for inflationary currency. To make this point clear, let us imagine a country that produces nothing and exports nothing. Could this country import products? At least in the long run, it could not, because there would be no demand for the currency on foreign exchange markets if there were no products which could be bought with it. The exchange rate between the national currency and foreign currencies would then be so weak that nobody could afford to buy products from abroad. The purchasing power of an economy's currency depends exclusively on the **productivity of its own people** in relation to the amount of its own currency in circulation.

4.4.4 Price Bubbles as Partial Inflation

Speculation and inflation are directly connected. In fact, price bubbles can be interpreted as "partial inflation" (Fuders & Max-Neef, 2014a, b: 348).[4] There are many theories which try to explain speculation as a psychological phenomenon, mass behavior in a panic situation, the flock following its leader, etc. However, here is a simpler explanation: the money supply is ever increasing. Individuals and institutions who wish to invest money will most likely not invest in food or consumer products, i.e., in the markets where inflation is measured, but in more durable assets: real estate, stocks, commodities, gold or bitcoin. The growing demand in asset markets drives up prices, but those rising prices in asset markets do not cause an increase in the consumer price index (CPI) since this commonly used measure of inflation only takes into account consumer goods.

4.4.5 The Inflation-Deflation Paradox

As we have just discussed, an ever-increasing money supply seeks forms of lucrative investments, and this causes the formation of asset price bubbles. The bull market in these markets is further supported by low interest rates. It may be so much money is withdrawn from the consumer goods markets to be invested in stocks or real estate that even deflationary tendencies prevail in the markets for consumer goods. That

[4]The idea to term price bubbles as "partial inflation" was inspired by a conversation with Karl Albrecht Schachtschneider.

means, even though we have an ever-increasing money supply, this will not automatically be reflected in the consumer price index (CPI) that measures inflation. This phenomenon, that might be termed the "*inflation-deflation paradox*" (Fuders, 2011: 36; Fuders et al., 2013: 51; Fuders & Belloy, 2013: 66 f.; Fuders, 2022) could be observed, e.g., in Chile in end of 2009 (see Fig. 4.3). The *inflation-deflation paradox* is the reason why people are not aware of how many monetary units truly exist, a situation that Irving Fisher (1928) once termed the "money illusion".

This paradox of measuring low inflation (or even deflation) alongside a growing money supply is fueled by low interest rates, which make investments in stocks and real estate (or gold and Bitcoin) relatively more attractive than investing in "virtual wealth" (Soddy, 1933; 1934: 24 ff.) as, e.g., bank accounts and debt securities, all the while steadily growing asset bubbles increase the opportunity costs of spending money on consumer goods. Yet, there is a secondary effect that very low interest rates have as well. Money that does not yield interest is likely to be hoarded in relatively large amounts under the mattress and thus will be out of circulation, a phenomenon we can corroborate in the eurozone. Figure 4.4 depicts the cash "circulation" in the eurozone. "Circulation" is put in inverted comas because as of 2021 only about 18% of all transactions worldwide were paid with cash (FIS, 2022). In other words, the growing cash in "circulation" is mainly used to hoard money physically at home and thus is, in fact, out of circulation. Money that is out of circulation decreases demand for goods and services (the velocity V in Irving Fisher's Eq. (4.2) declines) provoking again lower prices in the markets for consumer goods. Since interest rates tend to decrease in the long run (Sect. 5.5.1 discusses this point), it is more likely that the *inflation-deflation paradox* occurs the longer a monetary system exists.

Viewed from this perspective, asset bubbles may not be as negative as they are commonly made out to be. In fact, we might even consider them beneficial since they serve as an escape valve for the ever-increasing money supply. If all of the money in circulation were spent on food and consumer goods, prices would increase drastically, thus making everyday necessities unaffordable.

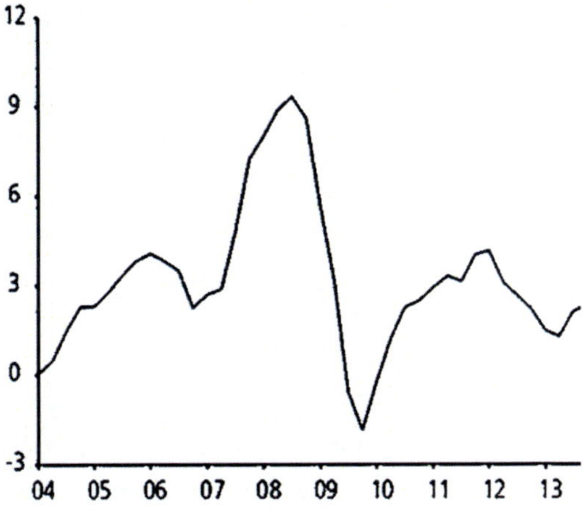

Fig. 4.3 CPI (in % from year ago) 2004–2013 in Chile. (Source: Banco Central de Chile (2014:10))

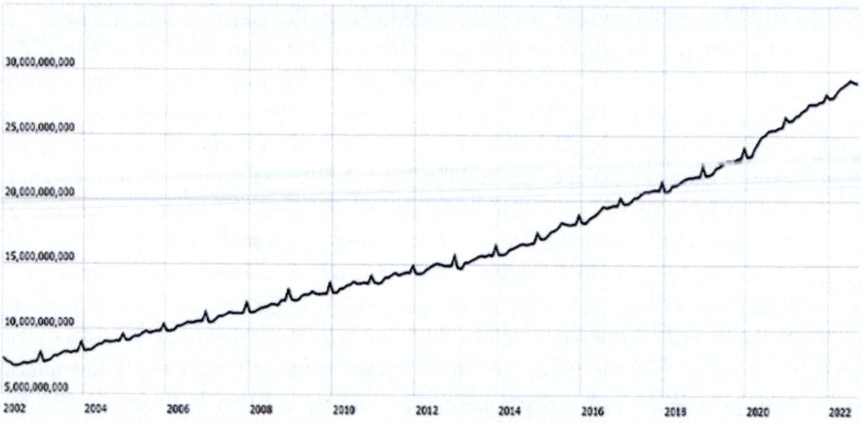

Fig. 4.4 Cash circulation in the eurozone. (Source: ECB (2022))

4.4.6 The Increase of All Prices Equals All Interest Payments

If we measured inflation not via price increases in consumer goods, where inflation is determined via arbitrarily chosen goods and weighting factors, but taking into account all price increases, including real estates, stocks, gold, cryptocurrencies, i.e., prices of whatever goods, services, assets we can buy, including secondhand prices, then this price increase would reveal Irving Fisher's (1928) *money illusion*. Sure, it is probably very difficult, not to say impossible to calculate all price increases of all goods and services, especially if we also include prices of second-hand goods. But it is not impossible to compute all interest payments in a year. What is the relevance of interest payments, some readers might ask. The answer is simple: the increase of all prices should be equal to all interest payments. Hence, if we take the sum of all interest payments and divide it by all persons (natural and legal) P in an economy then we obtain not only the per capita interest burden but also the per capita inflation burden (IB*c*) in Eq. (4.6).

$$IB_c = \frac{\sum \text{interets payments}}{P} \tag{4.6}$$

4.4.7 Foreign Investments Do Not Change the Money Supply

Readers trained in macroeconomics might now object that inflation could also be caused by the influence of the external sector, e.g., foreign direct investments or an increase in exportation. In this case, money supply would increase through money inflow coming from outside. However, **foreign investments do not change the money supply** of an economy; the domestic money supply does not increase as a result of foreign investment or growing exportation, in the same way as there is no such thing as "capital" outflows (better: currency outflows) if someone invests money abroad. For example, if a US citizen invests in Chile this does not have any effect on the supply neither of Chilean pesos, nor on the supply of US dollars. The sum of Chilean pesos circulating in Chile as well as the sum of US dollars circulating in the US keeps *ceteris paribus* unchanged. If someone makes a wire transfer from the US to Chile, the recipient in Chile will receive Chilean pesos on his bank account. There are no more, freshly produced Chilean pesos, but only those money units that have been already circulating. The banks in the US and Chile taking part in this transaction of transferring money will clear positions among each other. And even if someone holds a US dollar account in Chile (which for business clients and corporations in possible) then still the supply of Chilean pesos keeps unchanged, there is not one peso more exiting neither in Chile nor somewhere else in the world.

However, a person or business who receives US dollars in Chile will most likely sell the dollars at some point in order to pay their bills in Chile. If many US dollars are sold on the foreign exchange market this will cause the dollar to depreciate relative to the Chilean peso. Direct investments and increased exports thus do have an impact on the exchange rate, and this can in turn have an impact on inflation since many consumer goods are imported, and inflation is measured via prices of consumer goods. But this effect works in the exact opposite manner from what is usually taught (see, e.g., Duden Wirtschaft, 2016). The inflation rate will not increase but decrease due to exportation and direct investments since the repricing of the peso will make imports cheaper, not more expensive. It is noteworthy that this influence on the inflation rate only occurs because inflation is measured **solely via consumer goods**. If we considered all prices of all goods as proposed above, then direct investments or exports would not have any influence on inflation. Reduced spending on imported consumer goods will be offset by increased spending somewhere else in the economy, for example in real estate or stocks. In case the saved money would be deposited in a bank account then and the bank will put it back into circulation.

4.4.8 Heavy Deflation When the System Breaks Down

Deflation is the opposite of inflation and is defined as a decrease in the general price level of goods and services. In other words: deflation is when the inflation rate falls below zero. Deflation increases the exchange value of money, thereby allowing people to buy more goods with the same amount of money. Although this at first glance might sound like a desirable situation, it is not. It increases the real value of debt and induces a deflationary spiral that can plunge the economy into deep recession. In economics the cause of deflation is not well understood. Many texts and theories have been offered over the last two centuries regarding deflation. However, having discussed the view that our existing form of money has a basic design flaw that results in a constant need for the "drug" of interest to be administered in order to make money circulate, we can offer a corresponding explanation of the causes of deflation.

Deflation occurs when money does not circulate well. Due to the flawed design of our current form of money and its storable nature, the only way to keep it circulating and fulfilling its function as 'blood' of the economy, and thus to prevent deflation, is to pay interest as "reward for parting with liquidity" (Keynes, 1936: 167). But interest makes deposits and debt grow in an exponential manner, since there is no interest without debt, as described above. This system must break down eventually since it is impossible to infinitely expand the (credit) money supply. Banks are forced to continually expand the volume of loans because deposits also grow constantly. Over time it becomes ever more difficult to find enough new solvent borrowers to keep expanding the credit volume. Accordingly, banks start lending to borrowers with ever lower creditworthiness (subprime). They do so not because they are greedy but because they must expand the credit volume in order to survive. And when declining credit standards cause loans to go into default on a large scale this puts the solvency of the whole banking system at risk, because when one bank goes bankrupt, it can initiate a chain of bank bankruptcies, especially in the fractional reserve banking system (banks hold only a fraction of their customer's demand deposits as liquidity reserve physically at the bank or at the central bank account).

4.4.9 *Regular Financial Crises*

If the system collapses, money will not be kept circulating via new loan formation and people begin to store money physically at home thus causing deflation to set in. Sinking prices give an extra incentive to postpone consumption since tomorrow prices will be even lower. Eventually this affects even healthy businesses who are unable to pay their bills and their employees' salaries thus plunging the economy into recession or depression. This is why in the current system it is impossible to achieve a 0%-inflation rate in the long run. As we already observed, in the current system stable money does not exist. Either the money flows in accordance with the growth dynamic inherent in interest, in which case the money supply inflates, or the system breaks down and money stops flowing, resulting in deflation. As soon as our financial system stops inflating, i.e., the money supply stops expanding because the banks can no longer find enough solvent borrowers, the money cycle comes to a standstill, resulting in deflation. Since moderate inflation is considered to be better than deflation and the risk of sliding into deep economic depression central banks usually aim for an inflation rate of 2–3%.

Without constant expansion of the money supply deflation will set in, i.e., the only way to prevent deflation is to keep the system running. Therefore, bail outs become necessary (as the European Stability Mechanism). This is why the ECB, the Fed, the Bank of England and the central bank of Japan decided to keep the financial markets liquid by printing money (Steltzner, 2009; Frühauf, 2009). In the EU cash hoarding has grown since 2008 about 30.000%. This is the reason we see relatively moderate inflation rates in Europe. Money is kept under the mattress or in real estate, stocks or Bitcoin, where we see growing price bubbles, or money is kept in bank accounts, but banks are unable to make it circulate (*inflation-deflation paradox*). Of course, if at some point people lose confidence in the system, they will recognize the *money illusion* and start spending it on consumer goods thus causing hyperinflation. The risk of price bubbles and inflation is considered to be less bad than deflation. But, of course, keeping a sick system artificially alive is not a solution to the underlying problem.

That the system causes crises at regular intervals can be confirmed in any textbook on economic history. Later in this book in Chap. 16 when we discuss "economic resilience in presence of externals shocks" we will discuss the origins of financial crises and long-term economic cycles in greater detail, and we will see that all crises follow a similar pattern. We will also understand why the current chaos in the world (Covid-19, war in the Ukraine) is likely to trigger the next financial crisis, a crisis that would have occurred eventually anyway, even without the pandemic and war (Fuders & Max-Neef, 2014a, b; Fuders, 2010a, 2016a, 2017, 2021).

4.5 Interest Makes Growth in the Real Economy Mandatory (Microeconomic View)

Having now understood the reason for inflation we can now analyze the need for constant growth in the real economy. It is not (only) greed that makes us produce more. We are obliged to do so. There are four reasons that contribute to this growth imperative.

1. A steady increase of the total debt means that we (households, governments, companies) are ever more indebted, and this is the reason why we must continually do more just to maintain the *status quo* (for a similar argumentation see: Kennedy, 1990, 2011; Creutz, 1993; Binswanger, 1996, 2006; Costanza, 2012; Löhr, 2012; Lietaer et al., 2013). Looking at corporate balance sheet from 20 years ago and comparing them with those of today confirms this view. On average, corporations have much more borrowed "capital" (it should be called borrowed money, not capital) than they did 20 years ago.
2. Even businesses with no debt are not immune from the growth imperative. This is because governments are forced to pass on increased debt burdens to citizens and companies in the form of higher taxes. In fact, in most countries the government is the biggest customer of the banking industry and the volume of money loaned to the State is significantly higher than the amount loaned to private borrowers.
3. Workers require salary increases to compensate for inflation, which as outlined above is the result of the growing money supply. Inflation also affects the national budget which is why governments are prone to increase taxes even in the absence of public debt.

4. Even those fortunate companies who have no debt, who do not suffer from growing taxes and whose workers do not ask for pay increases to keep up with inflation are still not immune to the pressure to expand. This is because interest is the **opportunity cost** of any productive investment. The interest rate that banks pay is the benchmark against which the success of any productive investment is measured (e.g., Copeland et al., 2008; Common & Stagl, 2008). Any business that does not generate a return at least as high as what the business owner could receive by depositing money in a bank account is economically unviable (for a similar argumentation see Suhr, 1988, 1989). Hence, **interest is the rhythm to which the real (productive) economy has to dance** (Fuders, 2017), and not just a "fetish" as some believe (Hamilton, 2003).[5]

Some might object that instead of pursuing growth, businesses could instead try to cut costs as a way to maintain the status quo without needing to increase income. This is true and this helps to explain why we tend to see rising unemployment in economies which do not grow. Labor is typically the largest expense category for most companies. Chapter 7 is dedicated to investigating this problem in greater detail, specifically as it relates to SDG No 8. But first we need to analyze the effect of interest on money supply.

References

Alegre Saz, L., Berné Manero, C., & Fundame, C. G. (2010). *Fundamentos de economía de la empresa: perspectiva funcional* (2nd ed.). Ariel.

Aristotle. (1995). Politik 1st book. In *Aristoteles Philosophische Schriften in sechs Bänden. Band 4*. Felix Meiner Verlag.

Banco Central de Chile (BCCh). (2014). *Monetary Policy Report June 2014*. Banco Central de Chile, Santiago

Binswanger, H. C. (1996). Geld und Wachstumszwang. In B. Bierbert (Ed.), *Die Dynamik des Geldes: Über den Zusammenhang von Geld, Wachstum und Natur* (pp. 113–127). Campus.

Binswanger, H. C. (2006). *Die Wachstumsspirale: Geld, Energie und Imagination in der Dynamik des Marktprozesses* (2nd ed.). Metropolis.

Bundesbank. (2019). *Geld und Geldpolitik*. Bundebank.

Carver, T. (1883). The place of abstinence in the theory of interest. *The Quarterly Journal of Economics, 8*(1), 40–61.

Common, M., & Stagl, S. (2008). *Introducción a la Economía Ecológica*. Reverté.

Copeland, T., Weston, J., & Shastri, K. (2008). *Finanzirungstheorie und Unternehmenspolitik. Konzepte der Kapitalmarkorientierten Unternehmensfinanzierung* (4a ed.). Pearson.

Costanza, R. E. A. (2012). *Building a sustainable and desireable economy-in-society-in-nature*. United Nations Devision for Sustainable Development.

[5] This is true if the sole purpose of the business is to make money. However, entrepreneurs could, of course, also start their business because they wished to make use of their talents in order to produce products that contribute to the common good, i.e., to the quality of life of people, and not only for financial gains. In that case, entrepreneurs would care less about interest as an opportunity cost, and it might be sufficient for them to earn enough to live off it. Sect. 17.3 discusses this point.

Creutz, H. (1993). *Das Geldsyndrom. Wege zu einer krisenfreien Marktwirtschaft*. Wirtschaftsverlag Langen MÜller/Herbig.

Cuadrado, J. R., et al. (2015). *Política Económica – Elaboración, objetivos e instrumentos* (5th ed.). McGraw-Hill.

Deka. (2014). *Erklärfilm Zinseszinseffekt.* [Online] Available at: https://www.youtube.com/watch?v=81h5QpYzmto&list=PLF_jjVkf8BM2Gz7uuUUkHrLnAme3vrU1w&index=28. Accessed 11.10.2022.

Duden Wirtschaft. (2016). Importierte inflation. In *Duden Wirtschaft von A bis Z: Grundlagenwissen für Schule und Studium, Beruf und Alltag*. Bibliographisches Institut.

ECB. (2022). *Banknotes and coins circulation.* [Online] Available at: https://www.ecb.europa.eu/stats/policy_and_exchange_rates/banknotes+coins/circulation/html/index.en.html [Accessed 17 10 2022].

Farley, J. E. A. (2013). Monetary and fiscal policies for a finite planet. *Sustainability, 5*, 2802–2826.

Fernández Díaz, A., Parejo Gámir, J. A., & Rodríguez Sáiz, L. (2011). *Política Económica* (4a ed.). McGraw-Hill.

FIS. (2022). *The global payment report.* FIS (Fidelity National Information Services).

Fisher, I. (1912). *The purchasing power of money*. New York: Macmillan.

Fisher, I. (1928). *The money illusion*. Adelphi.

Fisher, I. (1932). *Booms and depressions – Some first principles*. Adelphi.

Frühauf, M. (2009). Die Verstaatlichung des Anleihemarktes. *F.A.Z.*, Issue 02.04.2009, p. 9.

Fuders, F. (2009). Die natürliche Wirtschaftsordnung als Option nach dem Zusammenbruch. *Aufklärung & Kritik, 16*(2), 128–145.

Fuders, F. (2010a). Alternative concepts for a global financial system – An answer to the present world financial crisis. *Estudios Internacionales, 166*, 45–56.

Fuders, F. (2010b). Warum der Zins auch moralisch nicht zu rechtfertigen ist. *Humane Wirtschaft, 2*, 26–29.

Fuders, F. (2011). Wie Zerstörung von Sachkapital dem Finanzsystem in die Hände spielt. *Humane Wirtschaft, 2/2011*, 34–38.

Fuders, F. (2016a). Smarter money for smarter cities: How regional currencies can help to promote a decentralised and sustainable regional development. In *Decentralisation and regional development – Experiences and lessons from four continents over three decades* (pp. 155–185). Springer.

Fuders, F. (2016b). Negativzinsen ja – aber richtig. Issue 4/2016, pp. 6–11.

Fuders, F. (2017). Neues Geld fÜr eine neue Ökonomie – Die Reform des Geldwesens als Voraussetzung fÜr eine Marktwirtschaft, die den Menschen dient. In *Finanzwirtschaft in ethischer Verantwortung – Erfolgskonzepte fÜr Social Banking und Social Finance* (pp. 121–183). Springer-Gabler.

Fuders, F. (2020a). *The 'tragedy of the commons' and the role of the money interest rate*. Springer.

Fuders, F. (2020b). Freigeld und Freiland: fÜr eine wirklich freie, nachhaltig funktionierende und gerechte Wirtschaftsordnung. *Humane Wirtschaft, 2/2020*, 16–24.

Fuders, F. (2021). The role of money for a healthy economy. In *Urban health and wellbeing programme – Policy briefs: Volume 2* (pp. 17–22). Springer.

Fuders, F. (2022). Das Inflation-Deflation-Paradoxon. *Fairconomy, 1/2022*, 10–14.

Fuders, F., & Belloy, P. (2013). Terremoto del 27.02.2010 en Chile – Cómo el sector financiero se beneficia de la destrucción de capital real. *Realidad Económica,* 61–75.

Fuders, F., & Max-Neef, M. (2014a). Local money as solution to a capitalistic global financial crisis. In *From capitalistic to humanistc business*. Palgrave Macmillan.

Fuders, F., & Max-Neef, M. (2014b). Dinero, deuda y crisis financieras. Propuestas teórico-prácticas en pos de la sostenibilidad del sistema financiero internacional. In *Economía Internacional. Claves teórica-prácticas sobre la inserción de Latinoamérica en el mundo*. LATIn.

Fuders, F., Mondaca, C., & Azungah Haruna, M. (2013). The central banks dilemma, the inflation-deflation paradox and a new interpretation of the Kondratieff waves. *Economía (U. de los Andes), 36*, 33–66.

Fullarton, J. (1845). *On the regulation of currencies* (2nd ed.). John Murray.

Gesell, S. (1949). *Die natürliche Wirtschaftsordnung durch Freiland und Freigeld*. Rudolf Zitzmann.

Hamilton, C. (2003). *Growth Fetish*. Allen & Unwin.

Kennedy, M. (1990). *Geld ohne Zinsen und Inflation – Ein Tauschmittel, das jedem dient*. Permakultur-Verlag.

Kennedy, M. (2006). *Geld ohne Zinsen und Inflation – Ein Tauschmittel, das jedem dient* (9th ed.). Goldmann.

Kennedy, M. (2011). *Occupy Money – Damit wir zukünftig alle die Gewinner sind*. J. Kamphausen.

Keynes, J. M. (1930). *A treatise on money – Vol. 1 The pure theory of money*. Cambridge University Press.

Keynes, J. M. (1936). *General theory of employment, interest and money*. Hartcourt.

Lietaer, B., Arnsperger, C., Goerner, S., & Brunnhuber, S. (2013). *Geld und Nachhaltigkeit – Von einem überholten Finanzsystem zu einem monetären Ökosystem. Ein Bericht des Club of Rome/EU-Chapter*. Europa Verlag.

Löhr, D. (2012). The euthanasia of the rentier – A way toward a steady-state economy? *Ecological Economics, 84*, 232–239.

Mankiw, G., & Taylor, M. (2014). *Economics*. Cengage Learning.

McCulloch, J. H. (1975). *Money and inflation – A monetarist approach*. Avademic Press.

OECD. (2023). *Main economic indicators (database)*. [Online] Available at: https://doi.org/10.1787/data-00052-en. Accessed 1.2.2023.

Sell, F., Sauer, B., & Engelkamp, P. (2020). *Einführung in die Volkswirtschaftslehre*. Springer Gabler.

Soddy, F. 1933 (Reprint 1961). *Wealth, virtual wealth and debt – The solution of the economic paradox*. Briton.

Soddy, F. (1934). *The role of money – What it should be contrasted with what it has become*. Routledge.

Statista. (2021). *Value of Euro banknotes in circulation in Europe from January 2008 to January 2021*. [Online] Available at: https://www.statista.com/statistics/1117483/value-of-euro-banknotes-in-circulation-in-europe/. Accessed 12.10.2022.

Steiner, R. (1979). *Aufgaben einer neuen Wirtschaftswissenschaft, Nationalökonomischer Kurs – 14 Vorträge, gehalten in Dornach vom 24. Juli bis 6. August 1922 für Studenten der Nationalökonomie*. Rudolf-Steiner-Verlag.

Steltzner, H. (2009). Auf dem Londoner Gipfel. *Frankfurter Allgemeine Zeitung (FAZ)*, Issue April 2nd, p. 1.

Süchting, J. (1995). *Finanzmanagement, Theorie und Politik der Unternehmensfinanzierung* (6th ed.). Gabler.

Suhr, D. (1988). *Alterndes Geld – Das Konzept Rudolf Steiners aus geldtheoretischer Sicht*. Novalis Verlag.

Suhr, D. (1989). *The capitalistic cost-benefit structure of money – An analysis of money's structural nonneutrality and its effects on the economy*. Springer.

Wienert, H. (2001). *Grundzüge der Volkswirtschaftslehre* (2nd ed.). Kohlhammer.

Wikipedia. (2022). *Entry "Money Supply"*. [Online] Available at: https://en.wikipedia.org/wiki/Money_supply [Accessed 12 10 2022].

Chapter 5
The Effect of Interest on the Money Supply, Demand and Growth

5.1 Interest Makes Growth Possible (Macroeconomic View)

On one hand, interest creates a requirement to grow, but on the other it also helps make that growth possible. To understand this point, we need to examine the effect that the interest rate has on the money supply and demand. Here a **completely new view** will be offered, contrary to conventional monetary theory. I would even go so far as to say that—at least in the long run—it is the opposite of what is normally taught. This misunderstanding could be an important contributing factor to the problems inherent in our monetary system being not yet broadly recognized.

According to conventional wisdom lowering interest rates fosters economic growth and *vice versa*. However, the following paragraphs will show that only in the very first moment lowering interest rates can foster economic growth while increasing interest rates can slow it down. In the medium and long run the effect of the interest rate on the money supply works the **other way around** and we see a clear positive correlation between interest rates, inflation, and GDP growth rate, when looking at the history (Fig. 5.1) of the US economy (for other economies see Lee & Werner, 2018): the higher the interest rate, the higher inflation and the GDP growth rate. That the interest rate and the growth of the money supply are **necessarily positively correlated** can be understood from three perspectives, which will be discussed in the next paragraph.

© The Author(s), under exclusive license to Springer Nature Switzerland AG 2023
F. Fuders, *How to Fulfil the UN Sustainability Goals*,
https://doi.org/10.1007/978-3-031-37768-6_5

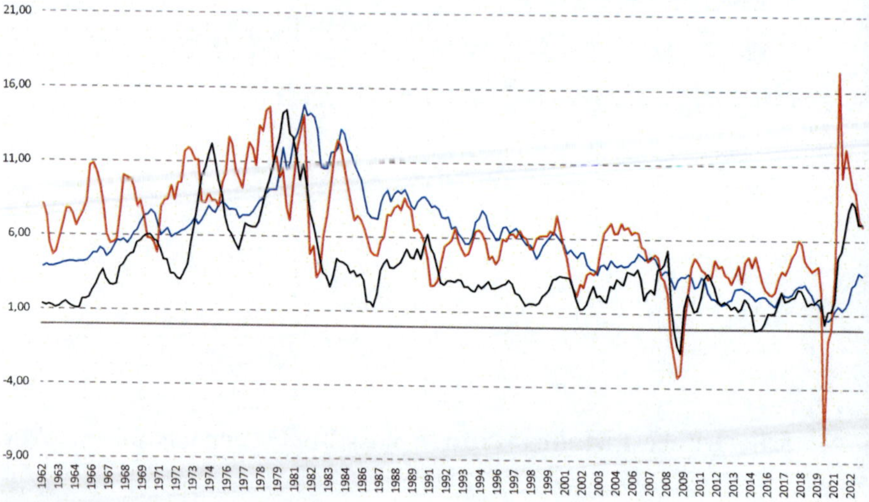

Fig. 5.1 Interest rate, inflation and GDP growth rate (US). Blue: market yield on 10-Year Treasury Constant Maturity Rate; black: Consumer Price Index (% change from year ago); red: GDP growth (% change from year ago). (Source: Own work. Data provided by Board of Governors of the Federal Reserve System, US Bureau of Economic Analysis and OECD (2023). Retrieved from Federal Reserve Bank of St. Louis (https://fred.stlouisfed.org))

5.2 Explanation of the Relationship Between Interest, GDP Growth and Inflation

5.2.1 *First Perspective: Interest Makes Deposits Grow*

As outlined above, interest makes deposits grow. The higher the interest rate, the faster deposits grow and therefore the higher will be the total money supply (composed of cash plus deposits). This effect might be called "*interest money creation*" (Fuders, 2011; Fuders et al., 2013). The role of the interest rate in the exponential growth of the money supply can be seen in the US (Fig. 4.2) as well as virtually any other country. A higher money supply means more monetary units circulating and, as more money is spent, aggregate demand grows. In other words, the interest rate not only pushes producers to maintain or increase profits, but it also makes economic growth possible by increasing the money supply and aggregate demand.

Here, it is crucial to understand that (i) as long as bank accounts earn interest the money supply will grow and (ii) once interest has led to the creating of monetary units, those units cannot be destroyed unless there is a 'reset' of the whole financial system. Money supply thus **always grows when there is a positive interest rate**. The higher the interest rate, the faster the money supply will grow. And as explained above interest makes deposits grow exponentially. This is true even for a very low interest rate since any positive interest rate will eventually lead to the amount of the original deposit doubling. And everything that experiences periodic doubling grows exponentially. The difference between higher and lower interest rates is simply that the flat part of the exponential growth curve is longer for lower interest rates. This also means that, from the moment the interest rate falls, the money supply curve *ceteris paribus* will show a disruption but will eventually start to again grow exponentially from that point on (dashed line in Fig. 5.2). Only a zero-interest-environment would create a stable money supply over the long term, as shown on the right-hand graph of Fig. 5.2, which depicts the growth of the money supply with positive interest rate until the interest falls to zero % (dashed line).

One might wonder why then the US money supply in Fig. 4.2 does not show such a disruption. Even though the interest rate has fallen sharply since 2009 the money supply has kept growing exponentially. Well, this is why I used the term *ceteris paribus* ("all other things being equal"). It is necessary to emphasize the *ceteris-paribus* constraint here, since of course money supply (cash plus deposits) also grows via central bank money printing. And this is precisely what the Fed and other central banks in the industrialized world have done since interest rates fell sharply in the aftermath of the 2008 financial crisis. The Fed has increased the monetary base since 2009. Figure 5.3 contrasts the US money supply M3 (which was depicted already in Fig. 4.2) with the US monetary base M0, i.e., the money directly produced by the central bank. As we can see, since 2009, the central bank has significantly expanded the monetary base in an environment of falling interest rates, thus compensating for the decreased *interest-money-creation* effect.

Already at this time the former chairman of the Fed Ben Bernanke was referred to as "helicopter Ben", for his policy of metaphorically letting money rain on the economy from a helicopter. In 2020 when interest rates had once again fallen to near zero and thus almost no money creation effect was occurring via interest, the Fed did what it never before had done in its history: it increased the monetary base from 3.2 to 6 trillion USD in just 1 year. The reason why we did not observe the disruption in money creation as depicted in Fig. 5.2 which would normally be caused by

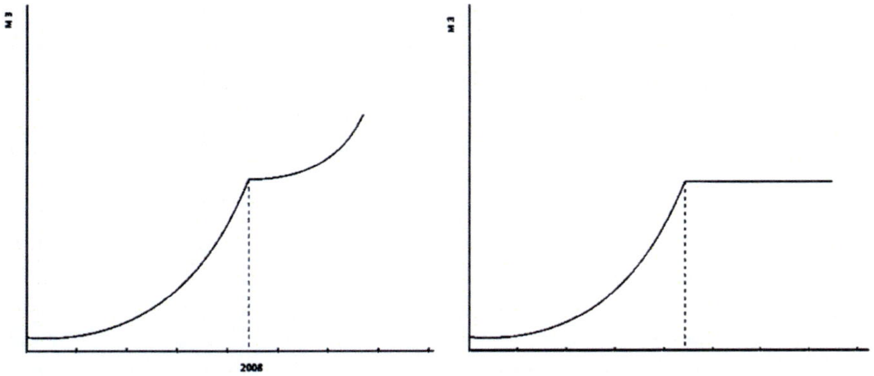

Fig. 5.2 Scheme money supply growth. Left: effect of lowering interest rates (but maintaining >0); right: effect of lowering interest rates to 0%. (Source: On work)

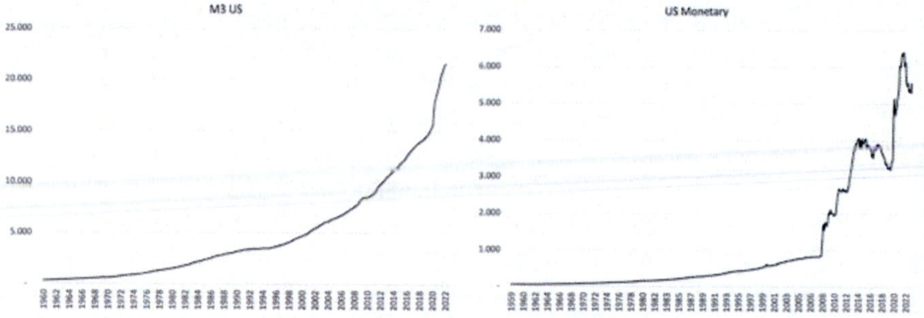

Fig. 5.3 Money supply M3 and M0 for the US. Left: US money supply M3 (in billions). Right: US monetary base M0 (in billions). (Source: Own work. Data provided by OECD (2023) and Board of Governors of the Federal Reserve System. Retrieved from Federal Reserve Bank of St. Louis (https://fred.stlouisfed.org))

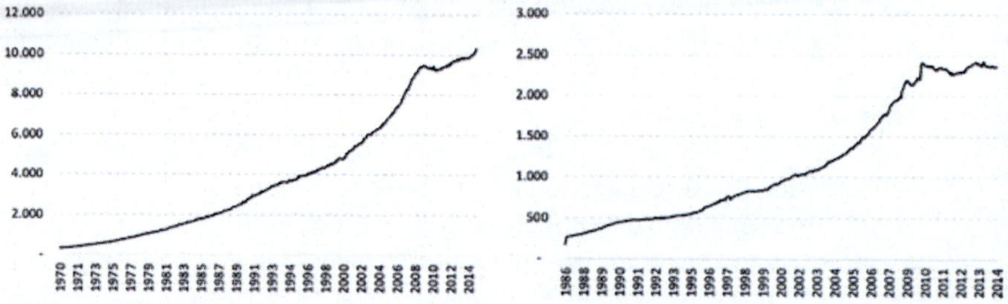

Fig. 5.4 Money supply M3 eurozone and UK (until 2014). Left: Euro money supply M3 (in billions). Right: British Pound money supply M3 (in billions). (Source: Own work. Data provided by OECD (2023). Retrieved from Federal Reserve Bank of St. Louis (https://fred.stlouisfed.org))

falling interest rates is that the central bank counteracted the decreased *interest-money-creation* effect by increasing the monetary base.

The effect of falling interest rates on the money supply can be better observed in the eurozone and the UK, because the ECB and Bank of England initiated a policy of money printing not directly after the crisis 2008 but in 2014. Figure 5.4 plots the growth of the money supply in the UK and the eurozone until 2014. The disruption in growth of the money supply beginning in 2008 when interest rates fell sharply can be clearly seen: the money supply in the UK and eurozone showed almost no growth whatsoever.

However, from 2014 on, the ECB and the Bank of England started copying the "helicopter Ben" policy and inaugurated a similar expansionary monetary policy injecting fresh money into the economy, i.e., printing money, nowadays referred to as "*Quantitative Easing*" (QE), a term coined by former ECB president Mario Draghi. The money supply (M3) began to once again increase from 2014 on, but in this case not because of deposit growth via interest but simply because the central banks printed money. Figure 5.5 depicts the Euro and British Pound money supply

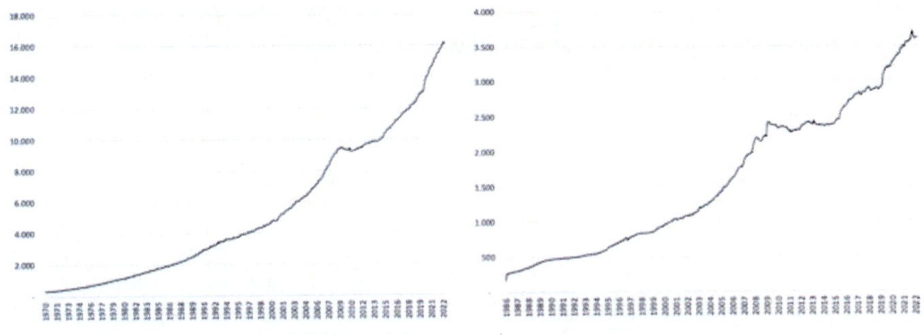

Fig. 5.5 Money supply M3 for eurozone and UK. Left: Euro money supply M3 (in billions). Right: British Pound money supply M3 (in billions). (Source: Own work. Data provided by OECD (2023). Retrieved from Federal Reserve Bank of St. Louis (https://fred.stlouisfed.org))

(M3) until 2022. We can observe how the QE policy compensated for the decreased interest money creation.

Note that even if all of the loans in an economy were repaid, the money supply would not return to its original level since deposits would have earned interest (Fuders et al., 2013; Fuders, 2017). Here we can already see why it is impossible for every debtor repay their debts (principal plus interest) (Kremer, 2009) and why it is unavoidable that financial crises occur at regular intervals. Both points will be discussed in more detail later in Sect. 5.6.4. However, according to classical monetary theory, the effect of the interest rate on the money supply and economic growth is thought to be exactly the other way around. Indeed, central banks lower interest rates if they wish to increase economic activity. This leads us to the second perspective.

5.2.2 *Second Perspective: Interest Requires Banks to Grant Loans*

So, why is it that central banks lower interest rates if they wish to increase economic activity and raise them if they wish to slow down economic activity and inflation? Central banks and conventional monetary theory nowadays (for approximately the last 70–100 years) usually apply the *money multiplier* model to explain the effect that the interest rate has on the money supply. As explained in Chap. 2 in my view this concept is correct and according to this concept money is created, and thus the money supply grows, through the expansion of credit. Central banks raise interest rates if they fear inflation and wish to slow down economic growth, as this is supposed to have the effect of disincentivizing the granting of loans (e.g., Cuadrado et al., 2015; Fernández Díaz et al., 2011). A brochure published by the German Bundesbank (2019: 156) asserts that:

> Higher interest rates strengthen the incentive to save and make borrowing more expensive. If less money is spent as a result, this tends to dampen aggregate demand and thus price developments. Conversely, lower interest rates tend to lead to rising demand (also financed by borrowing) and thus to a higher price level (author's translation).

The same argument is the standard explanation of the effect of monetary policy on inflation and economic growth which can be found in most textbooks. However, this explanation is based on the erroneous idea that money deposited in a bank is taken out of circulation, so as if a bank was a strongroom and money once deposited there would not circulate anymore.

However, the expansion of credit in most cases slows down **only immediately** after raising interest rates. In the medium and long term, higher interest rates result in more people depositing money into bank accounts, as correctly the quote from the brochure above describes. Yet, money deposited in bank accounts **does not stay out of circulation for long**. Growing deposits put enormous pressure on banks to increase the volume of money loaned out, since without income from loans, banks would be unable to pay interest on deposits without compromising their solvency. Remember that without debt there is no interest. Hence, money deposited in bank accounts will—as long as the monetary system is not in crisis—circulate in the economy via loans.[1] **The more money that is deposited, the more loans the bank will have to grant**. In other words: the *interest money creation* as explained in the previous paragraph obliges the expansion of loan volume. Hence, growing deposits is consistent, at least in the long run, with an increase and not a decrease of the loan volume.

Some might argue that if banks are unable to find enough creditworthy borrowers, they can generate revenue to pay depositors via other types of investments, such as the stock market. But even if they pursued such a strategy the money would remain in circulation, since depositor money would be passed on to the sellers of the stocks purchased and thus keeps circulating in the economy. Hence, even in this case, the money supply (cash plus deposits) continues to grow as long as banks put depositor money back into circulation via investments.

Therefore, even viewed from the perspective of the *money multiplier* explanation higher interest rates eventually lead to an increased money supply and thus higher aggregate demand and economic growth, and *vice versa* (Fuders et al., 2013; Fuders, 2017; Stehlík et al., 2017), which is why it should be obvious that conventional monetary policy should be completely reconsidered (Lee and Werner, 2018).

[1] Here we can also ask whether the so-called "Keynesian multiplier" really exists. In the view of post-Keynesian economics, the government can smooth out economic cycles via anticyclical spending and saving. That is, in times of high economic growth and increased tax revenue the government can save money that can later be spent if the economy weakens. However, money that is saved in bank accounts will keep circulating regardless via loans. So, what difference does it make if the money is put into circulation by the government or the banking system?

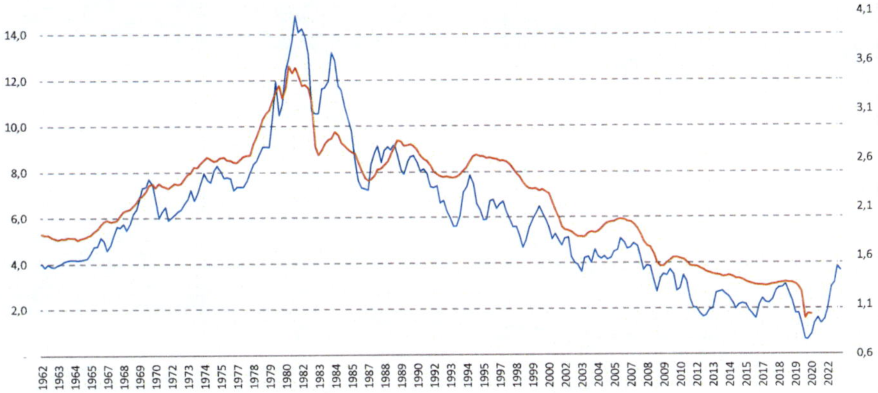

Fig. 5.6 Interest and velocity of circulation (US). Blue: market yield on 10-Year Treasury Constant Maturity Rate (in %, left axis); red: velocity zero maturity money (ratio, right axis). (Source: Own work. Data provided by Federal Reserve Bank of St. Louis and Board of Governors of the Federal Reserve System. Retrieved (2/2023) from Federal Reserve Bank of St. Louis (https://fred.stlouisfed.org))

5.2.3 *Interest Is a Reward for Parting with Liquidity*

The third and simplest explanation why higher and not lower interest rates spur aggregate demand is that the interest rate is a "*reward for parting with liquidity*" (Keynes, 1936: 167).[2] The higher the interest rate, the higher the incentive to lend out money (or to save it in a bank which lends it for us). Hence, the higher the interest rate, the more easily money circulates in the economy and *vice versa* (see Fig. 5.6). To come back to the above used metaphor of money as the blood of the economy and interest as the drug that makes the blood (money) circulate: the more of the drug you apply the greater will be the circulation (and the stronger will be the side effects, such as inflation, price bubbles and economic crises).

5.3 Empirical Evidence and Causality

5.3.1 *Empirical Evidence*

What has been said so far can be verified empirically. Notwithstanding the possibility of large short-term fluctuations, the interest rate and GDP growth rate are **clearly positively correlated** when viewed over long time periods, something that was already evident in Fig. 5.1 which illustrates this relationship for the US economy

[2] This is an excellent definition. However, many textbooks state that the interest rate was a premium to abstain from consumption. That this is wrong we have already discussed above in Sect. 4.2.4.

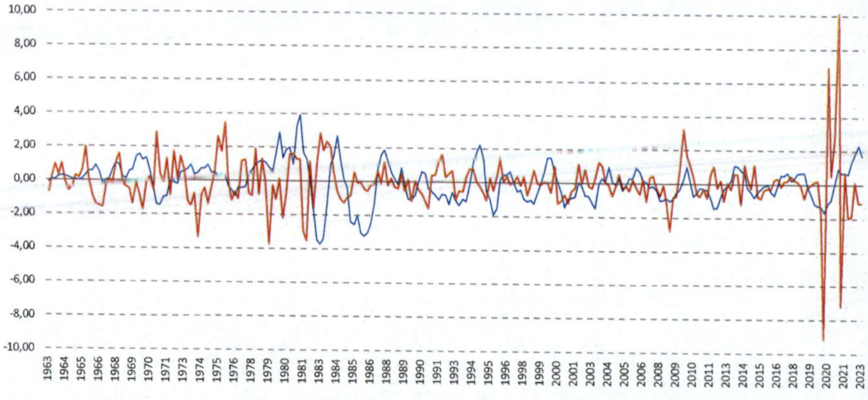

Fig. 5.7 Red line: real GDP growth rate (change from year ago, growth rate previous period); blue line: 10-Year Treasury Constant Maturity Rate (change from year ago in %). (Source: Own work. Data provided by Board of Governors of the Federal Reserve System and OECD (2023). Retrieved from Federal Reserve Bank of St. Louis (https://fred.stlouisfed.org))

(for other key economies see Lee and Werner, 2018). Even more clearly, we can observe this positive correlation Fig. 5.7 If the classical interpretation of the relationship between the interest rate and the money supply and GDP was correct, interest rates and GDP growth should be negatively correlated (Fuders et al., 2013; Stehlík et al., 2017).

Another way to describe the relationship is like this: in a freshly reset financial system, still solvent and far away from over-indebtedness and collapse, high interest rates can and will be charged for credits and paid on deposits. This leads people to deposit money at banks, which in turn increases the volume lent out. The growing money supply stimulates demand and production. The economy booms and companies as well as households are—because they find themselves in the boom phase—able to pay high interest rates. The more indebted households and businesses become, the more difficult it is for banks to find new debtors, which is why interest rates for loans (and those paid on deposits) go down. With falling interest rates, money supply still increases, but at a lower rhythm and so does economic activity. Therefore, in the long run, as we can observe from Fig. 5.1, interest rates, inflation and GDP growth rates develop in a synchronized manner (Fuders et al., 2013).

Since 2008 prime rates in most developed countries were approaching the zero-percent mark. The growth of the money supply of those countries had indeed slowed down as can be seen in Fig. 5.4 that show a moderate growth of the money supply of both the Euro and the British Pound since 2008 (and before ECB's QE policy started). The relationship can be seen even more clearly in Fig. 5.8 which plots the growth of Euro and Pound money supply M3 against the respective average interest rates.[3] As interest rates declined sharply, the money supply of both currencies

[3] Even a very low interest rate causes exponential growth of the money supply with the flat part of the exponential curve being longer than for higher interest rates. Only an interest rate of zero would

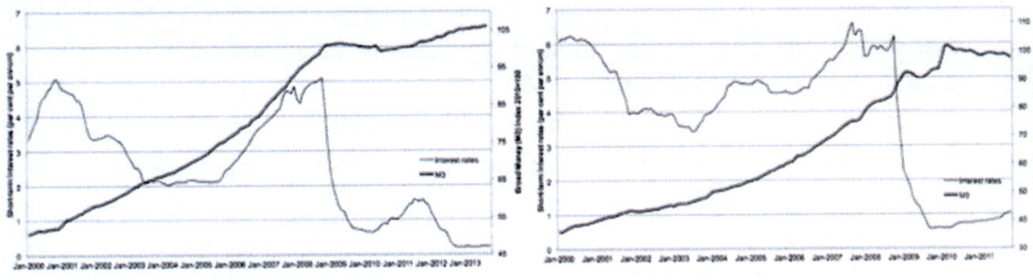

Fig. 5.8 Money supply M3 and Interest Rates. Left Euro, right British Pound. (Source: Own work. Data provided by OECD (http://stats.oecd.org), accessed 6/2013)

increased moderately. This is contrary to conventional monetary theory, and it verifies the explanation above. According to conventional wisdom, low interest rates should lead to an expansion of the money supply (e.g., Mankiw and Taylor, 2014; Dornbusch et al., 2009).

5.3.2 *Discussion*

Some reader trained in monetary theory might now object that the causality would go the other way around: a higher GDP growth rate goes along with higher interest rates because the central bank, fearing inflation, will raise rates to slow down the economy (and *vice versa*). Or lenders simply require higher rates to compensate for expected inflation. However, it is the purpose of this book to challenge this conventional understanding. As explained above, inflation is caused by the steady increase of money supply which is a result of the interest rate. It is due to the interest rate that the money supply grows consistently over time. Even though the real economy tries to keep pace, the money supply **starts to decouple** from GDP leading to an increase in the quantity of monetary units relative to the quantity of goods and services. Interest is not the consequence of inflation, but **inflation is the consequence of interest**. If nobody charged interest for loaning out money, Fullarton's *law of reflux* (see next Sect. 5.5) would finally work and there would be no inflation or deflation.

Another explanation for the positive correlation using conventional thinking could be that an increase in economic activity leads to higher demand for loans and therefore interest rates as the 'price' of renting money rises. According to this logic, interest rates follow the lead of the economy: first GDP grows, which is then followed by an increased demand for credit thus causing interest rates to rise as, e.g., Lee and Werner (2018) suggest. This interpretation can be deemed as partly correct and is not inconsistent with the analysis laid out in this book. Of course, the interest

prevent expansion of the money supply. However, in the current system an interest rate of zero has other devastating consequences: people hoard money under the mattress and money does not circulate. This has the potential to lead to deflation, which is what happened throughout the industrialized world beginning in 1929.

rate as the 'price' of renting money depends on supply and demand of money. And more credit (money) is demanded for investments when GDP grows, which then makes the money supply grow via the *money multiplier* effect. However, the necessity of paying interest means that the economy must grow **even more** than it would naturally do. And natural economic growth only occurs if people have not yet satisfied their fundamental human needs and therefore demand additional products and services. Naturally, if money was neutral and perishable (as real goods are), there would be no further economic growth once everyone had reached a living standard that satisfied their fundamental humans needs.[4]

Furthermore, we would not engage in excessive hoarding of goods because they perish. We would also not hoard excessive amounts of money since it too would be perishable and therefore not a suitable medium for hoarding surpluses. The natural rhythm of life would be more relaxed, and the economy would be able to truly find a long-term equilibrium. But in today's economy where money is easily storable, we not only have an incentive to produce more than is needed since surpluses are storable in the form of monetary units, but we are obliged to do so. **The interest rate requires excess growth**. The long-term equilibrium as described in microeconomics (see e.g., Frank, 2015) can therefore unfortunately never be achieved. And since usually the money supply grows faster than production the 'financial economy' starts to decouple from the real economy resulting in inflation and eventually crisis. This leads us to the next point.

5.4 Interest Destroys Fullarton's *Law of Reflux*

We have discussed the fact that the higher the interest rate, the faster the money supply grows (along with aggregate demand and GDP). A lower interest rate still causes the money supply to grow but at a slower pace, even though conventional monetary theory says the exact opposite. However, it is not solely the interest rate that has an expansionary effect on the money supply. If there is more demand for loans, for example because the economy is in a growth phase, this could also cause the money supply to grow, but only as long as there is an increase in demand for loans resulting from the increased economic activity. Conversely, if the economy is in a contractionary phase less loans will be demanded, and less money created. However, if money was neutral and the interest rate not necessary as non-hoarding premium to make it circulate, then the demand for loans would be the only variable determining an increase or decrease of the money supply. Then, as already mentioned above, we would most probably have stable money, we would have neither inflation nor deflation, and we would not experience recurring financial crises since the money supply would automatically adjust to economic cycles. And the long economic waves

[4] I will come back to this idea in Sect. 15.3.

(Kondratieff waves) would also be avoided since they can be interpreted as cycles of the financial system (see Chap. 16). This automatic adjustment of the money supply to economic growth is what Fullarton (1845: 67 ff.) once called the *law of "reflux"*, and which probably describes how things would work if money was truly neutral.

In the current system, however, as expressed in Eq. (5.1) the money supply M expands via the natural demand for Money M_{nd} (we could also say natural demand for credit) **plus** interest r.

$$\Delta M = \Delta M_{nd} + r \tag{5.1}$$

That means, the money supply M will increase as much as the natural demand for money M_{nd} increases plus the additional credit that banks have to grant in order to be able to pay interest while the interest rate *r* depends on the natural demand for money:

$$r = f(M_{nd}) \tag{5.2}$$

Banks therefore find themselves under constant pressure to grant more credit than would naturally be demanded and can only remain solvent as long as they are able to do so. Banks must therefore constantly increase their loan volume since deposits are also constantly growing due to interest. If at some point banks are unable to find enough borrowers to keep increasing their loan volume, they will not be able to keep up with the need to pay interest on deposits and will have to declare bankruptcy. When this happens on a large scale, we experience a financial crisis.[5]

Therefore, the money supply always runs a bit ahead of the productive economy. Even though we compulsively try to grow in order to keep pace with the steadily growing money supply, we will never 'reach' it. Here we see the reason why in the current system we always experience some inflation (except for the situation described as the *inflation-deflation paradox*), until the money supply has decoupled so much from the real economy that the interest burden cannot be met via an increase in economic output, which in turn leads to widespread credit defaults causing banks to fail and thus become unable to fulfill their most essential function of keeping money circulating and sending the economy into a deep economic depression.

[5] Before allowing this to happen, the bank will typically continue to expand lending by lowering credit standards and lending to clients with questionable solvency. If those (subprime) loans then default on a large scale banks go bankrupt.

5.5 Central Banks Are Trend Followers, Not Trend Setters

5.5.1 *In the Long Run Interest Rates Tend to Go Down*

We can now argue that the role of the central bank is **more that of a trend follower** than a trend setter. As we can see in Fig. 5.9 in the long run—and not withstanding

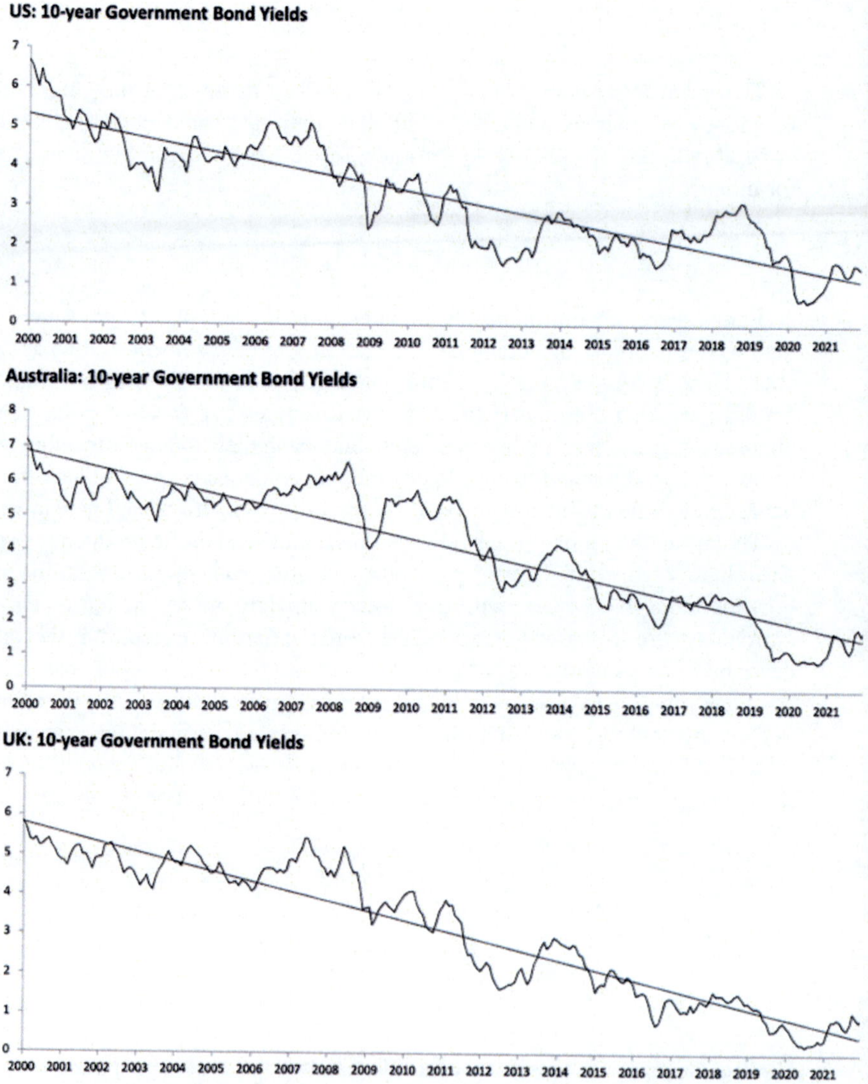

Fig. 5.9 10-year government bond yields and trendline for the US, Australia and UK. (Source: Own work. Data provided by OECD (2023). Retrieved from Federal Reserve Bank of St. Louis (https://fred.stlouisfed.org))

short and medium-term fluctuations such as economic stimulus packages to combat the COVID-19 crisis—interest rates seem to consistently trend downward. This is no accident (see also Stehlík et al., 2017). Low interest rates reduce the problems which banks face in the long run: of finding enough creditworthy borrowers to generate the revenue needed to pay interest on ever-growing deposit accounts. The amount of loans that must be issued does not grow quite as fast when interest rates are low, which gives banks some leeway in finding new borrowers. Additionally, it is easier to find new borrowers when loans are cheaper. It is worth noting that this is consistent with Irving Fisher's observation that interest rates fall whenever the economy enters a debt crisis (Fisher, 1933: 342; Dornbusch et al., 2009: 462). The closer the financial system gets to the collapse, a fate which invariably faces our monetary system sooner or later, the lower interest rates must go (Bichlmaier, 2010; Fuders, 2011; Fuders et al., 2013). This happens with or without central bank intervention.

To put it in other words: interest causes deposits to grow, thus increasing the money supply. Hence, higher interest rates cause the money supply to grow more quickly. In Sect. 5.2.1 I referred to this effect as *interest money creation*. Eventually the money supply grows to such an extent that the interest rate (the price of renting money) falls since the higher the money supply in relation to the demand for money (demand for loans) the lower is the price of renting money, the interest rate. Another way of saying this is that banks have ever-more difficulty finding new borrowers the more indebted people already are and thus have to lower interest rates. These views are two sides of the same coin. Since the money supply is the counterpart of debt, we can describe the phenomenon either from the perspective of the money supply or debt. If the money supply grows total debt also grows, thus making it harder for banks to continue increasing their lending, which then forces them to lower interest rates. The higher the money supply, the lower are (*ceteris paribus*) interest rates. Nobody doubts this relation. The new insight is that the money supply grows via interest.

The Fed and ECB have been criticized for their "cheap money" policies over the past decade. In reality, the central banks would probably have liked to increase interest rates. Nevertheless, the role of central banks is, as is well known, a passive one (Fernández Díaz et al., 2011; Cuadrado et al., 2015; Bundesbank, 2019). Central banks can only indirectly influence the market by setting the interest rate at which commercial banks can borrow from the central bank. However, it is not possible to raise interest rates if people are not willing to borrow money at higher rates. If the central bank forced commercial banks to increase interest rates without there being demand for loans at those rates, then money would not circulate via lending and there would be a risk of deflation and financial crisis. **Central banks mainly follow market trends** and not vice versa, a phenomenon that could be described as the central bank's dilemma (Fuders, 2011; Fuders et al., 2013). It is likely that this dilemma is the true reason for the low interest rate policy of central banks in most industrialized countries during the past 15 years, whereas many believe the main reason for this policy was to avoid a credit crunch.

Once we understand that interest is a premium that must be paid to holders of money in order to keep money circulating, this sheds new light on the past 10 years of deflationary pressures in the eurozone, where interest is now (until spring 2022) practically nonexistent on deposits. Yet central bankers seemed to be surprised at the low inflation rate despite low interest rates.[6] This may be because conventional economic theory teaches that low interest rates cause inflation to rise. The opposite is true, and the reason why is not difficult to understand. The higher the interest rate, the faster deposits grow and with it the quantity of money available for consumption. At a zero-interest rate level, the money supply stops growing unless the central bank prints money (Fuders et al., 2013). Of course, the central bank cannot simply raise interest rates if it wants to stimulate the economy. After all, interest rates are so low because banks cannot find enough willing borrowers. So, the central bank has **no choice but to follow the market trend**. The problem: in a close-to-zero interest rate environment money does not circulate. This leads us to the next point.

5.5.2 *In the Current System: Central Banks Are Obliged to Print Money*

As mentioned, the ECB, Fed and other central banks in the industrialized world have been heavily criticized for the politics of "cheap" money in the last decade. A new term has even been invented: so-called "Quantitative Easing" (QE). QE refers to buying government bonds or other sovereign debt obligations. It is questionable whether this policy is even consistent with the charters of the respective central banks.[7] Central banks are allowed to buy and sell bonds of other countries as part of their Open Market Policy (e.g., Fernández Díaz et al., 2011) but **not of the same country** whose monetary policy the central banks are managing. Buying bonds of the same country the central bank belongs to (e.g., ECB buying bonds of European countries) is the same as granting loans to the State, even though they do not buy the bonds directly from the respective State but on the market. This is a form of artificially enlarging the central bank's balance sheet, i.e., putting fresh money into circulation. This policy has been used aggressively by the ECB, Fed, Bank of England and Bank of Japan since 2008.

Why do central banks do this? The answer is simple. If they did not, there would be a risk of not enough money circulating, because ever more money is being hoarded literarily under the mattress. Recall Fig. 4.4 that depicts the cash circulation in the eurozone. The reason for this behavior is that the interest rates were (until March 2022) close to zero in most industrialized nations, and there is therefore no incentive to lend money out. Remember, the interest rate is a non-hoarding premium, a "*reward for parting with liquidity*" (Keynes, 1936: 167). Since money is (unnaturally) easily storable it will be hoarded at home if holders of money are not

[6] The ECB, together with the US Federal Reserve Bank, has been calling for submissions of scientific papers to explain the low interest rates and low inflation rate (Fed, 2016).

[7] However, the European Court of Justice (2018) ruled that this policy does not infringe EU law, cf. ECJ C-493/17, 2018.

offered an incentive to lend it out. In consequence, money does not flow properly in a near-zero interest rate environment.

So, low interest rates provide little incentive to lend money out (or to deposit it in banks which will then lend it out). If more and more money is hoarded at home under the mattress, the velocity of money circulation (V) in Irving Fisher's *quantity* Eq. (4.2) decreases. To counteract deflation and avoid the occurrence of a situation similar to that of 1929, central banks print money. The so-called QE policy practiced by the ECB, the Fed, and the Bank of England is nothing but a desperate attempt to avoid deflation. But the freshly printed money also does not flow and ends up under mattresses or back at the central bank in the accounts of commercial banks.

When the ECB first started its QE policy, the German journal Handelsblatt ran a noteworthy headline, which summarizes the point well: "Banks are drowning in money – For the first time in the history of the Federal Republic of Germany, deposits exceed loans granted (author's translation)" (Schreiber and Dohms, 2014). One might wonder why the ECB increases the monetary base if markets are already drowning in liquidity? The answer again is that money does not flow adequately. Banks have been repeatedly accused of preferring to park money at the ECB as so-called excess liquidity instead of granting loans (see Fig. 5.10). This is problematic because money deposited at the central bank is not available to the general economy. Parking money at the central bank has the same effect as hoarding cash under the mattress. The problem is that banks can no longer find sufficient creditworthy borrowers. If they nevertheless granted loans, there would be a significant risk of default. When customers default on a large scale, the financial system collapses. Banks would then be accused of greed and reckless lending, as was the case in the so-called subprime crisis of 2008.

In their desperation, central banks try to avoid deflation by artificially expanding the monetary base. That is, instead of changing the design of money so that it circulates by itself, ever more flawed money—the type of money that does not flow

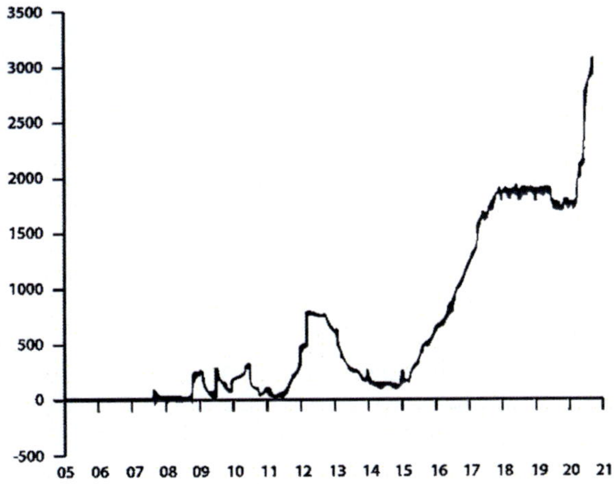

Fig. 5.10 Excess liquidity eurosystem (in billions of Euros). (Source: Own work. Based on data provided by ECB and data retrieved from Corradin et al., 2021)

well—is printed and ends up in ever larger amounts under the mattress or as excess liquidity at the central bank. Instead of injecting more and more of the non-flowing money into the economy in order to avoid deflation, we could produce money that flows better. This was the proposal of Silvio Gesell, who proposed a form of money that would naturally flow, without the need for interest to make it do so (see Chap. 10).

5.5.3 In the Current System: Central Banks Are Obliged to Lower Reserve Requirements over Time

In February 2012 the European central bank lowered the reserve requirement from 2% to 1%. The Fed and other central banks just recently even put the reserve requirement down to 0%. Why do central banks not only flood markets with "cheap" money but also tend to consistently lower reserve requirements? This question is even more significant in times of extreme excess liquidity and unstable financial markets. Should the central bank not increase reserve requirements instead of lowering them in order to reduce the risk of a bank run? The answer is, the longer the financial system exists the more the central bank is forced to not only lower interest rates and to print money but also to lower the reserve requirement.

The base money B, as long as it ends up in a bank account (which is the most likely scenario), grows by interest. And since the total money supply M is made up of cash plus short- and long-term deposits as well as other debt instruments that yield interest, we can define:

$$\mathrm{M} = \mathrm{B}(1+r) \tag{5.3}$$

We also know that money grows in the banking sector via the *money multiplier*, that is the ratio of base money to the reserve requirement, as explained above:

$$\mathrm{M} = \mathrm{B} / \mathrm{reserve} \tag{5.4}$$

That means:

$$\mathrm{B}(1+r) = \mathrm{B} / \mathrm{reserve} \tag{5.5}$$

We see that if the money supply grows by interest rate r (on the left side of the equation), the base money B (on the right side of the equation) must also grow if the central bank does not lower the reserve requirement. Hence, *interest money creation* effect (see Sect. 5.2) requires the central bank to increase base money B and/or lower the reserve requirements. This is what central banks all around the world are doing.

5.6 Our Money Is Not Neutral and Is Not Capital

5.6.1 The Fairytale of Monetary Neutrality

Textbooks on economics teach the theory of the "monetary neutrality" (e.g., Stiglitz, 1998: 187; Mankiw and Taylor, 2014: 587 ff.). According to this theory, the real (productive) economy is in the long run not influenced by money since money functions only as medium to facilitate the exchange of goods and services. In this context, a student with whom I discussed the topic once remarked that money could be compared to a knife: You can perform useful activities with it, such as cutting bread, or you can murder people with it. It depends on the user. The knife itself is, according to the opinion of my student, neutral. This metaphor is an expression of the misunderstanding of the effect of the interest rate. The fact that the interest rate induces the money supply and debt to constantly grow has been ignored. Money is not neutral (Suhr, 1988, 1989; Silva Moreira et al., 2016; Fuders, 2017; Sener, 2016). Rather, it generates a constant increase in a country's total money supply and debt and generates the need to grow. In fact, money behaves more like a knife—to stay with the metaphor—which has a life of its own. At first, it is small and useful. But then it begins to grow, first slowly and then ever faster, in an uncontrolled manner and, above all, it starts to stab and hurt people indiscriminately. Our money, as it is currently constructed, is not neutral. It **forces the real economy to grow** and **constantly expands the total money supply and total debt** and, thus, contributes to **growing inequality**.

And there is another problem, not commonly discussed in the mainstream economic literature, that is worth mentioning and that destroys the concept of monetary neutrality: the marginal utility of money (MU_M), that is the utility we gain from earning the next unit of money, depends on our "virtual wealth" (W_v) to use the expression coined by Frederick Soddy (1933), i.e., the units of money we already possess:

$$MU_M = f(W_v) \tag{5.6}$$

The more monetary units someone possess the less they will value earning the next unit (Fig. 5.11). This is a general problem that extreme wealth inequality produces and that makes any market equilibrium imperfect. The classical definition of efficiency, i.e., the cost of the last unit produced (marginal costs) equals the value the consumer attributes to this good (price), is perverted by the fact that the monetary units used to express costs and prices do not have the same perceived utility for each individual. A given price can be unaffordable for one person and extremely cheap for another. Hence, marginal production costs might equal price, but not necessarily the marginal utility of the consumer (Fuders and Pasten, 2020).

Fig. 5.11 $MU_M = f(W_v)$. (Source: Own work)

5.6.2 *Money Is Not Capital*

The reason why we find it difficult to grasp the problems inherent in our monetary (and especially the interest rate) system is that we do not distinguish the interest rate from real economic profits. This in turn could be due to the circumstance that since like 150 years we use the terms "money" and "capital" as synonyms, as if money was a factor of production like capital.[8] Many writings and books use the term "capital" (even in their title), but actually mean money (Marx, 1872; Bendixen, 1912; Hayek, [1950] 2009). Karl Marx, for example, explicitly asserts that money too is capital (Marx, 1872: 128 ff.; similar Marxist textbooks, e.g., Bogdanoff, 1925: 139 ff.). Many misunderstandings regarding the effects of interest on our economies and societies can be traced back to this basic misunderstanding. Capital goods (machines) serve in the production process along with labor, energy, and other inputs. Yet **money is not a factor of production**. It does not work (as banks sometimes would have us believe—"let your money work") nor does it produce offspring as Aristotle ([384 BC] 1995: 23) a little sarcastically pointed out. As outlined above money is not a real good and, thus, does not have any intrinsic value (Galbraith, 1983: 78; Keynes, 1983).

Money is only a medium to facilitate the exchange of real goods and services. With money we can buy capital, but money is not capital. If we believe that money is capital, then it is impossible to appreciate the difference between earning a profit from producing real goods and services and lending money. In economic models the two terms are often mixed up. Sometimes these models first distinguish "financial capital" (money) form real capital (see e.g., Frank, 2015) only later to mix them up, while the interest rate that banks pay is the benchmark used to evaluate any other productive investment. Many sophisticated models have been derived (e.g., the so-called CAPM—Capital Asset Pricing Modell) and Nobel prizes have been awarded

[8] While classical textbooks as that of Carl Menger still (correctly) emphasized that the "concept of money is entirely foreign to the concept of capital" (Menger [1871] 1976: 305).

Fig. 5.12 Plow and sack potato as "interest"

for theories on how to most profitably allocate investment portfolios, but in the end they all do the same thing: they equate profits earned from real productive investment with the interest rate.

The conflation of capital with money leads not only to the confusion of real economic production with interest received by lending money, but also to the wrong idea that interest was natural and had always existed. A classic example goes as follows. A blacksmith lends one of his plows to a farmer for a fixed period and receives a sack of potatoes as a reward from the harvest (Fig. 5.12). The sack of potatoes would represent the "interest" for lending the plow (Rösl, 2006). However, this reasoning is based on the conflation of capital with money. **The plow is real capital**. It is not money. The sack of potatoes is a real economic gain. A real good (with intrinsic value) was produced. In this example the blacksmith does not make money with money, which is why his gain is not interest but rather a lending fee for a real good (real capital). Note that this would be true even if the blacksmith was not paid with potatoes but with money that the farmer had earned by selling the potatoes. In this case, money is still not earned with money. Money supply will not grow, the financial system cannot decouple from real economy. If all 'loans' worked like this, there would be no money (and price) bubbles, and no need to maintain a certain economic growth rate.

5.6.3 Interest Is Not a Tax

Some may have difficulties recognizing the problems inherent in the money interest rate because they confuse them with taxes. The Bible says: "So give back to Caesar what is Caesar's, and to God what is God's" (Bible-NIV, 2015: Mathew 22:21). To be clear, interest and taxes are two **completely different things**. Conflating interest with taxes is similar to conflating interest with real economic production. Neither earning profits in the real (productive) economy, i.e., selling a product or service for profit, nor the paying of wages to labor nor the payment of taxes leads to an increase in money supply. Money merely flows from one hand to another. Interest, however, causes money (and debt) to multiply, as was already understood thousands of years ago by Aristotle ([384 BC] 1995), and this leads to the decoupling of the financial economy from the real economy.

5.6.4 Our Monetary System Is Comparable to a Ponzi Scheme

Some might view the paying of interest as comparable to paying for a good or service. Why would the money supply grow thereby? We can better understand this question by asking where the interest paid on deposits originates from if the monetary base (money issued directly by the central bank) has not been increased? Sure, the interest is ultimately paid by the borrowers, but where do they get the money to pay the interest? In fact, if interest is paid while the monetary base remains unchanged, someone else in the economy has to take out a loan so that this interest can be paid. This loan again requires the payment of interest, which makes a new loan necessary somewhere else in the economy. Figuratively, this effect could be described as a **snowball effect** that, once set in motion, grows larger and larger over time (until it crashes and causes a financial crisis).

This is the reason why it is mathematically impossible for everyone to repay their debts (Kremer, 2009). We could compare this system with a *Ponzi* scheme. A *Ponzi* scheme is a fraudulent (and illegal) pyramid investing scheme which generates returns for earlier investors with money taken from later investors without generating any real work, without producing real value, real goods or services (and without contributing to the well-being of society). A *Ponzi* scheme works as long as it continues to grow, i.e., while investors remain unaware of the fraud, and it is possible to find new investors from whose funds earlier investors can be repaid. The system unravels when the supply of new investors dries up. Our existing monetary system exhibits some similar dynamics. It breaks down when a significant amount of loans default.

To better understand the dynamics at work, let us imagine an economy consisting of only two people: "Happy Doña" owns all 100 monetary units in the economy and "Unlucky Fellow" who has no money at all (Fig. 5.13). Mr. Unlucky Fellow would also like to know how it feels to hold some money and borrows 10 money units from Happy Doña. He does not spend the money and at the end of the term pays it back to Happy Doña. But where does the interest come from that he now has to pay? The answer is, Mr. Unlucky must, in order to be able to pay the interest, take out a new loan from Happy Doña, that requires the payment of even more interest. After some time, Unlucky Fellow will be so indebted that, in order to be able to repay, he will have to sell off his belongings to Happy Doña so that he can, with the money receive, pay his debts. Happy Doña will thus end up with all money and some (or all) of Unlucky Fellow's belongings.

Fig. 5.13 Happy Doña and Unlucky Fellow

5.6.5 Clear Distinction Between Interest and Real Economic Profits

The reason why many believe that money is neutral might originate in a **missing distinction** between the interest charged for lending money and real economic profits. This relates to the failure to distinguish between money and capital as described above. If one believes that money is capital, i.e., a factor of production, then there is no clear distinction between profits driven out of lending money and lending real physical capital. The fact that in economics and in finance frequently any earnings are labelled as "interest",[9] leads to confuse returns generated out of productive investments with interest that is charged for loans.[10] Only the (money) interest rate generates an exponential growth of the money supply (and total debt) of an economy. It is possible to generate profits from productive investments even in a zero-interest environment. However, this profit is then generated solely from real economic output. Such production—free of the burden of interest—would result in an economy that is not compelled to constantly grow in order to keep up with the requirements of interest but would also lead to major differences from the current system from a moral-ethical perspective (see Chap. 9).

[9] Unfortunately, even in Silvio Gesell's work, whose monetary reform proposal will be discussed in Chap. 10 as the only truly workable solution, there is not always a clear distinction between interest from lending money and profits generated by productive investments.

[10] Interestingly, for Muslims the Koran prohibits to compare productive gains with interest for lending money, cf. Surah 2:275. The Holy Bible also prohibits lending for interest, see already references in Fuders (2017). Chapter 9 will examine this topic in more detail.

Drawing a clear distinction between interest and real economic profits is not always easy. As explained above, one of the characteristics of interest is that it must be paid **independently** of the success of the investment in real economic production financed with it. The distinction between interest from lending money and profits generated through real economic production can be illustrated by the following example:

A car mechanic asks his friend to invest in his business. The friends agree to do business together, and both invest a sum of money. The business consists of buying damaged vehicles, repairing them and then reselling them for a profit. The friends agree to share the profits equally. Although the mechanic's friend invested a larger sum of money, the mechanic commits to make up the difference in the form of labor. However, it turns out that the business doesn't go as well as anticipated, and not only are no profits generated, but there is a loss. The mechanic apologizes to his friend and says that he will pay him back the full amount that the friend had put in, but not the promised profit. The mechanic's friend was a bit taken aback since he had expected to earn a profit. He felt entitled to the promised profit. But when the latter slept on it for a night, he recognized that he was treated more than fairly in recouping the original sum invested. In fact, he reflected that it would have been fair if he had shared in the loss of the business, especially considering that his friend had done all the work. Why should the mechanic bear the loss alone? He therefore told his partner that he only wanted to receive the original sum minus half of the loss. In this way, both friends shared the loss.

This example (which took place in real life in the author's circle of friends), illustrates the difference between lending money (which requires the payment of interest regardless of the performance of the underlying business) and an equity investment, in which the investor shares in both profits and losses. If the friend who provided the money had demanded the repayment of his investment plus the promised profit, despite the fact that the business generated a loss, the investment would have been a classic loan (and the profit would have represented interest). However, the fact that he agreed to share in the loss meant that his participation in the business was ultimately not a loan but an equity investment. Thus, the key point to distinguish a loan from an equity investment, is whether a return is required **regardless** of the performance of the business financed with the loan. Unlike loans, equity investments participate in the risk and only yield a profit if the underlying business is profitable. And unlike loans equity investments do not cause growth of the money supply and debt.

5.7 The Growth Imperative Even Applies to Natural Resources

5.7.1 *The Tragedy-of-the-Commons Metaphor*

The in Chap. 4 described obligation to grow economically even applies to natural resources, regardless of whether they are publicly or privately owned (for an in-depth analysis see Fuders, 2020). In this context Herman Daly once remarked quite sarcastically but accurately that natural resources that do not reproduce at the pace of the interest rate are potentially in danger of extinction (Daly & Cobb, 1989).

Many books on environmental economics or green economics suggest that the destruction of the environment and the overexploitation of natural resources can be partly explained by the "Tragedy of Commons" (Hardin, 1968). In this metaphor, the overexploitation of nature results from natural resources (and as well as ecosystem services provided by them) being freely accessible goods (*Open Access Regimes*). Classical economic theory describes this as a "market failure" which can be corrected by assigning these goods private property rights (Demsetz, 1967; Cheung, 1970; Ault & Rutman, 1979; Barkley & Seckler, 1972; Dales, 1972; Smith, 1981; Welch, 1983; Stevenson, 1991; Common & Stagl, 2008; Daly & Farley, 2011). This argument assumes that only privately owned good can be efficiently allocated by the market. To put it more simply, private actors will take care of their investments in order to maximize their economic benefit. Generally speaking, this is correct. However, depending on the circumstances, especially given high interest rates, even private owners of natural resources may be incentivized to destroy them.

5.7.2 *The Goose that Lays Golden Eggs*

It is generally correct that we care more about what is our "own" than what is commonly owned, a fact that even (Aristotle, [384 BC] 1995) pointed out. This is the reason why communism does not deliver efficient results and communism is not a solution for sustainability or inequality. I will come back to this point later in Chap. 13. However, even a privately owned resource will not necessarily be immune to the pressure to grow and can be overexploited. Chile is an excellent example. Although the coastal areas are divided into exclusive management zones for aquaculture (*áreas de manejo)* and the forests are predominantly privately owned, neither aquaculture nor forestry are sustainably operated (regarding aquaculture see Sommer, 2009; Allsopp et al., 2008 regarding forestry: see Pastén et al., 2020; Manuschevich, 2020; Donoso & Romero, 2020).

That even private owners might overexploit and even completely destroy the natural resource has been pointed out also by Ostrom (1990). The reason is, again, the interest rate: the higher the interest rate, the smaller the Net Present Value (NPV) of future harvests. The NPV of a future cash flow is the value that would have to be deposited in a bank today, at the current interest rate, for the deposit to reach the expected future value of the cash flow. If the NPV of future harvests is less than the amount one can obtain by selling all of one's fish stock aquaculture or timber today, it is economically profitable to sell the entire stock today instead of managing aquaculture or harvests sustainably (Ackermann, 1994). In other words: if the interest rate paid by the bank is high enough, it might be worth it to harvest the entire stock and let the money "work" in the bank.

The problem can be explained through the children's **parable of the goose that lays golden eggs** (Daly & Farley, 2011). A lucky farmer one morning discovers that one of his geese started laying eggs of pure gold (Fig. 5.14). The farmer gets greedy and wants to get all of the eggs at once. To get the eggs he kills the goose and then finds no golden eggs inside. The moral of the story is that we shouldn't be greedy. But now let's change the story a little bit. Imagine that an angel appeared to the farmer and told him that the goose was immortal and would go on laying golden eggs forever. The farmer's children and grandchildren would continue receiving one golden egg per day. Should he sell the goose? Many would intuitively answer that of course he should not sell the magical goose. But from an economic perspective the question of whether or not the goose should be sold depends on two things—the price and the interest rate. If someone offers a price that is high enough that the money deposited in a bank would yield more interest than the value of the gold produced by the goose each day, then it would be a good decision to sell the goose. This is true even if the goose was immortal and would go on laying golden eggs forever.

The moral of this parable is that not even the owner of private property is free from the pressure of ensuring that his investment (the goose, a forest, or a stock of fish) generates a return that is at least equivalent to the rate of interest he would

Fig 5.14 Goose that lays golden eggs

receive if he sold the investment and deposited the proceeds in a bank. Here we find the reason for why clear cutting is still a common method of harvesting timber in many countries.

5.7.3 *Privatization Does Not Necessarily Avoid Unsustainable Resource Management*

That the interest rate is an opportunity cost for any productive investment places a value on time ("time is money"). This explains not only why clear cutting is the predominant method of harvesting timber (in countries where it is still legal), but also why **forest owners prefer to plant exotic, fast growing species** instead of native ones (For the case of Chile see chapters Pastén et al., 2020; Manuschevich, 2020; Donoso & Romero, 2020). Why do we see owners of forests in Chile, for example, planting exclusively eucalyptus or pine instead of a diversity of native species? A pine tree grows quickly, but it also sells relatively cheap. By contrast, a native tree grows more slowly but sells at a much higher price. It is not necessarily less profitable to plant native species. If, for example, a native tree takes twice as long as a pine tree (*Pinus radiata*) before it can be harvested, but would also sell for three times the price, as is the case with Chilean raulí - *Nothofagus alpina* one might think that a plantation owner would prefer to plant raulí, since the return in relation to the time needed until harvest is higher than that of a pine tree.

Yet in practice this turns out not to be the case. Instead, forestry companies prefer to plant exotic species, thereby causing loss of biodiversity, soil degradation and increasing the risk of forest fires (consider Chile in January 2016). The reason for this behavior is that profit calculations are distorted by the fact that **time is not a neutral factor**, because interest places a value on time. Accordingly, a species that grows faster often appears to be more profitable than a species that grows slowly, even if the latter sells at a much higher price relative to the time until harvesting. In other words, depending on the interest rate, the NPV of a tree that is harvested in 30 years might be less than the NPV of a species that is harvested in 15 years, even though it sells for twice the price (i.e., even though the proportion between the time to harvest and profit is the same). This 'time-is-money' pressure applies to both privately as well as publicly owned resources.

There are three ways to avoid exotic species monocultures. The first is for governments to restrict such actions, as is done in many parts of the world. The second is to implement subsidies (Pastén et al., 2020) or Payment-for-Ecosystem-Services (PES) Schemes (Schmitt Filho & Farley, 2020; Rosa Conceição & Börner, 2020). The third and probably most efficient is to create a monetary system in which interest does not distort how economic actors calculate profits (Fuders, 2020). This would allow for profitable long-term investments in natural resources, for example sustainable forestry practices such as planting of native species.

5.7.4 *Natural Resources in Open Access Regimes and the Role of Scale*

As we have just discussed, the interest rate and the resulting obligation to grow is a major factor behind the unsustainable management of even privately owned resources. That is to say, even though we usually take better care of our own property than of community property, and even though private owners have vested interests in maintaining their source of income, privately owned natural resources are at risk of unsustainable use. This is not (only) because of the greediness of the owners but also because of the obligation to grow at least enough to match the interest rate, which is the opportunity cost of any productive investment. On the other hand, resources are not automatically overused just because they are publicly owned. Many resources were **created in such abundance** that applying traditional technologies we would never be able to overuse them. For example, it would be essentially impossible to overexploit the oceans using small scale fishing boats, as they are still widely used in South America and Asia (Fig. 5.15). But unfortunately, the fishing industry is building ever larger boats, some of which catch 80 tons of fish or more in a day.

An argument can be made that the reason natural resources are overexploited is not due to the fact that they are freely accessible goods, but rather that it is a problem of **scale**. To continue with the fishing example, if we were still fishing with small artisanal boats, there would be no overexploitation **despite it being a free access regime**. There would be no danger of overfishing. The fact that Hardin's "Tragedy of the Commons" is today more relevant than ever is because the world has become relatively small in relation to exploitation technologies and not because the human population has grown, as Hardin suggested. Throughout human history, the ocean has been freely accessible, but in former times it was not "empty" or even close to

Fig. 5.15 Traditional Fisher boats in southern Chile (Los Molinos Bay near Valdivia). (Photo: Felix Fuders)

being overexploited as it is today. One might say, God foresaw the world with such an abundance that normally, if we used traditional fishing technologies, it is unlikely that we could ever exploit the ocean to the point of emptying it, even if the population of the world doubled. Rather, we contend that the ocean is being emptied by "**gigantism**", to use the words of Max-Neef (1986: 58, 63, 136, 149, 151 f., 184).[11] If we fished at a "**human scale**" (Max-Neef et al., 1991), then we would probably never empty the ocean.

One could ask what, then, is the reason for gigantism? Is it rooted in human greed? The answer is that it does not (only) have to do with greed. The owner of a modern fishing ship that currently exploits the ocean by catching large hauls of fish does not necessarily earn more than a fisherman 100 years ago. This is because the costs associated with the modern ship are enormous and therefore a relatively large number of fish must be caught just to recover those costs. It might be that the fishing company is so indebted that it does not even have enough income to pay down its debt. Why, then, are bigger and bigger fishing ships built? The answer is that the building of bigger ships is rooted in the **obligation to grow economically**, stemming from the interest rate that exerts a constant pressure to produce more and more, a fact that, unfortunately, is poorly understood in economic thinking. Our current financial system has existed for approximately 700 years, and during that time the economy has only done well if it was able to constantly increase production.[12]

5.7.5 *Worst Case Scenario: Open Access Combined with the Obligation to Grow*

We concluded that the ocean is not over exploited because it is an open access regime but because of the obligation to grow and the "gigantism" that this causes. Of course, open access combined with the obligation to grow and given modern harvesting technologies is the worst case scenario for the natural resource. And if the stock decreases and fish become scarce, fish prices will tend to skyrocket, a fact that we can observe by comparing fish prices today with those of two decades ago. This, in turn, generates an even stronger incentive to take out as fast as many fish as possible and let the money "work" on a bank account.

[11] The first author to conceptualize the term "gigantism" was probably Charbonneau (1969).

[12] Approximately 700 years ago, the "Bracteate," a coin that was difficult to hoard, was replaced by the "*denarius perpetuus*," the money that is used today (see: Walker, 1959; Azkarraga et al., 2011: 54; Fuders & Max-Neef, 2014b).

5.7.6 *Risks Are Consequently Underestimated*

Not only is the net present value (NPV) of future cash flows lower the higher the interest rate is and the further into the future the expected cash flow will occur. Also, the inherent risks of new technologies, such as nuclear power, genetic engineering or radiofrequencies are weighted less heavily in the present the higher the interest rate is since the present value of future damage is also discounted. Imagine a nuclear super-disaster taking place 200 years in the future. Although the costs would be enormous, the present value of these costs would be relatively small. If we were to discount costs equivalent to the current world GDP in 200 years with a discount rate of 5%, we would obtain a present value equivalent to that of a good apartment (Heal, 1998; Azqueta, 2007). Here we see again the effect of the exponential nature of compound interest, which turns even the largest future cash flows into small NPVs, and thus results in a systematic underestimation of future risks.

5.7.7 *Summary*

This chapter aims to demonstrate that the overexploitation of natural resources cannot be attributed solely to the existence of free access regimes. Rather, it originates from the **obligation to grow**. The root cause of this obligation is the existence of interest as an opportunity cost of every productive investment. Consequently, and contrary to conventional economic understanding, the privatization of natural resources that were previously freely accessible will not solve the underlying problem or increase economic efficiency. We also observed that the granting of private property rights to natural resources that were previously freely accessible does not necessarily prevent unsustainable management of those resources, and in fact does not even ensure that those resources will not be exploited to the point of total exhaustion.

5.8 Discussion

Mainstream economic thinking still doesn't recognize the existence of the growth imperative or explain why it exists. Money is generally thought of as neutral, at least in the long term (e.g., Mankiw & Taylor, 2014). If this were true, money would not generate pressure to grow. A similar reasoning as brought forth in this book in support of a growth imperative inherent in our monetary system we can find in Daly and Cobb (1989), Kennedy (1990, 2011), Creutz (1993, 2018), Douthwaite (1999), Lietaer (2002), Binswanger (1996, 2006), Sorrell (2010), Löhr (2012), Costanza et al. (2012), Lietaer et al. (2012), Eisenstein (2012, 2013), Farley et al. (2013), Heinsohn and Steiger (2013), Kuzminski (2013), Fuders and Max-Neef (2014a, b),

Boyle and Simms (2015), and Henke (2017). However, there are also scholars who outright deny the existence of a growth imperative inherent in our monetary system. In the following section, these critical views will be discussed and discarded.

5.8.1 The Circular-Flow Argument

The most current argument against the notion of a growth imperative inherent in our monetary system is what could be termed the "circular-flow" argument. We find it in papers published by Freydorf et al. (2012), Wenzlaff et al. (2014), Berg et al. (2015), Jackson and Victor (2015), Richters and Siemoneit (2017, 2019), and Larue (2020). Regarding the argument (see argument No 1 in Sect. 4.5 of this book) in support of the idea that ever-growing debt requires us to constantly work harder in order to keep up with the interest burden, thus creating a permanent necessity to grow Freydorf et al. (2012), Wenzlaff et al. (2014) as well as Berg et al. (2015) argue that, if interest income is fully consumed, a stationary redistributive economy would be possible. To cite Wenzlaff et al. (2014: 50):

> The need for growth can only be justified by the incomplete consumption of interest income (…). Only then must the economy grow (at least) to the extent of the unconsumed portion, (...) The growth constraint is thus not a consequence of money creation, but a consequence of income-dependent saving combined with an interest rate determined by liquidity preference (author's translation).

In a similar way Jackson & Victor (2015: 45) conclude:

> The fact that the charging of interest on its own does not lead to a growth imperative could perhaps have been inferred from the realisation that the only interest payments which contribute directly to the GDP are the net interest payments of firms. All other interest payments turn out to be transfers between sectors and neither restrict nor enhance aggregate demand in themselves.

Larue makes an akin argument (2020). He draws on an example used by Lietaer et al. (2012) to illustrate the growth imperative inherent in our interest based monetary system. This example goes like this: A foreigner proposes to lend money to villagers but asks them to repay it with interest at the end of the year. Accepting that offer allows the villagers to produce more but also generates an obligation to grow in order to repay the debt.[13] Larue goes on to argue that if we modify this example only slightly then the growth imperative would not result. Specifically, Larue asks the reader to assume that the money lender is not a foreigner as in the original example but a member of the community. If that person consumes all his revenue, for instance by buying the villagers' goods, money would flow back to them, so they could repay their loans without being required to produce more (Larue, 2020).

The key argument in the above examples is that interest earnings are spent and thus represent income for someone else in the economy. Hence, the money keeps

---

[13] A similar example we find in Eisenstein (2013).

circulating. The authors conclude that even though the individual who borrows money has to pay back principal plus interest and therefore is indeed obliged to grow, for the economy as a whole this is not the case, as long as money keeps circulating. Some authors such as Wenzlaff et al. (2014) and Berg et al. (2015) admit that saving could be a problem which would result in a growth imperative, but the true cause of the growth imperative would then be the "liquidity-preference" as Keynes (1936: 165 ff., 194 ff.) termed it, not interest.

Here we again encounter the common misunderstanding that money saved is withheld from circulation, which is not the case. Money saved in banks is kept in circulation via loans or other investments since the bank must pay interest on deposits. With the exception of money saved physically at home, money keeps circulating in any case.

The problem arises not from saving but from interest paid on loans (or bank deposits, which can be thought of as loans granted to banks). Any interest paid (or demanded) creates new money units in bank accounts (or loan contracts) **that did not previously exist**. I referred to this in Sect. 5.2 as "interest money creation". This means, if somewhere in the economy (or the village in Larue's (2020) example) interest is paid and the central bank has not increased the money base, then the interest payment requires necessarily a new (and again interest bearing) loan somewhere else in the economy (see Unlucky fellow's parable in Sect. 5.6.4).

A circular flow of all monetary units **is not possible** without constant increase of debt. It is not easy to see this in an economy consisting of millions of people rather than just two. If interest causes bank accounts to grow, total debt must also grow. Another way to think of this is to keep in mind that money is the counterpart of debt, as one person's money is another person's debt. If deposits grow via interest, the money supply (cash plus deposits) also grows. But where does this extra money come from if the central bank does not increase the monetary base? Some of the cited papers use sophisticated mathematic models. This gives the impression of rigorousness. However, any model is only as good as its assumptions. Depending on how we define the assumptions, we can influence the outcome of any mathematical model. The models used in the cited papers neglect the effect of "interest money creation".

5.8.2 The Inverted-Causality Argument

Another current argument against a growth imperative inherent in our monetary system is what could be termed the "inverted-causality" argument. We find it in papers published by Jackson and Victor (2015), Cahen-Fourot and Lavoie (2016), Lee and Werner (2018), and Larue (2020). Jackson and Victor (2015) construct a mathematical model in which they allegedly demonstrate that changes in the money supply are driven by what is happening in the real economy, rather than the other way around. Increased output results in increased loans demanded by firms, creating a higher level of deposits from households. They conclude (p. 40) that

> (...) with a suitable choice of initial values, a stationary state economy is possible. More interesting for our purposes in this paper is that this result is obtained from an economic model with interest-bearing debt.

According to Larue (2020), following the reasoning of the post-Keynesians, an important argument against the existence of a growth imperative would be that money creation is in their eyes an endogenous process. Citing Cahen-Fourot and Lavoie (2016), Larue (2020: 7) argues "Money is created through bank credit when economic agents have a credit-worthy demand for it". In other words, this would mean that "the rise in production takes shape in the mind of producers before money is created and is effectively realized when credit is granted and money is created to finance it". According to this reasoning, growth in production drives growth of the money supply, therefore "the money supply can in no way be held responsible for the growth of output, which entails that there cannot be a money-led growth imperative" (Larue, 2020: 7). A similar reasoning is advanced in a paper by Lee and Werner (2018).

In order to respond to this viewpoint, I would first point out that banks cannot create money endogenously (out of nothing), as we discussed in Chap. 2. However, it is true that a growing economy can lead to more demand for credit and thus the money multiplier effect can lead to an increase in money supply within the banking sector (but not via a single bank). As discussed above, this can be viewed positively since it could cause Fullarton's *law of reflux* to finally work. The money supply would automatically adjust to the needs of the real economy. In an interest-free monetary system, this would cause the money supply to adjust automatically to economic cycles, resulting in no (or much less) inflation/deflation.

What post-Keynesians overlook is the role of interest rates. Interest puts pressure on banks to **increase the volume of debt** since it causes deposits to grow. As long as interest is paid on deposits (even if the rate is very low), banks **must** continuously increase loan volume or else they will go bankrupt. Seen from this perspective, demand for debt does not depend on productive output, but **the other way around**. Banks will employ ever more aggressive marketing practices in order to expand loan volume and will even lower credit standards and make loans to riskier borrowers if necessary. Or they can try to generate returns through other forms of investments such as the stock market or investment banking. When such behavior leads to inevitable losses, there will invariably be accusations of greediness or unlawful behavior.[14]

In their model, Jackson and Victor (2015) also neglect the *interest-money-creation* effect (see Sect. 5.2). Again, if the requirement for the payment of new

[14] If this is not enough the public sector usually serves as a 'good' (loan demanding) client of banks, a sort of 'borrower of last resort'. Also, as we currently can observe in world politics pandemics or wars seem to serve as a perfect pretext for government to be a good (loan demanding) client for banks and to serve as borrower of last resort. Chapter 16 will take a deeper look at interest as the true cause of recurring financial crises.

interest has been created (be it through a bank account or a loan contract), where does this money come from if the central bank has not increased the money base? The answer is it can only come from the creation of new loans.

5.8.3 *Neglecting Compound Interest*

Another current argument against a growth imperative inherent in our monetary system is the denial of the compound interest effect on money supply. Berg et al. (2015) as well as Richters and Siemoneit (2017, 2019) argue that proponents of the monetary growth imperative model would neglect the fact that creditors can consume or invest their interest income, so that money flows back into the economy and thus no growth of the money supply is caused. They assert that a growth imperative only results if creditors choose to keep their money in banks, since only then is compound interest paid, thus leading to exponential growth. However, as explained above (Sect. 4.3), interest income that is spent will most likely end up in another bank account somewhere else in the economy and thus generate interest there. Hence, all bank accounts taken together (and thus the total debt and money supply) always grow exponentially due to interest.

5.8.4 *The Profit-Maximization Argument*

Some authors argue that the aim of steadily increasing profits and in particular the pursuit of labor productivity growth is the primary cause of the growth imperative (Jackson and Victor, 2015). However, this perspective is backwards and conflates cause with effect. Why should we unceasingly attempt to increase output and productivity when we are already doing well? Conversely, interest-driven growth creates a constant pressure to improve efficiency and especially to improve labor productivity. This is because if firms fail to keep pace with the growth imperative the second-best strategy is cutting costs. We will come back to this topic in Chap. 7.

And even if we accept the notion that there is a greed-driven growth imperative, this would also be solved by the solution presented in Chap. 10. Indeed, in the natural money system that will be proposed, the Keynesian preference for liquidity (Keynes, 1936), which ultimately gives rise to the existence of interest as a non-hoarding premium, is nullified by the fact that money in this proposal perishes like real goods and is thus not permanently accumulable. The desire to earn more and more, i.e., the desire to grow, would be significantly curbed since profits could not be accumulated and preserved without risk.

5.8.5 *The Ambiguous-Role-of-Interest Argument*

Diedrich (1994) studies the case of forests (as an example of renewable natural resource management) and argues that the interest rate has a dual role, one to act as "discount rate" and one to act as the "price of capital". In his view, both roles exert opposite effects on the exploitation of renewable natural resources, and "the overall outcome then depends on the relative powers of the two tendencies" (Diedrich, 1994: 4). On the one side, as also is argued in this book, interest as discount rate for future revenues artificially overvalues time, thus creating pressure to grow fast, i.e., to heavily exploit the natural resource. On the other side, money as the "price of capital" makes capital intensive production more expensive since the money market subjects producers to the interest rate as the price of capital ("capital" is defined in Diedrich's (1994) model correctly as the value of physical capital stocks).

The author admits in agreement with Ackermann (1994) that the long-term use of forests is more intensive in high interest rate environments than under low ones. But in the short term he argues that high interest rates lead to a shift from capital intensive to labor-intensive production models that result in less exploitation of timber. This is because "in its role as discount rate, the higher interest rate makes rising future production costs less relevant compared with current costs" (Diedrich, 1994: 13). Similar arguments have previously been made by Farzin (1984).

However, this is, as the author correctly states, only true in the short-term. And even in the short-term, it is not necessarily correct. The assumption made in Diedrich's model (1994) that high capital costs lead to more labor-intensive production is questionable. Business owners won't simply discard existing productive capital and replace it with labor as a result of rising capital costs. And even if they did, this would not necessarily result in reduced throughput of timber. Why would the owner of a wood chip processing company, for example, demand less timber? Diedrich argues that as labor-intensive production is relatively less dependent on a certain minimum level of timber utilization, reduced timber supplies would lead to dramatic increases in the uses of other factors of production. This argument is weak. It is not clear that a more labor-intensive production process necessarily changes the "timber sensitivity" as Diedrich (1994: 13) argues.

Some argue that higher interest rates reduce overall production (e.g., Musu, 1994) and thereby counteract the effect of interest as a discount rate for future cash flows (or as opportunity cost of any productive investment). However, as discussed in Sects. 5.1 and 5.2, there are powerful arguments that the effect of interest on the money supply is not correctly understood yet. As argued above, contrary to conventional wisdom, higher rates lead in the long run to increased GDP growth. Hence, interest not only incentivizes growth but also makes that growth possible. Irrespective of strong short-term fluctuations, empirically we can observe a clear long-term positive correlation between interest rates and GDP growth rates (while the GDP growth rate follows the interest rate, see again Fig. 5.1).

Finally, regarding the case of natural resources, some might argue that higher interest rates make harvesting more costly, and this counteracts any pressure to grow (e.g., Farzin, 1984). This is true. But there is no reason to assume that higher harvesting costs can fully counteract the growth imperative resulting from interest as the benchmark against which the returns of all productive investments are measured. Investors take harvesting costs into account. If they decide to invest they do so because their calculations indicate that a profit can be made. And these profits must be at least as high as the interest rate that can be earned risk-free via a bank account or other financial asset. We could even argue the other way around that higher harvesting costs **create additional pressure** to increase productive output in order to counteract those higher costs.

5.8.6 *Summary*

We discussed and discarded five arguments which are often brought forth to argue against the existence of a growth imperative induced by our monetary system. All of these arguments are rooted in a misunderstanding of the effect interest exerts on money supply (and debt). Lending creates new money, monetary units that were not created by the central bank and which only can be repaid throughout the creation of new interest-bearing debt. From a macroeconomic perspective, this is the most important consideration. However, the most important point from a microeconomic perspective (supply side)—and in my opinion the most important point of all—is that the interest rate is the opportunity cost of any productive investment. Any investment must yield at least what the investor could earn by lending money to a bank. Therefore, the economy must move at least at the rhythm of the interest rate. As far as I know, none of the authors who deny the existence of a growth imperative inherent in our monetary system address this argument thoroughly.

5.9 Conclusions

While Chap. 4 discussed the growth imperative from a microeconomic (*supply side*) point of view, Chap. 5 explained the growth imperative from a macroeconomic (*demand side*) point of view. Both viewpoints look at the problem from different perspectives and combine to create a **completely new understanding** of the growth imperative. This perspective explains why the growth imperative applies to privately owned resources as well as resources in open access regimes (Fuders, 2020).

This chapter advances the hypothesis that environmental degradation is not solely a product of classic market failure but also results from the fact that the economy is obliged to grow. This is the **primary failure** of our economic model, though it is not recognized as such in conventional economic texts. The growth imperative is rooted not only in the greed of economic agents seeking to increase profits but is also driven by a financial system in which money is the counterpart of debt and both the money supply and total debt grow constantly through the mathematical logic of interest. This dynamic exerts a constant pressure on the economy to grow at least at the same rate. In other words, regardless of whether or not natural resources are freely accessible goods, they will be overexploited in the long term because of this logic of growth. The next Chap. 6 discusses the connection between economic growth and the degradation of our natural environment in greater detail.

References

Ackermann, F. (1994). The natural interest rate of the forest: Macroeconomic requirements for sustainable development. *Ecological Economics, 10*, 21–26.

Allsopp, M., Johnston, P., & Santillo, D. (2008). *Challenging the aquaculture industry on sustainability*. Greenpeace/University of Exeter.

Aristotle. (1995). Politik 1st Book. In *Aristoteles Philosophische Schriften in sechs Bänden. Band 4*. Felix Meiner Verlag.

Ault, D., & Rutman, G. (1979). The development of individual rights to property in tribal Africa. *Journal of Law and Economics, 22*(1), 163–182.

Azkarraga, J., Max-Neef, M., Fuders, F., & Altuna, L. (2011). *La evolución sostenible II*. Eskoriatza.

Azqueta, D. (2007). *Introducción a la economía ambiental*. McGraw Hill.

Barkley, P. W., & Seckler, D. W. (1972). *Economic growth and environmental decay – The solution becomes the problem*. Hartcourt Brace.

Bendixen, F. (1912). *Geld und Kapital*. Duncker&Humblot.

Berg, M., Hartley, B., & Richters, O. (2015). A stock-flow consistent input–output model with applications to energy price shocks, interest rates, and heat emissions. *New Journal Physics, 17*, 015011.

Bible-NIV. (2015). *NIV Zondervan study Bible: New international version*. Grand Rapids: Zondervan.

Bichlmaier, S. (2010). Umlaufsicherung vs. Niedrig-Zins-Politik und Inflationsstrategie der Notenbanken. *Humane Wirtschaft, 2*, 6.

Binswanger, H. C. (1996). Geld und Wachstumszwang. In B. Bierbert (Ed.), *Die Dynamik des Geldes: Über den Zusammenhang von Geld, Wachstum und Natur* (pp. 113–127). Frankfurt.

Binswanger, H. C. (2006). *Die Wachstumsspirale: Geld, Energie und Imagination in der Dynamik des Marktprozesses* (2nd ed.). Metropolis.

Bogdanoff, A. (1925). *A short course of economic science* (2nd ed.). Communist Party of Great Britain.

Boyle, D., & Simms, A. (2015). *The new economics a bigger picture*. Earthscan.

Bundesbank. (2019). *Geld und Geldpolitik*. Bundebank.

Cahen-Fourot, L., & Lavoie, M. (2016). Ecological monetary economics: A post-Keynesian critique. *Ecological Economics, 126*, 163–168.

Charbonneau, B. (1969). *Le jardin de Babylone*. Gallimard.

Cheung, S. (1970). The structure of a contract and theory of a non-exclusive resource. *Journal of Law and Economics, 13*(1), 49–70.

Common, M., & Stagl, S. (2008). *Introducción a la Economía Ecológica*. Reverté.

Corradin, S., Hoerova, M., & Schepens, G. (2021). *Euro area money markets over the past 15 years: Changes, driving factors and implications for monetary policy.* [Online] Available at: https://www.ecb.europa.eu/pub/economic-research/resbull/2021/html/ecb.rb210318~fb6634551a.en.html. Accessed 20 Oct 2022.

Costanza, R., et al. (2012). *Building a sustainable and desirable economy-in-society-in-nature*. United Nations Division for Sustainable Development.

Court of Justice of the European Union. (2018). *Judgment in Case C-493/17*. Court of Justice of the European Union.

Creutz, H. (1993). *Das Geldsyndrom. Wege zu einer krisenfreien Marktwirtschaft*. Wirtschaftsverlag Langen Müller/Herbig.

Creutz, H. (2018). *Das Geldsyndrom* (4th ed.). Kubo.

Cuadrado, J. R., et al. (2015). *Política Económica – Elaboración, objetivos e instrumentos* (5th ed.). McGraw Hill.

Dales, J. H. (1972). *Pollutions property and prices* (4th ed.). University of Toronto Press.

Daly, H., & Cobb, J. (1989). *For the common good: Redirecting the economy toward community, the environment, and a sustainable future*. Beacon Press.

Daly, H., & Farley, J. (2011). *Ecological economics – Principles and applications* (2nd ed.). Island Press.

Demsetz, H. (1967). Toward a theory of property rights. *The American Economic Review, 57*(2), 347–359.

Diedrich, M. (1994). Environmental resources and the interest rate. *Economie Appliquée, 2*, 161–180.

Donoso, P. J., & Romero, J. E. (2020). Towards a new forest model for Chile: Managing forest ecosystems to increase their social, ecological and economic benefits. In F. Fuders & P. Donoso (Eds.), *Ecological economic and socio ecological strategies for forest conservation* (pp. 159–170). Cham.

Dornbusch, R., Fischer, S., & Startz, R. (2009). *Macroeconomía* (10th ed.). McGraw-Hill.

Douthwaite, R. (1999). *The ecology of money*. Green Books.

Eisenstein, C. (2012, March). We can't grow ourselves out of debt, no matter what the Federal Reserve does. *The Guardian*. https://www.theguardian.com/commentisfree/2012/sep/03/debt-federal-reserve-fixation-on-growth

Eisenstein, C. (2013). *Ökonomie der Verbundenheit wie das Geld die Welt an den Abgrund fÜhrte – und sie dennoch jetzt retten kann*. Scorpio.

Farley, J., et al. (2013). Monetary and fiscal policies for a finite planet. *Sustainability, 5*, 2802–2826.

Farzin, Y. H. (1984). The effect of the discount rate on depletion of exhaustible resources. *Journal of Political Economy, 92*(5), 841–851.

Federal Reserve Bank (Fed). (2016). *International research forum on monetary policy*. [Online] Available at: https://www.federalreserve.gov/newsevents/conferences/irfmp2016-call-for-papers.htm. Accessed 20 Oct 2022.

Fernández Díaz, A., Parejo Gámir, J. A., & Rodríguez Sáiz, L. (2011). *Política Económica* (4a ed.). McGraw-Hill.

Fisher, I. (1933). The debt deflation theory of great depressions. *Econometrica, 1*(4), 339–357.

Frank, R. (2015). *Microeconomics and behaviour*. McGraw-Hill/Irwin.

Freydorf, C., et al. (2012). *Wachstumszwang in der Geldwirtschaft? Theoretische Erwägungen*. Denkwerk Zukunft.

Fuders, F. (2011). Wie ZerstÖrung von Sachkapital dem Finanzsystem in die Hände spielt. *Humane Wirtschaft, 2*, 34–38.

Fuders, F. (2017). Neues Geld fÜr eine neue Ökonomie – Die Reform des Geldwesens als Voraussetzung fÜr eine Marktwirtschaft, die den Menschen dient. In *Finanzwirtschaft in ethischer Verantwortung – Erfolgskonzepte fÜr Social Banking und Social Finance* (pp. 121–183). Springer-Gabler.

Fuders, F. (2020). *The 'tragedy of the commons' and the role of the money interest rate*. Springer.

Fuders, F., & Max-Neef, M. (2014a). Local money as solution to a capitalistic global financial crisis. In *From capitalistic to humanistic business*. Palgrave Macmillan.

Fuders, F., & Max-Neef, M. (2014b). Dinero, deuda y crisis financieras. Propuestas teórico-prácticas en pos de la sostenibilidad del sistema financiero internacional. In *Economía Internacional. Claves teórica-prácticas sobre la inserción de Latinoamérica en el mundo*. LATIn.

Fuders, F., & Pasten, R. (2020). Allocative efficiency and property rights in ecological economics: Why we need to distinguish between man-made capital and natural resources. In *Ecological economic and socio ecological strategies for Forest conservation – A transdisciplinary approach focused on Chile and Brazil* (pp. 43–56). Springer Nature.

Fuders, F., Mondaca, C., & Azungah Haruna, M. (2013). The central bank's dilemma, the inflation-deflation paradox and a new interpretation of the Kondratieff waves. *Economía (U. de los Andes), 36*, 33–66.

Fullarton, J. (1845). *On the regulation of currencies* (2nd ed.). London: John Murray.

Galbraith, J. K. (1983). *El Dinero – De dónde vino/Adónde fue* (2nd ed.). Hyspamerica.

Hardin, G. (1968). The tragedy of the commons. *Science, 162*, 1243–1248.

Hayek, F. A. (2009). *The pure theory of capital (Reprint form second impression 1950)*. Ludwig van Mises Institute.

Heal, G. (1998). *Valuing the future: Economic theory and sustainability*. Columbia University Press.

Heinsohn, G., & Steiger, O. (2013). *Ownership economics: On the foundations of interest, money, markets, business cycles and economic development*. Routledge.

Henke, S. (2017). *Fließendes Geld für eine gerechtere Welt: Warum wir ein alternatives Geldsystem brauchen, wie es funktioniert und welche Auswirkungen es hat*. Tectum Wissenschaftsverlag.

Jackson, T., & Victor, P. A. (2015). Does credit create a 'growth imperative'? A quasi-stationary economy with interest-bearing debt. *Ecological Economics, 120*, 32–48.

Kennedy, M. (1990). *Geld ohne Zinsen und Inflation – Ein Tauschmittel, das jedem dient*. Permakultur-Verlag.

Kennedy, M. (2011). *Occupy money – Damit wir zukünftig alle die Gewinner sind*. J. Kamphausen.

Keynes, J. M. (1936). *General theory of employment, interest and money*. Hartcourt.

Keynes, J. M. (1983). Economic articles and academic correspondence. In D. Moggrigde (Ed.), *The collected writings of John Maynard Keynes* (p. 402). Macmillan.

Kremer, J. (2009). Eine andere unsichtbare Hand des Marktes: Von den blinden Flecken der Volkswirtschaftslehre. *Humane Wirtschaft, 1*, 2–12.

Kuzminski, A. (2013). *The ecology of money: Debt, growth, and sustainability*. Lexington Books.

Larue, L. (2020). The ecology of money: A critical assessment. *Ecological Economics, 178*, 106823.

Lee, K. S., & Werner, R. (2018). Reconsidering monetary policy: An empirical examination of the relationship between interest rates and nominal GDP growth in the U.S., U.K., Germany and Japan. *Ecological Economics, 146*, 26–34.

Lietaer, B. (2002). *The future of money – Creating new wealth, work and a wiser world*. Random House.

Lietaer, B., Arnsperger, C., Goerner, S., & Brunnhuber, S. (2012). *Money and sustainability: The missing link – A report from the Club of Rome – EU Chapter to the Finance Watch and the World Business Academy*. Triarchy Press.

Löhr, D. (2012). The euthanasia of the rentier—A way toward a steady-state economy? *Ecological Economics, 84*, 232–239.

Mankiw, G., & Taylor, M. (2014). *Economics*. Cengage Learning.

Manuschevich, D. (2020). Land use as a socio-ecological system: Developing a transdisciplinary approach to studies of land use change in South-Central Chile. In F. Fuders & P. Donoso (Eds.), *Ecological economic and socio ecological strategies for forest conservation* (pp. 79–97). Cham.

Marx, K. (1872). *Das Kapital – Kritik der politischen Oekonomie. Erster Band, Buch I: Der Produktionsprozess des Kapitals* (4th ed.). Meissner.

Max-Neef, M. (1986). *La Economía Descalza – Señales desde el mundo invisible*. Nordan.

Max-Neef, M., Elizalde, A., & Hopenhayn, M. (1991). *Human scale development – Conception, application and further reflections*. Apex.

Menger, C. (1976). *Principles of economics (Translation of the German version "Grundsätze der Volkswirtschaftslehre" Vienna 1871)*. Ludwig von Mises Institute.

Musu, I. (1994). Sustainable economy and time preference. *Structural Change and Economic Dynamics, 5*(1), 81–86.

OECD. (2023). *Main economic indicators (database)*. [Online] Available at: https://doi.org/10.1787/data-00052-en. Accessed 1 Feb 2023.

Ostrom, E. (1990). *Governing the commons: The evolution of institutions for collective action*. Cambridge University Press.
Pastén, R., Nazal, N., & Fuders, F. (2020). Subsidizing green deserts in southern Chile: Between fast growth and sustainability of forest management. In F. Fuders & P. Donoso (Eds.), *Ecological economic and socio ecological strategies for forest conservation* (pp. 59–78). Cham.
Richters, O., & Siemoneit, A. (2017). Consistency and stability analysis of models of a monetary growth imperative. *Ecological Economics, 136*, 114–125.
Richters, O., & Siemoneit, A. (2019). Growth imperatives: Substantiating a contested concept. *Structural Change and Economic Dynamics, 51*, 126–137.
Rosa Conceição, H. D., & Börner, J. (2020). Understanding adoption and design of incentive-based forest conservation policies: A case study of the SISA Program in Acre, Brazil. In F. Fuders & P. Donoso (Eds.), *Ecological economic and socio ecological strategies for forest conservation* (pp. 241–258). Cham.
Rösl, G. (2006). *Regionalwährungen in Deutschland – Lokale Konkurrenz für den Euro? – Diskussionspapier Reihe 1: Volkswirtschaftliche Studien (No 43/2006)*. Deutsche Bundesbank.
Schmitt Filho, A., & Farley, J. (2020). Transdisciplinary case study approaches to the ecological restoration of rainforest ecosystems. In F. Fuders & P. Donoso (Eds.), *Ecological economic and socio ecological strategies for forest conservation* (pp. 185–212). Cham.
Schreiber, M., & Dohms, H. (2014). Banken ertrinken im Geld – Zum ersten Mal in der Geschichte der Bundesrepublik übersteigen die Einlagen die ausgereichten Kredite. *Handelsblatt, 17*(10), 2014.
Sener, U. (2016). *Die Neutralitätstheorie des Geldes Ein kritischer Überblick* (2nd ed.). Universitätsverlag Potsdam.
Silva Moreira, T. B., Miranda Tabak, B., Mendonça, M. J., & Sachsida, A. S. (2016). An evaluation of the non-neutrality of money. *PLoS, 11*(3), 1–20.
Smith, R. (1981). Resolving the tragedy of the commons by creating private propeny rights in wildlife. *CATO, 1*, 439–468.
Soddy, F. (1933) (Reprint 1961). *Wealth, virtual wealth and debt - The solution of the economic paradox*. Briton.
Sommer, M. (2009). Acuicultura Insostenible en Chile. *Revista electrónica de Veterinaria, 10*(3), 1–23.
Sorrell, S. (2010). Energy, economic growth and environmental sustainability: Five propositions. *Sustainability, 2*, 1784–1809.
Stehlík, M., et al. (2017). Financial and risk modelling with semicontinuous covariances. *Information Science, 394–395*, 1–6.
Stevenson, G. (1991). *Common property economics – A general theory and land use applications*. Cambridge University Press.
Stiglitz, J. E. (1998). *Macroeconomía* (2nd ed.). Ariel.
Suhr, D. (1988). *Alterndes Geld – Das Konzept Rudolf Steiners aus geldtheoretischer Sicht*. Novalis Verlag.
Suhr, D. (1989). *The capitalistic cost-benefit structure of money – An analysis of money's structural nonneutrality and its effects on the economy*. Springer.
Walker, K. (1959). *Das Geld in der Geschichte*. Rudolf Zitzmann Verlag.
Welch, W. (1983). The political feasibility of full ownership property right: The cases of pollution and fisheries. *Policy Sciences, 16*, 165–180.
Wenzlaff, F., Kimmich, C., & Richters, O. (2014). *Theoretische Zugänge eines Wachstumszwangs in der Geldwirtschaft*. Universität Hamburg.

Chapter 6
Economic Growth in the Long Run Is Unsustainable

Anyone who believes that exponential growth can go on forever in a finite world is either a madman or an economist.

Kenneth E. Boulding (1973) (quoted from Jackson & Victor, 2019: 950).

6.1 What Is Sustainability?

6.1.1 Four Lines of Thought

We have now understood that given our current monetary system we must grow. At the online encyclopedia "Investopedia" we even read: "A healthy economy is one where both exports and imports are experiencing growth" (Investopedia, 2021). The entry does not further explain this affirmation. How can an economy that is ever growing be healthy? Is it possible to ever grow? Can a constant growth be sustainable? This chapter will investigate this question. We will understand that constant growth cannot be healthy (see already Fuders, 2021), nor sustainable.

Since the word 'sustainable' is often used, misused and abused, we should start by clarifying our definition of the word. We can distinguish at least four lines of thought relating to the concept of sustainability (Azkarraga et al., 2011).

- **Sustainable economic growth:** We keep on growing while trusting in scientific-technical innovation to solve our ecological problems.
- **Weak sustainability**: Still based on an anthropocentric vision it gives emphasis on economic growth, but takes into account the limits of nature and, as a consequence, proposes 'controlling' the market as a solution to the environmental problem. The so-called green economics would be along these lines.
- **Strong sustainability:** Strong commitment to the preservation of all natural resources, without consideration of economic criteria. To solve the ecological

© The Author(s), under exclusive license to Springer Nature Switzerland AG 2023

F. Fuders, *How to Fulfil the UN Sustainability Goals*,
https://doi.org/10.1007/978-3-031-37768-6_6

crisis, technological and economic solutions are not enough; measures must be taken at the educational and social levels.

- **Deep Ecology:** Biocentric vision inspired by Arne Naess' (1973) that considers every living organism with an intrinsic value, whose protection is a moral and political duty. The solution to the problem requires a profound change in society.

These concepts could be arranged on a spectrum running from 'free markets' to 'consciousness' (see Fig. 6.1) as if market economy and consciousness were two opposed concepts and the more sustainable we want our lifestyle to become, the more conscious (and the less market orientated) we should behave. A key point is that the concepts of market economics and economic growth are often conflated. We are so conditioned to view economic growth as a beneficial and necessary force that it is hard for us to even imagine an economy that does not grow. As outlined in the previous chapter, economic growth and market economy are two completely different things. Changing our form of money would enable us to have a working market economy without the need for growth. The growth imperative emanates from our monetary system, not from market economy.

We could even argue that the truth is the other way around. If we truly want markets to function correctly, i.e., if we want an economy that serves people by helping them satisfy their fundamental human needs, we need to get as close as possible to the ideal of so-called "perfect competition" (Sect. 10.4 will further elaborate on this aspect). But we need to change our monetary system to one that is more consistent with nature, and thereby get rid of the **most important and at**

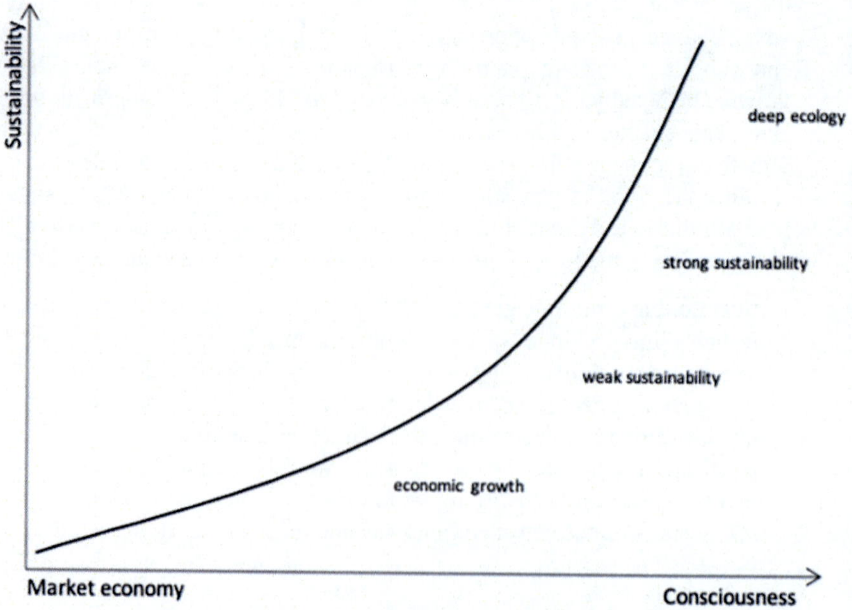

Fig. 6.1 Alleged connection between sustainability and consciousness. (Source: own work)

the same time least recognized reason for market failure (Fuders, 2017). The reason why we are so far away from achieving what in economics is termed "perfect competition" (and the herewith connected just remuneration of factors of production), and the reason why we are obliged to grow and engage in unsustainable behavior is our form of money. The problem is not the market economy but the economic growth imperative that perverts the market economy (Fuders & Max-Neef, 2014).

6.1.2 Sustainability as Marketing Tool to Promote Growth

Today the term "sustainability" has turned into a sort of marketing tool for businesses, enabling them to sell more and to grow. For example, energy-saving light bulbs are often sold with some kind of green stamp on it, supposedly signifying that they are energy efficient and less harmful to the environment. However, that these light bulbs are less harmful is not true, since they contain mercury, the production of which is, in fact, very harmful to the environment (and to workers employed in manufacturing them). When they break, they create a health hazard. Furthermore, they produce artificial light with an unnatural frequency, and they are suspected to contribute to depression and skin cancer. They also produce electro-magnetic pollution since they emit heavy electro-magnetic fields (Maes, 2007). Reports of headaches are frequent among those working under such light sources.

Another example is the promotion of modern automobiles using the argument that they are more efficient in the use of gasoline and therefore less harmful to the environment. Some countries for example Germany even promote the purchase of such vehicles via tax incentives. Such incentives can be viewed as hidden subsidies for the automobile industry which actually do the exact opposite of promoting sustainability. The energy required to produce a new automobile is hardly compensated by the more efficient use of gasoline. And many of these new automobiles are actually not more efficient than older models since they include all sorts of electrical devices which offset any gains in energy efficiency.

It is not just that sustainability is used as marketing tool to sell more products, but it sometimes even leads to the burning of more fossil fuels. Diesel particulate filters, which are now mandatory in many countries, make cars up to 3–13% less efficient (Abendblatt, 2005; Huanga et al., 2022). In the 1980s we were told that diesel emissions are less harmful to our health than those of gasoline engines. Suddenly the narrative changed, and they are now considered even more harmful than gas powered cars.[1] If we consider who owns the automobile industry (see Figs. 6.2 and 6.3),

[1] Similarly, it is questionable whether it makes sense to use headlights during the day. In many countries today this is mandatory in order to decrease the number of accidents while until some years ago it was on the contrary recommended to turn off the headlights during the day in order to save energy. Using headlights makes a car approx. 1–2% less efficient (Greco, 2013; Fuhrmann, 2013).

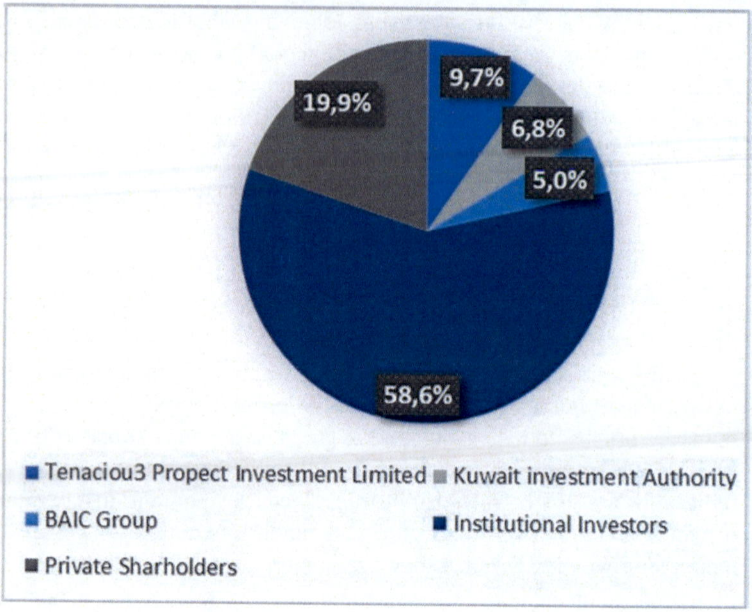

Fig. 6.2 Shareholder structure Daimler AG. (Source: own work. Based on data provided by Daimler AG)

we might gain a better understanding of why the automobile industry does not really wish to produce more efficient cars. A significant percentage of the automobile company stock is owned by oil producing countries and institutional investors who also hold investments in the oil industry.

Another point worth noting is that electrical cars are not an improvement on gas powered vehicles if the energy used to power them is not generated via clean methods. No country in the world currently generates electricity mainly through alternative sources like hydro, wind and solar. Even in Germany—which is one of the industrialized countries with the highest percentage of clean energy—only 44% of the total electricity is generated using clean methods (a high percentage compared to most other countries). Using electricity produced by conventional methods like fossil fuels and nuclear energy to power electric cars does not promote sustainability. Burning fossil fuels to move a turbine in order to generate electricity, i.e., transforming kinetic into electric energy and then later transforming the electric energy back into kinetic energy is probably even less efficient than directly burning fuel in the car.

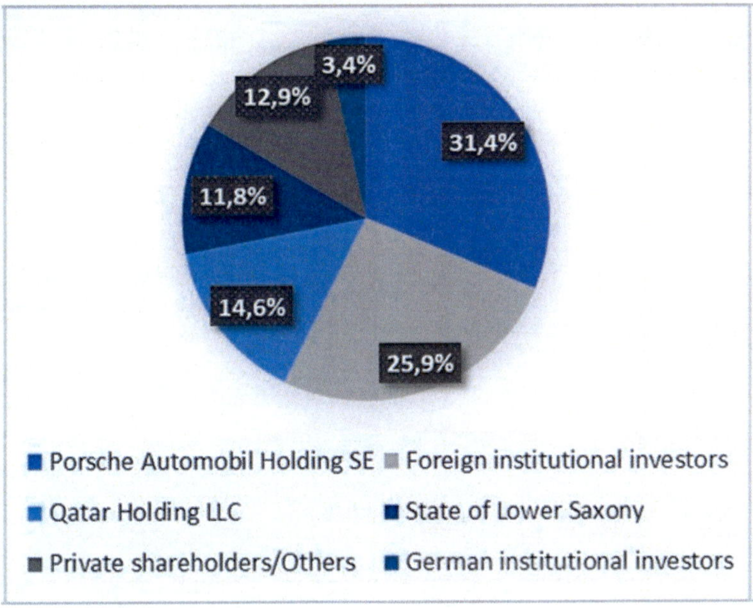

Fig. 6.3 Shareholder structure VW AG 2021. (Source: own work. Based on data provided by VW AG)

6.2 We Are an Integral Part of Nature

6.2.1 Living Within Limits Established by Ecosystems

In the ancient view, which prevailed until about 200 years ago, nature was something to be fought or overcome. Us against nature. This view has changed. Nature is not perceived as an enemy anymore. Thanks to the profound eco-social crisis and a growing awareness of environmental degradation, we have recently witnessed a flourishing of non-anthropocentric visions; that is, visions of a biocentric nature that, in different ways, displace the human being from the center of the scene and put all living organisms in their place.

This implies a change of worldview, beyond the political, economic, cultural, technological and social changes that the new paradigm of sustainability demands. According to this biocentric vision, along the lines of Arne Næss' Deep Ecology (Naess, 1973), human beings must view themselves as an integral part of nature, and that everything that deviates from this divine equilibrium, must be considered unnatural (including our monetary system).

Unfortunately, despite the spread of this kind of perspective, we find ourselves trying to find a balance between, on the one hand, the growing need to take care of the environment and, on the other, the increasing utilization of materials and energy consumed by today's world. Both sensitivity and unsustainability are growing

(Azkarraga et al., 2011). It seems that we are still far from understanding what it really means to build a sustainable society and a sustainable way of life. We are still far from grasping what the great socio-ecological transition really means. For many there is no incompatibility between economic growth and ecological balance. According to an orthodox view more of the former is required to accomplish the latter (the hypothesis known as the 'environmental' Kuznets Curve; we will discuss this idea in Sect. 6.5.4).

Put as simply as possible, sustainability is the durability of a system: A socioeconomic system is sustainable when, in its transformations and developments, it does not undermine the ecological bases on which it rests—and can therefore last over time (Azkarraga et al., 2011). Sustainability means living well (being able to satisfy fundamental human needs) within the limits established by ecosystems.

6.2.2 Three Defining Elements

According to (Riechmann, 2015: 181) true sustainability requires a substantial cultural transformation (or revolution). We can highlight three defining elements of such cultural transformation:

- A rediscovery of the cultural notion of **sufficiency**. Sustainability is not only about increasing efficiency, because it is possible for societies to be both efficient and unsustainable. In fact, the so-called "Jevons paradox" or rebound effect means that increases in efficiency are accompanied by increases in overall demand for resources.
- Sustainability is basically defined as respect for **biophysical limits**. This implies human self-restraint or self-limitation.
- Sustainability requires substantial **socio-political transformation**. This point includes the reform of our monetary system to achieve true monetary neutrality.

Recovering sufficiency and promoting the required socio-political transformation does **not necessarily** have to be a painful process. Rather, it could be accomplished automatically via changing our money such that it is truly neutral and in so doing eliminating the growth imperative. Since truly neutral money is not infinitely storable it is more difficult to hoard surpluses, thus putting automatic limits on our *liquidity-preference*, as Keynes (1936: 168) called our penchant for saving and hoarding. Also, since interest rates will not be positive the Net Present Value of future cash flows will be higher. Thus, we will automatically give more priority to the protection of natural resources so that they will still be productive for future generations.

6.2.3 Evolution of the Concept of Sustainability

Probably the most widely used definition of sustainability, the one provided by the United Nations through the Report of the World Commission on Environment and Development (known as the Brundtland Report), emphasizes responsibility towards future generations:

> Sustainable development is development that meets the needs of the present without compromising the ability of future generations to meet their own needs (Brundtland, 1987: 37).

Since that report, sustainable development has been interpreted in multiple ways, according to different ideological perspectives, values or interests, and many of these interpretations have little to do with sustainability (Azkarraga et al., 2011). In fact, it has become a catch-all concept, used alike by the profit-seeking entrepreneur as well as the farmer pursuing a risk-minimizing existence, the equity-seeking social worker, the first-worlder and nature lover concerned about pollution and the growth-maximizing public manager, the technocrat and the vote-counting politician (Bermejo, 2010). The Brundtland Report was the first widely recognized document to introduce the concept of sustainability to a broader public. Nevertheless, as far back as 1975, the prestigious Hammarskjöld Foundation (1975) published a report on development: "What now?" The report contained a fundamental idea: development must at the same time respect inner and outer limits:

- Inner limits: the satisfaction of fundamental human needs
- Outer limits: physical limits established by the biosphere or the carrying capacity of the planet.

If the latter are not respected, we will be faced with a problem of ecological unsustainability, and if the former are not respected, we will be faced with social unsustainability. The challenge for the future is how to guarantee well-being (social, economic, ecological, and psychic), and how to carry out sustainable development in social and ecological terms.

Perhaps the most important precursor to the Brundtland Report was the 1972 MIT study "Limits to growth" (Meadows, 1972), commissioned by the Club of Rome, which argued for the first time using a mathematical model that the dynamics of exponential growth on a limited planet cannot be sustainable. In short, a sustainable society is socially just, ecologically compatible (the biophysical limits of the planet) and economically viable. However, as long as we do not change our monetary system, we will neither be able to combine ecological compatibility with economic viability nor will there be social justice, as will be explained in the next chapter.

6.3 Economic Activity and the Biosphere

6.3.1 The Economy as a Sub-system of the Ecological System

While in the old vision the economy, society and ecology were thought of as different systems and sustainability was seen as a point of intersection between the different systems (Fig. 6.4), nowadays we understand that the economy is a sub system of society (Fig. 6.5). In fact, the economy is part of society. There is no economy without people. It is we, the consumers and producers who compose the markets. And society itself is just one part of the ecological system.

6.3.2 We Live on a Finite Planet

We live on a finite planet with finite resources: soil, minerals, fossil fuels, etc. One might argue that renewable resources are replenished. However, renewable resources are finite if we exploit them at a rate faster than that at which they are able to renew themselves. Even the extent to which we can make use of solar energy is not infinite. This is because capturing the sun's energy requires organisms to photosynthesize it, and these are finite. Likewise, photovoltaic technology requires space and materials that are not inexhaustible. It is necessary to emphasize this point because economic growth is the main goal of most economists and politicians. However, notwithstanding the possibility of making the production of goods and services more efficient, steady economic growth (increasing the productive output of the economy) requires more energy and more resources over the long run.

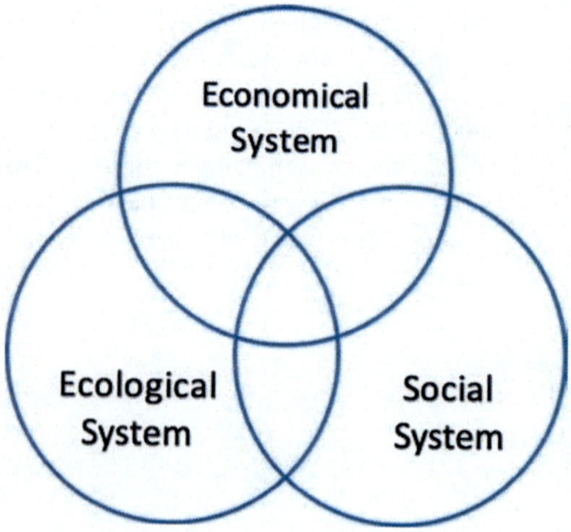

Fig. 6.4 Sustainability as intersection set. (Source own work)

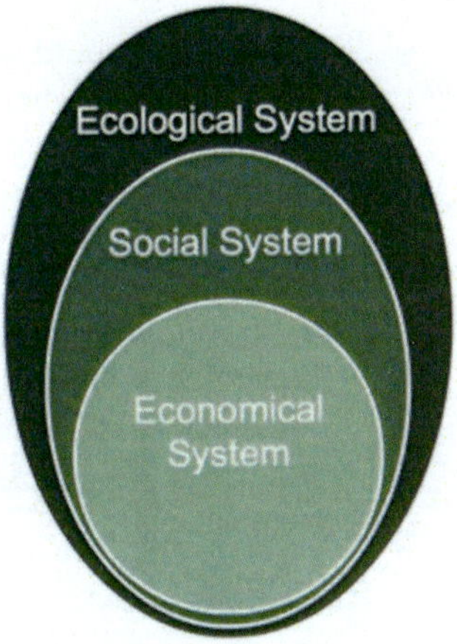

Fig. 6.5 Economy as part of society and society is part of the ecology. (Source: own work)

6.3.3 We Cannot Create Something Out of Nothing

We cannot create something out of nothing. Everything we produce requires energy and resources. This insight is crucial to understanding that a constant increase in the economic activity will eventually use up all our resources. This is what we can learn from the laws of thermodynamics (Clausius, 1854, 1867). The first law tells us that energy can neither be created nor destroyed. The second law tells us that energy moves towards greater homogeneity (law of entropy).

The second law means that all kinds of energy are eventually transformed into heat which is dissipated and no longer available for use. Entropy is a measure of the unavailable energy in a thermodynamic system. Example: Wood can be burned to heat a house or to power a steam engine. However, not all of the energy contained in the fuel used to power the engine is available as mechanical output. Some of the energy is dissipated in the form of heat and smoke. That heat and smoke are not useful to us and can in fact be harmful.

6.3.4 The Nature of the Economic System Is Entropic

If we understand that the laws of thermodynamics apply to both energy and matter, then the nature of the economic process is entropic:

- Transformation (the production of goods and services) requires energy. We can only make use of low entropy energy by transforming it into high entropy energy that we can no longer use.
- Everything that we produce returns to nature as waste or garbage (high entropy matter).

Therefore, we could define our economy as a system that transforms low entropy matter and energy into waste and unusable energy (energy with high entropy) supplying us with a flow of perceived satisfaction (Daly & Farley, 2011). I like this definition since the satisfaction of fundamental human needs should be the **ultimate goal of all economic activity**. In Chap. 15, I will come back to this idea. We will conclude that it is truly possible to transform our economy so as to get close to a "*Development at Human Scale*" (Max-Neef et al., 1991). That is a development which optimizes the satisfaction of fundamental human needs. While we would not be obliged to grow economically anymore, economic growth would still possible where necessary to further improve the satisfaction of fundamental human needs.

6.4 There Is No Such Thing as a 'Sustainable' Growth

We have just established that we cannot create something out of nothing. This is physics. Production requires physical resources provided by nature. Everything, even the most sophisticated productive process is ultimately sourced from nature. Thus, endless economic growth must inevitably use up all available resources sooner or later. This is true even for renewable resources such as fish populations or forests, since steady growth means that at some point, we will use resources at a faster rate than they can be replenished (for an in-depth analysis of this topic and further references see Fuders, 2020). Hence, economic activity cannot grow indefinitely. The limited supply of resources limits the size of the economy as Daly & Farley (2011) correctly put it. This is important to highlight since many are convinced that "sustainable growth" is possible. For example, the European Commission adopted an ambitious "Circular Economy Package", including legislative proposals aimed at reducing waste, boosting productivity and competitiveness, creating new jobs and fostering "sustainable economic growth" (European Commission, 2020).

But, as we have already discussed, if we do not change our monetary system, we will be perpetually obliged to pursue economic growth simply to maintain the *status quo of prosperity*. Thus, economic models have been developed arguing that it is possible to continue to grow without destroying the environment or depleting our resources. In the following section we will examine some of these models and explain the fallacies upon which they are based.

6.5 Discussion

6.5.1 Growth Models with Unrealistic Assumptions

Some argue that it is possible to increase production without using more resources through increasing efficiency. If true, this would make unlimited economic growth theoretically possible. Unfortunately, such arguments are based on unrealistic assumptions. While it is true that automobiles today use one-fifth the amount of steel used to make cars 50 years ago and that we can manufacture aluminum cans far more efficiently than in the past, it is also true that we use more steel in the manufacture of cars and more aluminum in the production of cans than ever before (Daly & Farley, 2011). More economic efficiency does not mean that fewer resources are used (Jevons Paradox).

In a similar way some models fall for the error to unrealistically assume that we were able to constantly improve technology so that emissions per tons or unit produced can be reduced.[2] Even though we can (and hopefully will) use cleaner technologies, apply filters, use green (renewable) energy resources, it is **physically impossible** to produce products with no input and no emissions at all. Although new technology may make it possible to produce a given product with fewer resources and less emissions, it will never be possible to produce it without any resources at all. A continuous growth of physical production even at a small rate would one day exhaust the last resource (Daly & Farley, 2011).

6.5.2 Growth Models with Unrealistic Production Functions

Another often used 'trick' to show that we could endlessly grow without destroying nature is to apply classic production functions using only capital (K) and labor (L) as production factors, i.e., where natural resources are not even considered, as if we could produce products by employing only capital and labor (see Eq. 6.1).[3] Of course, if natural resources do not appear in the model, they cannot be over exploited as Daly (1997) correctly pointed out.

$$Q = f(K,L) \tag{6.1}$$

[2] Texts that belong to this category are, e.g.: Chen (2002); Oh (2010); Mohapatra et al. (2016); Jardon et al. (2017); Cao and Wang (2017); Guo et al. (2018); Wolf et al. (2019); Deng et al. (2019); Du & Li (2019); Wang et al. (2019); Yao et al. (2019); Halkos and Bampatsou (2019); Jimenez et al. (2019); Gavrilyeva et al. (2020); Nasrollahi et al. (2020); Peng et al. (2020); Zhang et al. (2021).

[3] Texts that belong to this category are, e.g.: Chen (2002); Oh (2010); Cao and Wang (2017); Moran Molina and Gonzaga Añazco (2017); Vasyl'yeva (2021).

These classic production functions can be understood as a strong simplification and a relic of the romantic view of the old times (like 200 years or more ago) when only a couple million persons habited the world and natural resources were not scarce compared to the then existing possibilities of humans to exploit them. The damaging and exploitive effects to nature of human productive processes in those times were meaningless compared to the overwhelming abundance of nature. The reason why we still use such simplified production functions could also be a practical one. In economics we love to draw graphs and a production function with only capital and labor as production factors can be drawn on a two-dimensional graph.

Some more modern production functions do consider nature, or natural resources, (r in Eq. 6.2) but in an erroneous manner, e.g., as if the natural (stock-flow) resources and capital or labor (which are fund service resources) were substitutes: the relationship between the production factors is described as multiplicative (e.g., Cao & Wang, 2017; Vasyl'yeva, 2021). As an example, we could take a classic Cobb-Douglas production function with natural resource r:

$$Q = f(K,L) = mK^{\alpha}L^{\beta}r \tag{6.2}$$

If we want to produce more, e.g., cookies, we need not only more labor, but also more flour. It is not possible to increase the quantity of cookies by just increasing the number of people working in the kitchen or by buying a bigger oven. The relationship between stock-flow resources and fund-service resources is complementary and not substitutive (Daly, 1997; Daly & Farley, 2011).

6.5.3 *Clarification: Economic ≠ Physical Values and Development ≠ Growth*

What has been said does not mean that indefinite economic growth measured in nominal terms is not possible. Prices are measured in monetary units. If the quantity of monetary units in circulation (money supply) grows at a faster pace than the quantity of products, then we will experience inflation. Economic output measured in nominal monetary units can theoretically grow forever. What cannot grow indefinitely is the quantity of physical products, i.e., real (inflation adjusted) GDP, a fact that was understood as far back as Aristotle.

Our use of language can also cause confusion if we fail to differentiate between 'growth' and 'development'. Both terms are often used as synonyms. In Chile the government has used slogans such as "Chile grows with you" (*Chile crece contigo*). What they really mean is: "Chile develops with you". What is the difference between growth and development, one might ask? The answer is that while growth is quantitative, development is qualitative (Max-Neef, 1992). For example, GDP is a quantitative measure of all of the goods and services produced in the course of a year. But increasing the quantity of products and services does not necessarily mean that we are better off. One of the basic (and true) insights of microeconomics is the *law of diminishing marginal utility*, which says that the more we have consumed of a certain good, the less utility (satisfaction) we will get from consuming an additional unit. We might, in fact, be even worse off, if the production process depletes or destroys our natural environment.

As we have already observed, we cannot grow indefinitely, but perhaps we can 'develop' indefinitely (Daly & Farley, 2011). If we were to invent a way to measure development that was truly different than measuring GDP—maybe by measuring the subjective perception of the satisfaction of fundamental human needs (Fuders, 2011; Fuders et al., 2016; Fuders & Mora-Motta, 2021)—then perhaps it would be possible for this measure to grow indefinitely. We can constantly improve products and services to better satisfy human needs without necessarily increasing the quantity of products and services sold. Achieving sustainability **does not necessarily mean an end to development**, but it does mean halting the growth of overall physical production.[4]

6.5.4 Environmental Kuznets Curve

Some even argue that economic growth is not only compatible with a clean environment but a prerequisite. They argue that a clean environment is a luxury. If people are struggling to survive, they have more urgent problems to worry about than the protection of the environment. For example, someone who is struggling to survive is likely to accept a job in a factory that pollutes the air and water. Seen from this perspective, only rich nations have the freedom to worry about a clean environment, to abstain from consumption or production and let industries move to other parts of the world if they pollute too much.

Originally, the Kuznets' (1955) curve was about income inequality. According to Simon Kuznets, as economic growth progresses, income inequality initially increases and then later decreases (see Fig. 6.6).[5] In the 1990s this logic was applied to the environmental problem (Grossman & Krueger, 1995; for an overview of literature see Pastén & Figueroa, 2012). According to this idea as GDP grows, at first there is an increase in pollution, but a point is then reached at which people begin to demand stricter environmental protections, thus reversing the destruction of the environment. Indeed, in many 'first world' countries the air quality improved significantly during the last 40 years, e.g., in Germany. But, has air quality improved worldwide or only in some regions? The answer is that the industries which produce the most pollution simply moved to other parts of the world such as China or India, where there are less rigorous environmental protection laws (Common & Stagl, 2008). In addition, a constant increase in the output of industrial production will

[4] But even the satisfaction or fulfilments of fundamental human needs will probably not endlessly improve, for our needs are either satisfied or they are not. It is impossible to be more than 100% satisfied.

[5] This theory can be questioned if we understand that our monetary system constantly increases income inequality by mathematical logic the longer the system exists.

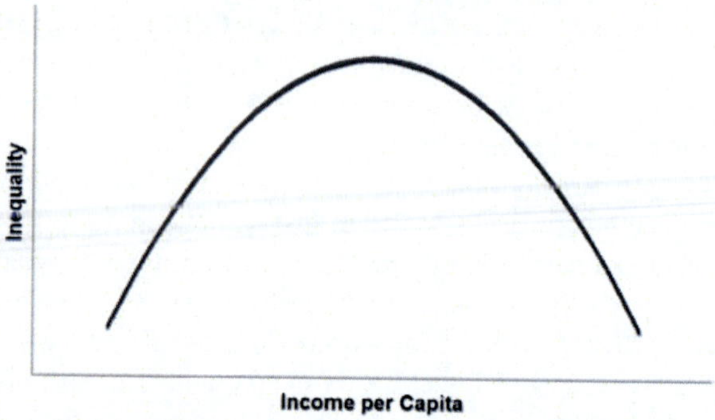

Fig. 6.6 "Kuznets curve". (Source: own work)

lead someday to the extinction of the last resource even using cleaner production methods and ever increased efficiency.

6.5.5 Circular Economy and Decarbonizing the Economy

Some argue that we could grow without further damaging the environment if we managed to come close to a 100% circular economy, that is an economy where all waste is completely recycled and reused in the production of new goods. We might ask if such a thing is even possible. But even if it is,[6] it will not solve the sustainability problem, as long as we are continually obliged to grow. As long as we continue to grow, **we will use more resources**, more inputs, more energy (first and second law of thermodynamics).

The same is true for the idea of "decarbonizing" the economy as strongly promoted by international organizations such as the UN, the WEF and the OECD (UNECE, 2021; WEF, 2021; D'Arcangelo et al., 2022). No doubt, replacing carbon fuel-based energy generation by renewable and clean energy sources such as hydro, wind or solar energy is a desirable goal. Nature gives us abundant resources of clean and free energy, so we should use them. But even if we had a 100% circular economy with 100% clean energy, we would still need physical resources (inputs) to produce goods and services. It should be noted that the production of services also requires physical resources and often large amounts of energy. The latter is important to point out since some would argue that increasing the production of services instead of physical goods would also be a way to decarbonize the economy and allow for sustainable growth. However, steady expansion of economic activity will eventually deplete all available resources. And even clean, renewable energy sources

[6] Prohibiting plastic and aluminum would be a big step towards a 100% circular economy since then all products would be made of wood, steel, leather, cotton. All waste would be organic, recycling would almost not be necessary anymore since nature would handle this for us.

will eventually be unable to keep up with the demand, since their production requires space, and there are only so many rivers on the planet.

6.6 Conclusions

Since economic growth (real GDP growth) means that we are always producing more and more over time, and since the first **law of thermodynamics** tells us that we cannot produce something out of nothing (Clausius, 1876; Radi & Rasmussen, 2013), a steady GDP growth rate in the long run **must** result in an increased use of natural resources. Even if we continue to improve efficiency, it will never be possible to produce products or services without any inputs. Therefore, steady economic growth in the long run must cause an increased use of natural resources and energy. Even renewable resources will eventually prove to be insufficient, since if we keep growing indefinitely at some point the extraction rate will be higher than the reproduction rate of the resource (on this argumentation see e.g., Daly & Farley, 2011). No matter how you analyze the facts, we cannot grow endlessly on a finite planet. Sustainable economic growth therefore cannot exist.

References

Abendblatt. (2005). *Rußpartikelfilter erhöhen den Sprittverbrauch.* [Online] Available at: https://www.abendblatt.de/ratgeber/auto-motor/article107044852/Russpartikelfilter-erhoehen-den-Spritverbrauch.html. Accessed 10.11.2022.

Azkarraga, J., Max-Neef, M., Fuders, F., & Altuna, L. (2011). *La evolución sostenible II*. Lanki (Mondragon Unibertsitatea).

Bermejo, R. (2010). *Menos es más: del desarrollo sostenible al decrecimiento sostenible*. Hegoa.

Brundtland, G. H. (1987). *Our common future*. Oxford University Press.

Cao, B., & Wang, S. (2017). Opening up, international trade, and green technology progress. *Journal of Cleaner Production, 142*, 1002–1012.

Chen, E. (2002). The total factor productivity debate: Determinants of economic growth in East Asia. *Asian-Pacific Economic Literature, 11*(1), 18–38.

Clausius, R. (1854). Über eine veränderte Form des zweiten Hauptsatzes der mechanischen Wärmetheoriein. *Annalen der Physik und Chemie, 93*(12), 481–506.

Clausius, R. (1867). *The mechanical theory of heat – With its applications to the steam engine and to physical properties of bodies*. John van Voorst.

Clausius, R. (1876). *Die mechanische Wärmetheorie*. Friedrich Vieweg.

Common, M., & Stagl, S. (2008). *Introducción a la Economía Ecológica*. Reverté.

D'Arcangelo, F., et al. (2022). *OECD economic policy papers no 31 – A framework to decarbonise the economy*. OECD.

Daly, H. (1997). Georgescu-Roegen versus Solow/Stiglitz. *Ecological Economics, 22*, 261–266.

Daly, H., & Farley, J. (2011). *Ecological economics – Principles and applications* (2nd ed.). Island Press.

Deng, Y., You, D., & Wang, J. (2019). Optimal strategy for enterprises' green technology innovation from the perspective of political competition. *Journal of Cleaner Production, 235*, 930–942. https://doi.org/10.1016/j.jclepro.2019.06.248

Du, K., & Li, J. (2019). Towards a green world: How do green technology innovations affect total-factor carbon productivity. *Energy Policy, 131*, 240–250. https://doi.org/10.1016/j.enpol.2019.04.033

European Commission. (2020). *Circular economy action plan.* [Online] Available at: https://ec.europa.eu/environment/strategy/circular-economy-action-plan_en. Accessed 15.10.2021.

Fuders, F. (2011). *Política Económica – Class manuscript*. Valdivia: Universidad Austral de Chile.

Fuders, F. (2017). Neues Geld für eine neue Ökonomie – Die Reform des Geldwesens als Voraussetzung für eine Marktwirtschaft, die den Menschen dient. In *Finanzwirtschaft in ethischer Verantwortung – Erfolgskonzepte für Social Banking und Social Finance* (pp. 121–183). Springer-Gabler.

Fuders, F. (2020). *The 'tragedy of the commons' and the role of the money interest rate*. Springer.

Fuders, F. (2021). The role of money for a healthy economy. In: Urban health and wellbeing programme – Policy briefs*: Volume 2.* : Springer Nature, pp. 17–22.

Fuders, F., & Mora Motta, A. (2021). El Índice de Desarrollo a Escala Humana: una primera evaluación. In C. Oyarzún, P. Donoso, & J. J. Nuñez (Eds.), *Historia natural, Servicios Ecosistémicos y Perspectivas de Desarrollo de la Cuenca de Llancahue* (pp. 173–194). Valdivia: Kultrun.

Fuders, F., & Max-Neef, M. (2014). Dinero, deuda y crisis financieras. Propuestas teórico-prácticas en pos de la sostenibilidad del sistema financiero internacional. In *Economía internacional. Claves teórica-prácticas sobre la inserción de Latinoamérica en el mundo*. LATIn.

Fuders, F., Mengel, N., & Barrera, M. (2016). Índice de desarrollo a escala humana: propuesta para un indicador de desarrollo endógeno basado en la satisfacción de las necesidades humanas fundamentales. In R. Pérez Garcés, E. Espinosa Ayala, & O. Terán Varela (Eds.), *Seguridad Alimentaria, actores Terretoriales y Desarrollo Endógeno* (pp. 63–106). Iztapalapa: Laberinto

Fuhrmann, T. (2013). *Zeit online: Tagfahrlicht – Spritverschwendung durch falsches Licht.* [Online] Available at: https://www.zeit.de/auto/2013-04/leserartikel-tagfahrlicht-spritverbrauch. Accessed 10.11.2022.

Gavrilyeva, T., et al. (2020). Economy-related green-house gases emissions and validation of the environmental Kuznets curve for Sakha Republic (Yakutia). *Polar Science, 23*, 100507. https://doi.org/10.1016/j.polar.2020.100507

Greco, F. (2013). *Blick – Wieso steigt der Verbrauch, wenn man mit Licht fährt?*. [Online] Available at: https://www.blick.ch/auto/service/tcs-ratgeber-wieso-steigt-der-verbrauch-wenn-man-mit-licht-faehrt-id2369761.html. Accessed 10.11.2022.

Grossman, G., & Krueger, A. (1995). Economic growth and the environment. *Quarterly Journal of Economics, 110*(2), 352–377.

Guo, Y., Xia, X., Zhang, S., & Zhang, D. (2018). Environmental regulation, government R&D funding and green technology innovation: Evidence from China provincial data. *Sustainability, 10*, 940.

Halkos, G., & Bampatsou, C. (2019). Economic growth and environmental degradation: A conditional nonparametric frontier analysis. *Environmental Economics and Policy Studies, 21*, 325–347.

Hammarskjöld Foundation (1975). *What now: another development, the 1975 Dag Hammarskjold report.* Dag Hammarskjold Foundation, Uppsala, 1975

Huanga, Y., et al. (2022). Effect of diesel particulate filter regeneration on fuel consumption and emissions performance under real-driving conditions. *Fuel, 320*, 123937.

Investopedia. (2021). *Investopedia – How importing and exporting impacts the economy.* [Online] Available at: https://www.investopedia.com/articles/investing/100813/interesting-facts-about-imports-and-exports.asp. Accessed 16.11.2021.

Jackson, T., & Victor, P. (2019). Unraveling the claims for (and against) green growth – Can the global economy grow indefinitely, decoupled from Earth's limitations? *Science, 366*(6468), 950–951.

Jardon, A., Kuik, O., & Tol, R. (2017). Economic growth and carbon dioxide emissions: An analysis of Latin America and the Caribbean. *Atmosfera, 30*(2), 87–100.

Jimenez, C., Moncada, L., & Ochoa-Jimene, D. (2019). Kuznets environmental curve for Ecuador: An analysis of the impact of economic growth on the environment. *Sustainability, 11*, 5896. https://doi.org/10.3390/su11215896

Keynes, J. M. (1936). *General theory of employment, interest and money*. Hartcourt.

Kuznets, S. (1955). Economic growth and income inequality. *The American Economic Review, XLV*(1), 1–28.

Maes, W. (2007). Glühbirne raus – Energiesparlampe rein? – Wir alle müssen was tun. Bei den Energiesparlampen spiele ich nicht mit. *Wohnung + Gesundheit, 9*, 124.

Max-Neef, M. (1992). *From the outside looking in – Experiences in barefood economics* (2nd ed.). Zed.

Max-Neef, M., Elizalde, A., & Hopenhayn, M. (1991). *Human scale development – Conception, application and further reflections*. Apex.

Meadows, D. L. (1972). *The limits to growth*. New American Library.

Mohapatra, S., Adamowicz, W., & Boxall, P. (2016). Dynamic technique and scale effects of economic growth on the environment. *Energy Economics, 57*, 256–264.

Moran Molina, G., & Gonzaga Añazco, S. (2017). Análisis de la medición del impacto ambiental como producto del crecimiento económico. *Revista Universidad y Sociedad*, Volume online: http://scielo.sld.cu/scielo.php?script=sci_arttext&pid=S2218-36202017000100012

Naess, A. (1973). The shallow and the deep, long-range ecology movement. A summary. *Inquiry, 16*(1–4), 95–100.

Nasrollahi, Z., Hashemi, M.-S., Bameri, S., & Taghvaee, V. M. (2020). Environmental pollution, economic growth, population, industrialization, and technology in weak and strong sustainability: Using STIRPAT model. *Environment, Development and Sustainability, 22*, 1105–1122.

Oh, D.-H. (2010). A metafrontier approach for measuring an environmentally sensitive productivity growth index. *Energy Economics, 32*, 146–157.

Pastén, R., & Figueroa, E. (2012). The environmental Kuznets curve: A survey of the theoretical literature. *International Review of Environmental and Resource Economics, 6*, 195–224.

Peng, B., Zheng, C., Wei, G., & Elahi, E. (2020). The cultivation mechanism of green technology innovation in manufacturing industry: From the perspective of ecological niche. *Journal of Cleaner Production, 252*, 119711. https://doi.org/10.1016/j.jclepro.2019.119711

Radi, H. A., & Rasmussen, J. O. (2013). *Principles of physics for scientists and engineer*. Springer.

Riechmann, J. (2015). *Autoconstrucción. La transformación cultural que necesitamos*. Catarata.

UNECE. (2021). *Technology brief – Hydrogen*. UNECE.

Vasyl'yeva, O. (2021). Assessment of factors of sustainable development of the agricultural sector using the Cobb-Douglas production function. *Baltic Journal of Economic Studies, 7*(2), 37–49.

Wang, Q., et al. (2019). Green technology innovation development in China in 1990-2015. *Science of the Total Environment, 696*, 134008. https://doi.org/10.1016/j.scitotenv.2019.134008

WEF. (2021). *How to build smart, zero carbon buildings – And why it matters.* [Online] Available at: https://www.weforum.org/agenda/2021/09/how-to-build-zero-carbon-buildings/. Accessed 28.11.2022.

Wolf, R., Benitez Rodriguez, A. M., & Medeiros, D. (2019). Estimativa da Curva de Kuznets ambiental para o Estado de Mato Grosso do Sul. *Investig. Agrar, 21*(2), 124–135.

Yao, S., Zhang, S., & Zhang, X. (2019). Renewable energy, carbon emission and economic growth: A revised environmental Kuznets curve perspective. *Journal of Cleaner Production, 235*, 1338–1352.

Zhang, L., et al. (2021). Caring for the environment: How human capital, natural resources, and economic growth interact with environmental degradation in Pakistan? A dynamic ARDL approach. *Science of the Total Environment, 774*, 145553.

Chapter 7
Economic Growth or Unemployment

Check for updates

7.1 The Force Behind Okun's Law

We have now analyzed the origin of the growth imperative and the destructive effect it has on our environment. Some might argue that rather than increasing output, businesses could use cost cutting as a way to maintain the status quo. Indeed, if a company in unable to grow, it will typically try to reduce costs in order to maintain a Return on Investment (ROI) at least as high as its opportunity costs, i.e., the interest rate. Human labor is usually the most significant category of costs, which is why there is a constant pressure to either grow or to replace humans by machines, i.e., to improve productivity.[1] This is the reason why the labor share in value added and income is constantly decreasing (Karabarbounis & Neiman, 2014; Dvorkin & Bharadwaj, 2019), and why we can empirically observe a negative correlation between growth and unemployment in all economies. In economics this negative correlation is known as "Okun's law".

According to Arthur Okun's (1962) "3-to-1" rule of thumb unemployment tends to fall by 1 percentage point for every 3-percentage- point rise in gross national product (see also Neely, 2010; Mankiw & Taylor, 2014). Only if we manage to grow at a faster pace than productivity increases, will we not see increased unemployment. But, again, the pressure to constantly improve productivity is driven not only by the greed of business owners but also by interest as the opportunity cost of any productive investment (see already Suhr, 1989). We can now understand why our financial system is the reason for both neoclassical as well as Keynesian models always seeking economic growth. In political economics textbooks, the panacea for unemployment is economic growth (see, e.g.: Fernández Díaz et al., 2011; Cuadrado

[1] If we improve "productivity", then this is usually not because workers become more efficient but because more and more efficient machines are employed replacing human labor. The same quantity of products is manufactured using ever less human labor.

© The Author(s), under exclusive license to Springer Nature Switzerland AG 2023

F. Fuders, *How to Fulfil the UN Sustainability Goals*,
https://doi.org/10.1007/978-3-031-37768-6_7

et al., 2015). Since all politicians, regardless of their political orientation, fear unemployment and the associated social unrest, everything is done to foster economic growth.

7.2 SDG No. 8 Tries to "Square the Circle"

That we either grow or otherwise unemployment increases, must be the reason why SDG No. 8 demands economic growth next to decent work. As outlined above, the laws of physics tell us that "sustainable" growth is an impossibility, akin to trying to "square the circle" (Azkarraga et al., 2011: 10). Nevertheless, as long as we maintain our current financial system we have to keep growing unless we want unemployment and precarious working conditions to increase. Despite the enormous export surplus in Germany in the last 22 years unemployment is still around 5%. Yet Germany can consider itself lucky. The EU-wide unemployment is much higher. Those who still have jobs have to work more and more and constantly increase their productivity, not only in Europe but also elsewhere. This results in stress-related illnesses, such as burnout syndrome, which hardly existed in the past.

Therefore, **degrowth strategies**—even though highly necessary to avoid further overexploitation of the environment—will cause unemployment and deep economic depression if the existing financial system is maintained. The current system forces a choice between unsustainable growth and unemployment and misery (Costanza et al., 2012; Henke, 2017). Since economic growth is the reason for the unsustainability of our lifestyles, a solution requires that we eliminate the need to choose between growth and unemployment. **This is only possible if we change our financial system.**

In this context, I would like to mention the experience of a European television team in 2006 while filming a report on the remaining indigenous population in Brazil.[2] In return for being allowed to film, the tribe members asked to see a film about Europe. They were shown a video of Europe's big cities and industrial production. The indigenous people were surprised that we work so much in Europe, even though we have the power over fire and all these machines. They, the natives, on the other hand, have no such technology and yet work only a few hours a day. Unemployment, on the other hand, is nonexistent in their villages. Later we will see that in large parts of Europe there was a time (about 1000–1300) when unemployment was not a problem and people only worked 4 days a week for only a couple hours each day. Nevertheless, this period is considered by historians as a time of great productivity, during which most of the castles and cathedrals were built. At that time, a monetary system existed in Europe that was similar to the one that will be proposed in Chap. 10.

---

[2] I do not remember the title of the report. It was broadcasted in June 2007 on the German radio station "B5 aktuell".

7.3 Discussion

7.3.1 *Divergence Between Savings and Investment as Cause of Unemployment?*

John Maynard Keynes gave us highly valuable insights, especially the most accurate definition of the interest rate as a "*reward for parting with liquidity*" (Keynes, 1936: 167). He was also convinced that the "most sensible way of gradually getting rid of many of the objectionable features of capitalism" would be by pushing interest to zero (Keynes, 1936: 221).[3] Nevertheless, it seems that he failed to understand that it is precisely the interest rate as a benchmark for the evaluation of the profitability, and thus feasibility, of any investment that puts enormous pressure on the productive economy to either grow or to cut costs and, in doing the latter, to generate unemployment. Instead, Keynes believed—as do post Keynesian economists up to the present day—unemployment to be caused by a lack of aggregate demand. This in turn is believed to be caused by a divergence between savings and investment (see any textbook e.g., Cuadrado et al., 2015).

However, as long as we don't save huge amounts of cash (bills and coins) literally at home 'under the mattress', this divergence between savings and investment can hardly exist. Keynes falls into the same misinterpretation of the role of banks as conventional monetary theory does when it comes to interpreting the effect of the interest rate on the money supply. Money deposited in a bank account will be kept in circulation via loans. With the exception of the amount of money that the bank holds to meet the reserve requirement, any money deposited in a bank account will be loaned out. As already mentioned, the reserve requirement is currently 0% in many industrialized countries, in others 1–4%. And keep in mind that the reserve requirement usually only applies to demand deposits, not on time deposits or inflows form selling obligations or other debt securities, which can be lent out without holding any cash reserve. Hence, and as already discussed in Sect. 5.2 money deposited in bank accounts **does not remain out of circulation for long**. Ever-growing deposits put enormous pressure on banks to increase their loan volume proportionally, since they are required to pay interest on those deposits. As long as the financial system functions correctly, money is kept in circulation.

Nevertheless, we must give some credit to this idea of the divergence between savings and investment as cause of unemployment, considering that Keynes wrote his masterpiece during the Great Depression. This deep economic crisis was caused

[3] It is noteworthy that Keynes here used the term "marginal efficiency of capital" not "interest rate". That "marginal efficiency of capital" refers to the interest rate can be surmised based on the context in which he uses term. Unfortunately, like many economists, Keynes uses the words "money" and "capital" as synonyms and mistakenly states that the money interest rate has "no uniqueness compared with other rates of interest" (Keynes, 1936: 225), despite the fact that he correctly recognized in a prior publication that money has no intrinsic value (Keynes, 1983), and thus cannot be a factor of production.

by a chain of bank failures. Households and businesses lost faith in banks and therefore indeed hoarded money at home 'under the mattress'. Money failed to circulate and to perform its function as a medium to facilitate the exchange of goods and services.

7.3.2 *The Keynesian Economic Perpetual Motion Machine*

We have just discussed the fact that as long as the monetary system functions correctly, and people save money in banks rather than underneath the mattress, and banks fulfil their function of keeping money in circulation, a major divergence between savings and investment cannot occur. Therefore, the so-called Keynesian multiplier effect is not an accurate description of reality. According to the Keynesian economic perspective a lack of aggregate demand can be compensated for by increased government spending (expansionary fiscal policy). The idea is simple. The demand for labor N^d in Eq. (7.1) is a function of Y (GDP):

$$N^d = f(Y) \tag{7.1}$$

If households and businesses do not create sufficient demand for GDP growth to reach a level of full employment, then we need to increase aggregate demand by increasing government spending (moving aggregate demand AD_1 to AD_2 in Fig. 7.1). Keynesian economists even believe that this positive effect could cause a self-sustained economic upturn, the so-called *Keynesian multiplier*. The reactivation of economic activity provoked by expansionary fiscal policy initiated by government spending would cause people to spend more since an increased GDP means higher income for households and businesses. Higher spending and investments by households and businesses would then again push the next round of GDP growth and so aggregate demand (AD_3 in Fig. 7.1). The second round of GDP growth may be less strong as the first one since a part of the extra income in not spent but saved. Some are even convinced that this economic upturn enabled for more taxes to be levied so that the initial spending could be refunded (Ackley, 1978). If this was true, then we could generate wealth out of nothing, a true economic perpetual motion machine.[4]

However, where does the money which is spent by the government come from? If the government increases taxes to finance increased spending, then **aggregate demand will not increase**. Money (potential demand) is withdrawn from one part and inserted into another part of the economy. If the government instead uses

[4]A perpetual motion machine (in Latin *perpetuum mobile*) generates energy out of nothing. To generate energy out of nothing is not possible for it contradicts the laws of thermodynamics. The concept is based on the false assumption that without government spending money would be kept out of circulation. What is not understood is that money saved by the government on a bank account will circulate via bank loans if it is not withdrawn and spent by the government.

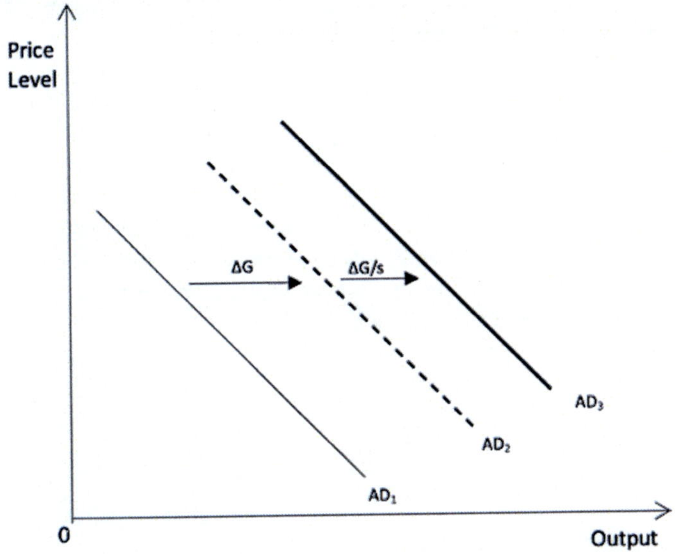

Fig. 7.1 Supposed Keynesian multiplier. Government spending (ΔG) increases aggregate demand from AD_1 to AD_2. The improved performance of the economy permits households and business to consume more, while only a part of the increased income is saved (s). (Source: Own work)

taxation to withdraw money from the economy during periods of economic growth, saves it on a bank account and then injects those funds back into the economy during periods of weakness (so-called *anticyclical policy*) this too is likely to have **no effect on aggregate demand**. Because it does not matter if money is kept in circulation via bank loans or via government spending. So, the money is kept in circulation either way. By the same logic, restrictive fiscal policy does not reduce aggregate demand. If we understand that money deposited in bank accounts will (as long as the monetary system has not yet plunged into crisis) always be loaned out and kept in circulation, then it follows that **fiscal policy is powerless to affect aggregate demand** and thus GDP.

7.3.3 *Is the "Keynesian Multiplier" Being Confused with the "Money Multiplier"?*

So why do so many subscribe to the Keynesian idea that government spending can increase aggregate demand? Entire economic doctrines—Keynesian and post-Keynesian—have been built on this concept. In fact, even the strongest critics of Keynes such as Friedrich A. von Hayek or Milton Friedman did not question the notion that government spending can be used to stimulate GDP growth (Hayek, 1976).

The reason why the idea of the Keynesian multiplier effect is so widely accepted may be because most countries do not accumulate funds to finance expansionary fiscal policy but instead engage in what Keynesians refer to as "deficit spending". In the event that expansionary fiscal policy is financed by public debt, it will indeed lead to an increase of aggregate demand. This is due to the fact that—as we have discussed above—money is the counterpart of debt, and any new credit creation will create money; an expansion of the economy's total debt volume will increase the money supply. **Deficit spending thus creates new, fresh money** that did not previously exist; thus, aggregate demand grows. It doesn't grow because of the Keynesian multiplier but because of the money multiplier effect.

This helps us understand why the Keynesian multiplier is supposed to diminish over time. The expansionary effect of government spending only lasts for a certain period of time. The explanation usually offered is that people do not spend all additional income caused by the expansionary fiscal policy but save some of it (Ackley, 1978; Cuadrado et al., 2015). According to this idea, Y (GDP) in Eq. 7.2 will increase as much as the government increases spending G divided by the savings rate s (defined as the inverse of the propensity to consume c).

$$\Delta Y = \frac{\Delta G}{1-\mathrm{c}} = \frac{\Delta G}{\mathrm{s}} \tag{7.2}$$

However, as long as the economy is not in a depression this so-called "leakage" from the circular flow of money **cannot exist**. As we already have discussed, as long as banks are performing their normal functions people generally do not save money under the mattress but at banks, which then lend that money out. Hence, money saved in banks will, for the most part, keep circulating in the economy. In the same way money paid as taxes (another alleged reason for leakage from the circular flow) is either spent elsewhere or saved in banks and thus returned to circulation via loans. So, why should the effect diminish?

The true reason for why the multiplier effect diminishes is that it is not the Keynesian multiplier but the money multiplier that expands aggregate demand. As outlined above, an original sum of 100 € can multiply to a maximum of 1.000 € if reserve requirements are 10%. That is to say, the multiplier effect does not decrease because of "leakage" effects as Keynesians believe but because the money multiplier does not increase the money supply indefinitely because banks are required to hold reserves.[5]

[5] Note that even without a legal reserve requirement, banks would still hold some reserves for safety reasons. Banks today hold even higher reserves than they are required, because they cannot find enough borrowers, which is an indicator that the world financial system is approaching the next serious crisis.

7.3.4 *Empirical Evidence That the Keynesian Multiplier Does Not Exist*

Since practically all countries pursue expansionary fiscal policy via deficit financing, it is very difficult to prove that the stimulative effect is caused by the monetary expansion that results from increased debt, rather than from government spending. The only way to prove that what is attributed to the Keynesian multiplier is actually due to the money multiplier is to find a country that pursues Keynesian expansionary policy financed by public savings rather than deficit spending (i.e., without increasing national debt).

One such country is Chile. Prior to the Covid-19 crisis Chile had very fiscally responsible governments that balanced its budgets and accumulated almost no national debt over the course of 30 years. Expansionary fiscal policy was, for the most part, financed by funds, which allowed for increased spending without adding to the national debt. So, if the Keynesian multiplier does not actually exist, this spending should have had no effect on aggregate demand or GDP growth. And this is precisely what was found in three studies conducted at Universidad Austral de Chile (Owuso et al., 2017; Mondaca et al., 2017; Ataway et al., 2019). The studies applied different methodologies, such as auto regression models or wavelet analysis, but the results of all three were the same: there is **no clear positive correlation** between government spending and GDP growth, either on a national level or on a regional level.

Some readers might object that it is well known that expansionary fiscal policy does not always translate into increased output and employment. According to Keynesian theory, this depends on the price elasticity of the aggregate supply curve, as depicted in Fig. 7.2. If aggregate supply is elastic, nothing prevents an increase in demand from being converted into an increase in output. However, if GDP is already high and we are on the inelastic (steep) part of the aggregate supply curve then

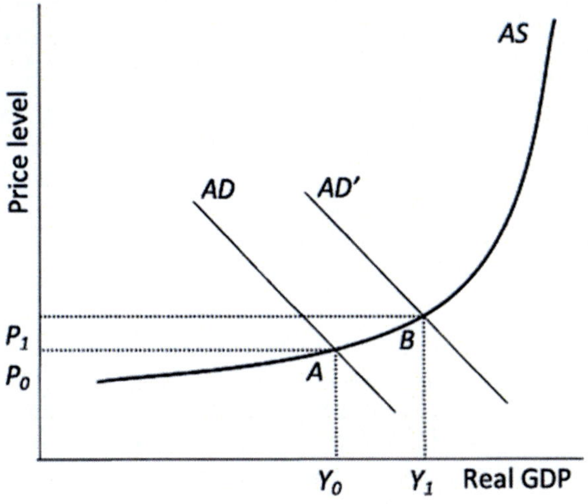

Fig. 7.2 Supposed effect of an increase of aggregate demand on GDP and inflation. (Source: Own work)

prices will rise (Peacock & Shaw, 1976) thus increasing the profits of producers without creating employment (Fernández Díaz et al., 2011; Cuadrado et al., 2015). The idea represented in Fig. 7.2 is that, in times of severe economic depression with low levels of national income (Y_0 in Fig. 7.2), people are worried about the future and tend to save more money, which constrains the economy. Under such circumstances, injecting money into the economy has little or no effect on inflation since people save most of the surpluses, and that money therefore doesn't circulate.

But this view in based on the incorrect classical interpretation of the role of banks as places where money is physically hoarded and thus taken out of circulation. In normal times, i.e., when the monetary system is working as it should and banks perform their function of making money circulate, this does not happen. If households and businesses deposit their money in banks, those banks keep the money in circulation. So, it doesn't matter whether the savings rate is high or low. Either the owners spend the money themselves or borrowers do. Any increase in aggregate demand—which as described above, is only possible by increasing national debt or by printing money—thus results in an increase of inflation and GDP growth.

Only if people literally hoard money under the mattress, would injecting money into the economy fail to generate inflation. But then it would also have little effect on economic growth and employment. People hoarding money at home rather than in banks is something that occurred during the Great Depression when a chain reaction of bank failures caused people to lose faith in the system. A similar thing was happening until recently in the US and in the Euro zone, both of which find themselves in what Keynes (1936: 169 ff.) referred to as a "liquidity trap". Due to the difficulty of finding new creditworthy borrowers, banks have lowered interest rates to close to zero. This incentivized people to hoard money at home, since without interest there is less reason to deposit money in banks.

7.3.5 Structural Deficiencies as Cause of Unemployment?

Many would argue that structural problems are the cause of unemployment, and thus neither labor market reforms nor macroeconomic policy are sufficient to solve the unemployment problem (see, e.g., Fernández Díaz et al., 2011). Fiscal policy should be complemented by other measures, especially structural reforms in different sectors, such as modernization, improving flexibility, efficiency, and supply-side stimulus. But what is the rationale for demanding such reforms? The answer is that these measures would improve productivity.

However, the driver behind the desire of business owners and investors to continuously increase productivity is, as explained above, the interest rate as the opportunity cost of any productive investment and the resulting requirement to either grow at this pace or to cut costs elsewhere so as to keep the Return on Investment (ROI) at least as high. Our monetary system is the main structural cause of this steady push to increase productivity (replacing human labor by machines) and thus increasing unemployment.

7.4 Does Population Growth Require Economic Growth?

Some might argue that population growth requires economic growth in order to maintain living standards. It is true that people need food and other goods and services in order to satisfy their fundamental human needs. However, the world has sufficient resources to support 3 times as many people as currently live on earth if we lived a lifestyle similar to that of a 100 years ago. Furthermore, as we have already discussed, the main driver of the growth imperative is not the satisfaction of fundamental human needs but the dynamics of our monetary system. It is the interest rate as opportunity cost of any productive investment that requires steady growth. No economy is immune to this obligation to grow. Growth has therefore become an end in itself.

Even countries that do not have growing populations and where people already enjoy abundant material wealth still pursue steady growth. These economies do not grow because their populations are growing or to improve the fulfilment of fundamental human needs, but for the sake of growth. And, as outlined above, this compulsion to grow is not (at least not solely) rooted in greediness. We must grow at least at the rhythm of interest as opportunity cost, otherwise our business model doesn't work (Suhr, 1989). Politicians of all political persuasions pursue public policies designed to foster growth via deregulation, investment incentives or fiscal policy, and they usually finance these policies by increasing public debt, because they fear unemployment.

Poverty is not a problem of having too many people in the world or a lack of resources but is primarily a problem of distribution, both within countries as well as between them. This brings us to the next topic. Steadily growing income inequality is predominantly caused by our monetary system. Of course, there are other reasons too. Unfortunately, just as the causes of the growth imperative are not yet well understood in mainstream economic analysis the problem of income inequality is generally not associated with the non-neutrality of money, which is why the solutions offered address the symptoms rather than the disease. Instead of redistributing wealth via taxes and transfers, it would be far better to eliminate the root cause of the problem.

References

Ackley, G. (1978). *Macroeconomics: Theory and policy*. Macmillan.

Ataway, S., Fuders, F., & Monti, G. (2019). *Deficit spending and aggregate demand in Chile: Validation of the Keynesian multiplier and its implication for development planning*. Universidad Austral de Chile.

Azkarraga, J., Max-Neef, M., Fuders, F., & Altuna, L. (2011). *La evolución sostenible II*. Lanki (Mondragon Unibertsitatea).

Costanza, R., et al. (2012). *Building a sustainable and desirable economy-in-society-in-nature*. United Nations Devision for Sustainable Development.

Cuadrado, J. R., et al. (2015). *Política Económica – Elaboración, objetivos e instrumentos* (5th ed.). McGraw-Hill.

Dvorkin, M., & Bharadwaj, A. (2019). *Declining labor share and U.S. industries*. [Online] Available at: https://www.stlouisfed.org/on-the-economy/2019/december/declining-labor-share-us-industries. Accessed 29 Nov 2022.

Fernández Díaz, A., Parejo Gámir, J. A., & Rodríguez Sáiz, L. (2011). *Política Económica* (4a ed.). McGraw-Hill.

Hayek, F. A. (1976). *Inflación o pleno empleo?* Unión Editorial.

Henke, S. (2017). *Fließendes Geld für eine gerechtere Welt: Warum wir ein alternatives Geldsystem brauchen, wie es funktioniert und welche Auswirkungen es hat*. Tectum Wissenschaftsverlag.

Karabarbounis, L., & Neiman, B. (2014). The global decline of the labor share. *The Quarterly Journal of Economics, 129*(1), 61–104.

Keynes, J. M. (1936). *General theory of employment, interest and money*. Hartcourt.

Keynes, J. M. (1983). Economic articles and academic correspondence. In D. Moggrigde (Ed.), *The collected writings of John Maynard Keynes* (p. 402). Macmillan.

Mankiw, G., & Taylor, M. (2014). *Economics*. Cengage Learning.

Mondaca, C., et al. (2017). *Estructura y comportamiento de las transferencias fiscales en Chile*. IDB (Inter-Amerikan Development Bank).

Neely, C. (2010). Okun's law: Output and unemployment. *Economic SYNOPSES, 4*, 1–2.

Okun, A. (1962). Potential GNP: Its measurement and significance. In *Proceedings of the business and economic statistics section*. American Statistical Association.

Owuso, C., Fuders, F., & Mondaca, C. (2017). *The role of decentralized public expenditure in the promotion of regional development in Chile*. Universidad Austral de Chile.

Peacock, A. T., & Shaw, G. K. (1976). *The economic theory of fiscal policy*. Allen & Unwin.

Suhr, D. (1989). *The capitalistic cost-benefit structure of money – An analysis of money's structural nonneutrality and its effects on the economy*. Springer.

Part III
Money and Inequality

Chapter 8
Our Current Monetary System Produces Income Inequality

8.1 Introduction

Income inequality is a major topic everywhere in the world. After the 2008 financial crisis we saw the emergence of movements like "Occupy Wallstreet" in the U.S., "Los Indignados" in Spain or "Blockupy" in Germany. In Chile the ongoing student protests culminated in the 2019 "Estallido Social" (Social Breakout), triggered by an increase in public transportation prices, which led to the current process whereby the country is drawing up a new constitution. A process that could be seen as unique in the world, for this is the first time since the French Revolution that ordinary people, are being allowed to actively participate in the formulation of a new constitution.

And even in countries that are supposed to have more balanced income distribution, such as Germany or the Scandinavian countries, growing inequality is becoming apparent. About 10% of the German population own 60% of the nation's wealth (OECD, 2015: 35). The situation is similar in many industrialized countries. In developing and emerging countries, the inequality of income distribution is often even more pronounced. It is not uncommon for just a few families to control most of the wealth of an entire country, as e.g., in Chile (Pérez Villamil, 2012; Fuders & Michaelis, 2016; Mieres Brevis, 2020). Even to those who do not support egalitarianism and who believe a certain amount of inequality is natural and beneficial, the magnitude of the current wealth inequality seems unjust. In this chapter we will see why our current monetary system is a powerful—probably the most powerful—driver of steadily growing income inequality. Chapter 9 will be dedicated to the moral-ethical aspects of the effects of interest on wealth redistribution.

© The Author(s), under exclusive license to Springer Nature Switzerland AG 2023

F. Fuders, *How to Fulfil the UN Sustainability Goals*,
https://doi.org/10.1007/978-3-031-37768-6_8

8.2 Exponentially Growing Deposits and Debt

As explained in Chap. 2, money is the counterpart of debt, which is why an economy's money supply and total debt **must** grow in tandem. We also concluded that this is not the actual problem. The problem is that both deposits and debt grow incessantly and exponentially via interest. Hence, while on the one side we find ever more monetary units in bank accounts, on the other side people are ever more indebted. How can this go hand in hand, one might ask. How is it possible that we become simultaneously richer (in terms of monetary units) and more indebted? The answer is an increasing number of people become ever more indebted while a smaller and smaller group becomes richer and richer. In other words: income inequality grows exponentially.

We can visualize this by plotting money supply and debt on the same graph. In Fig. 8.1 using data from the US, we can see that monetary assets mirror total debt. This is no surprise, since as was outlined above, there is no interest without debt. The gap between the upper and the lower points on the graph represent the (exponentially growing) inequality. The same gap can be observed in any country if a long enough period of time is observed. That means that anywhere and everywhere income inequality grows **exponentially according to mathematical logic**. And there is nothing that can change this, because "money is a credit-debt relation from which none can effectually escape" as Frederick Soddy (1934: 25) correctly put it, and as long as we have positive[1] interest rates both grow exponentially.

The Gini-index that measures income and wealth inequality shows similarly high values for almost all OECD countries as well as a significant increase in inequality over the last 30 years (OECD, 2015; Bárcena et al., 2018). Some countries try to mitigate these problems via redistributive economic policies, which is why the Gini-index after tax and transfers varies greatly from country to country depending on the extent of redistributive policies being implemented. However, the Gini-index before taxes and transfers depicts a similar picture in all OECD countries (OECD, 2015).

We can now understand why the richest nations are at the same time also the most indebted ones (counted in absolute figures): U.S., UK, Japan, France and Germany (CIA, 2022). Is the entire sophisticated infrastructure (houses, public buildings, streets, etc.) worth nothing, does it even have a negative value? The answer is that wealth, measured in money units, is the counterpart of debt. It is not that the infrastructure is worth nothing but that the economies with the biggest money supplies must also be those with the highest amounts of total debt.

[1] The term "positive" here does not refer to something good, but to an interest rate above zero (> 0).

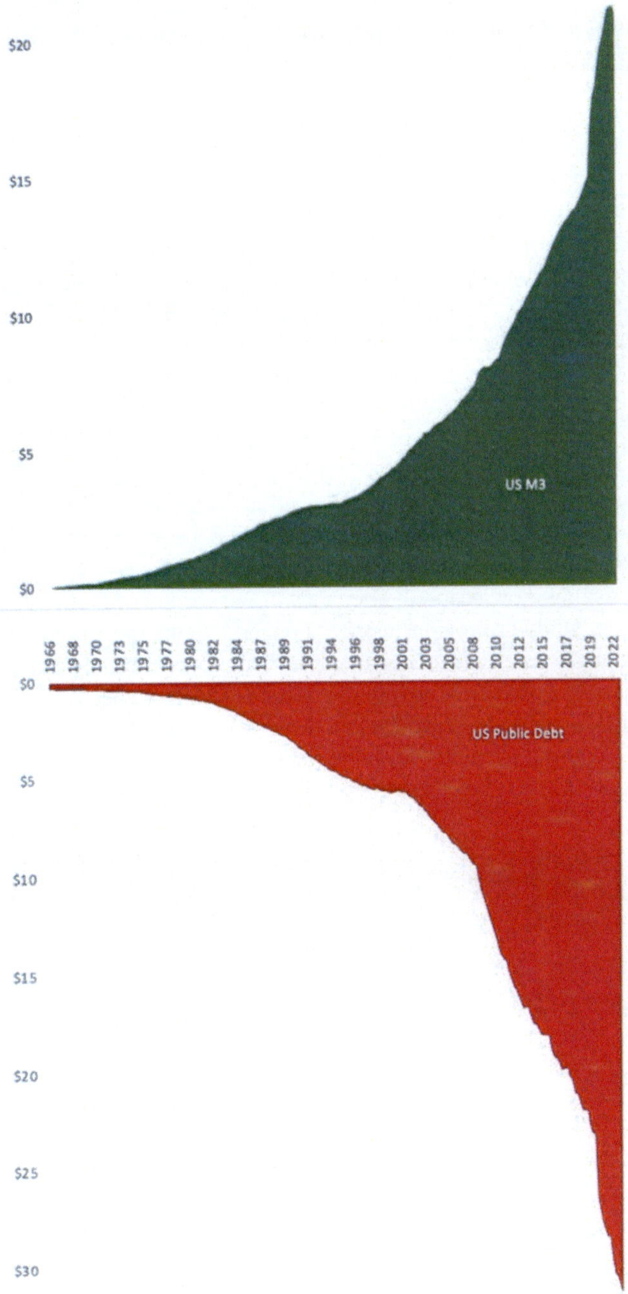

Fig. 8.1 US money supply M3 and total public debt (in trillions of USD). (Source: Own work. Data provided by OECD (2023) and US Department of Treasury. Retrieved from Federal Reserve Bank of St. Louis. (https://fred.stlouisfed.org))

8.3 In Perfect Competition Unjust Inequality Should Not Exist

8.3.1 Everyone Is Supposed to Gain According to Their Effort

It is quite interesting what we can learn here from economic theory. In the idealized microeconomic model of so-called "perfect competition", each market actor earns exactly the amount that equals the opportunity costs of the production factors employed (e.g., Frank, 2015, p. 361; Daly & Farley, 2011), i.e., the amount they could have earned elsewhere with the same input of capital, time, diligence, and knowledge. Competition between producers causes prices to go down (and quality to go up). In the long-term equilibrium prices reach a minimum, that is the value that equals the minimum point of Short-term Average Total Costs curve [SATC] and Long-term Average Total Cost curve [LATC] in Fig. 8.2.

In other words, in perfect competition company owners do not necessarily earn more than their employees. According to the theory, if they earn more, it is because they have invested more. For example, maybe they work on weekends, or maybe they risk their own capital. They may have specialized knowledge acquired through years of study or previous work experience. In the model of perfect competition, if company owners earn more than their employees, it is because they have made a greater investment of production factors. However, according to the theory, their profits should not exceed the opportunity costs of the production factors invested. So, according to the theory of perfect competition, income differentials can exist, but only to the extent that they reflect greater or lesser investment of production

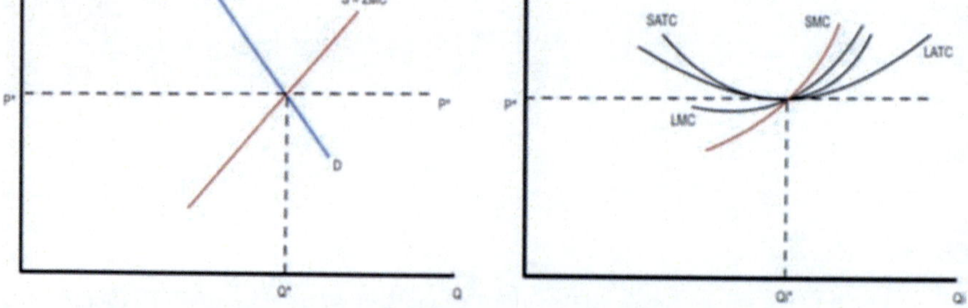

Fig. 8.2 Long-term equilibrium Perfect Competition. (Source: Own work. Explanation: The market price P* is at the lowest possible level (minimum Short-term Average Total Costs curve [SATC] and Long-term Average Total Cost curve [LATC]))

factors. Perfect competition thus results in an income distribution that reflects the concept known as "*suum cuique*" which goes back to Plato, Aristotle and Cicero and says that people's rewards should correspond to their efforts.

8.3.2 The Money Lender Earns an Income Without Creating Value

However, the principle that everyone gains according to their effort **does not apply to the money lender**. The latter earns without having contributed any production factors, because money has no intrinsic value and can therefore not be regarded as a production factor, as discussed above in Sect. 8.3.3. In classical economics—until around 150 years ago—production factors were correctly defined as: labor, land and physical capital. Money correctly was not considered a production factor. Money does not work. It is people combined with capital and land who create value. However, since today money and capital are often used interchangeably—as if money were a resource, a production factor—few notice the fact that the **money lender earns an income without creating value**. Hence, our monetary system does not only generate exponentially growing inequality, but even worse: it is **unjust inequality**.

Someone might now argue that perfect competition is only a model based on assumptions that seldom can be found in the real world. That is true. Nevertheless, the model does serve as a kind of 'economic compass' that indicates to us in which direction we can and should move. Even though we will never achieve perfect competition, it is still worth trying to get as close as possible. The reason why we are so far away from the two ideal outcomes that (i) everyone gains according to their effort and that (ii) there is no squandering of resources (allocative efficiency),[2] is **precisely because of our monetary system**. Notwithstanding that there are many other well-recognized reasons for so-called market failure, our monetary system can be regarded as the **most powerful and at the same time least recognized driver** behind income inequality and the obligation to grow. If we change our monetary system to one in which money is truly neutral (as will be proposed in Chap. 10) our economy would take a huge step towards the ideal of perfect competition, and in doing so, we would move towards "development at human scale" (Max-Neef et al., 1991) in which people can best satisfy their fundamental human needs. In Chap. 15 this aspect will be discussed.

[2] The marginal cost, i.e., the cost of the last unit produced equals the price, i.e., value the consumer assigns to the product.

8.3.3 The Marginal Productivity of the Money Lender Is Zero

What just has been analyzed from the point of view of production factors employed, can also be viewed through the lens of productivity. In the model of perfect competition, everyone earns an amount equal to their respective marginal productivity (Frank, 2015). Again, **this does not apply to the money lender**. Money lenders earn even though their marginal productivity is zero. After all, money does not produce. Rather it is borrowers who are productively active. Many people educated in economics will probably not immediately agree with this point of view. Of course, money has marginal productivity, they will say. For example, Keynes speaks of the "*marginal efficiency of capital*" (Keynes, 1936: 221) and refers to the possibility of earning interest on money as well as capital. The fact that in economics we believe money yields "marginal productivity" or "marginal efficiency", however, is precisely due to the aforementioned misunderstanding whereby we fail to distinguish between interest from money lending and profit resulting from real productive investments. This, in turn, may be due to the fact that no distinction is made between goods that have intrinsic value and money that represents such goods. The conflation of real capital and money subsequently leads to the conflation of interest from money lending with the return on real economic investment. Again: money is not productive. **It is people together with machines (capital) who create value**.

8.3.4 No Pareto Efficiency

Apropos of "efficiency", as long as positive interest rates exist, we will never reach Pareto efficiency. Pareto efficiency is a frequently used definition of efficiency in economics. According to this definition (based on Italian engineer and economist Vilfredo Pareto) a society is Pareto efficient if it is impossible to reallocate resources so as to make any individual better off without making at least one individual worse off (Frank, 2015). In other words, a Pareto efficient state is one in which it is not possible to improve one's well-being except at the expense of someone else. However, the latter is precisely occurring charging interest for money lending. As explained, the marginal productivity of the money lender is zero, which means gaining interest from money lending is an income at the expense of others. In Sect. 9.7 we will take this analysis even further and argue that gaining interest from money lending is comparable to the income of a blackmailer.

8.4 Is the Economic Theory Wrong?

So, is the economic theory wrong in stating that competition in the long term produces just incomes, ("just" in terms of everyone gaining according to their effort)? The answer is: **it is not wrong, but it is completely distorted** by our monetary system (Fuders & Max-Neef, 2014a, b). It would probably work the way the theory says it should if we replaced our existing irrational monetary system with a rational one, one in which it is not possible to generate profits simply by lending money.

This would not mean that everyone would earn the same amount or that wealth would be automatically evenly distributed. However, the kind of extreme and growing wealth inequality that characterizes our current system would not exist. And differences of economic outcomes would be based on differences in the amounts of investment of production factors performed by individuals. Of course, we could argue that such a system still wouldn't be completely fair since not everyone has the same access to production factors, and people have different God given capabilities. However, in the existing system the greatest cause of inequity is not due to different capabilities or access to production factors. Instead, it is due to the distorting influence of our current form of money.

References

Bárcena, A., Cimoli, M., García-Buchaca, R., & Pérez, R. (2018). *La ineficiencia de la desigualdad*. Cepal.

CIA. (2022). *Factbook, Country Comparisons: Debt—external*. [Online] Available at: https://www.cia.gov/the-world-factbook/field/debt-external/country-comparison. Accessed 25 May 2022.

Daly, H., & Farley, J. (2011). *Ecological economics—Prcinciples and applications* (2nd ed.). Island Press.

Frank, R. (2015). *Microeconomics and behaviour*. McGraw-Hill/Irwin.

Fuders, F., & Max-Neef, M. (2014a). Local money as solution to a capitalistic global financial crisis. In: *From capitalistic to humanistc business*. Palgrave Macmillan.

Fuders, F., & Max-Neef, M. (2014b). Dinero, deuda y crisis financieras. Propuestas teórico-prácticas en pos de la sostenibilidad del sistema financiero internacional. In: *Economía Internacional. Claves teórica-prácticas sobre la inserción de Latinoamérica en el mundo*. LATIn.

Fuders, F., & Michaelis, N. (2016). Marktkonzentration: Ein großes Hindernis für eine Nachhaltige Ökonomie. In: *Jahrbuch Nachhaltige Ökonomie 2016/2017* (pp. 271–294). Metropolis.

Keynes, J. M. (1936). *General theory of employment, interest and money*. Hartcourt.

Max-Neef, M., Elizalde, A., & Hopenhayn, M. (1991). *Human scale development—Conception, application and further reflections*. Apex.

Mieres Brevis, M. (2020). The dynamics of inequality in Chile: A regional look. *Revista de análisis económico, 35*(2). https://doi.org/10.4067/S0718-88702020000200091

OECD. (2015). *In it together: Why less inequality benefits all*. OECD Publishing.

OECD. (2023). *Main Economic Indicators (database)*. [Online] https://doi.org/10.1787/data-00052-en. Accessed 1 Feb 2023.

Pérez Villamil, X. (2012). *El mapa de la concentración en Chile*. [Online] Available at: https://www.elmostrador.cl/mercados/2012/02/15/el-mapa-de-la-concentracion-en-chile/. Accessed 2 Dec 2022.

Soddy, F. (1934). *The role of money—What it should be contrasted with what it has become*. Routledge.

Chapter 9
Moral-Ethical Aspects of Interest-Induced Inequality

... do not lend money at interest,
and do not take a bribe against the innocent.

—Holy Bible-NRS (2011: Psalm 15:5)

9.1 The Perfect Mechanism for Wealth Redistribution (from the Poor to the Rich)[1]

We have seen that the wealth gap (Fig. 8.2) results from the nature of our financial system, in which one person's debt represents another person's wealth and both multiply through interest (see also: Suhr & Godschalk, 1986; Kennedy, 1990; Creutz, 1993, 2018; Kremer, 2009). Unfortunately, the wealth redistributing dynamic of our existing monetary system is not well understood and is not generally discussed in mainstream economics. In fact, the prevailing (but erroneous) view is that savers are somehow being shortchanged when interest rates fall to very low levels, as is currently the case in many industrialized countries. In an article in the newspaper "Die Zeit" (Schieritz & Uchatius, 2014: 13–15), the situation was described as follows: "The interest rate also enabled cab drivers and saleswomen to build up a small fortune. The interest made millions of people rich and helped to create a broad middle class in the Federal Republic of Germany".[2]

This view fails to take into account that it is the savers themselves who ultimately pay the interest they receive on deposits. This is true even if they do not borrow money themselves, since interest is a hidden component in the prices of all products

[1] The following considerations on the moral-ethical problems inherent in our monetary system have already been discussed in Fuders (2017).

[2] Author's translation. Similarly, the former head of the German IFO Institute Sinn said that the low interest rate would cost Germans 300 billion euros (Reuters, 2014).

© The Author(s), under exclusive license to Springer Nature Switzerland AG 2023

F. Fuders, *How to Fulfil the UN Sustainability Goals*,
https://doi.org/10.1007/978-3-031-37768-6_9

(Kennedy, 1990; Creutz, 1993; Kremer, 2009)[3] because they and their suppliers finance themselves through debt. In fact, the broad middle class in Germany has developed not because of, but in spite of, our current monetary system. Especially in the first decades after World War II, the gap between the exponentially growing financial assets and the exponentially growing debt was not yet as large as it is today. The longer the financial system exists, the stronger are these redistributive effects.

And yes, of course, there are other causes of income inequality, such as market concentration (monopolized markets), collusion (price cartels) unfair competition or expensive education systems, which may reinforce the privileges of the elites. A monopoly is the contrary of perfect competition. Owners of a monopoly can charge higher prices than they would be able to in a state of perfect competition, which enables them to obtain unusually high profits. This extra profit (the so-called monopoly rent) is thus not based on the owner's effort but on the exploitation of a dominant market position. Here, a single person (the owner of the monopoly) gains at the cost of everyone else. Market concentration (the monopolization of markets) therefore exacerbates income inequality. The same effect have price cartels. Also, unfair competition can lead to unjustified profits, i.e., income that exceeds the normal remuneration of the invested production factors. Examples of unfair competition are cheating the consumer lying on the label or in advertisements.

Nevertheless, it is easy to demonstrate empirically that the gap between financial assets and total debt is growing, both in developed and developing economies.[4] Of course, not all income inequality should be considered unjust. If someone works hard it is reasonable for that person to earn more than someone who works less hard. The following sections will examine this aspect in greater depth. We will see that charging interest for lending money is comparable to the exploitation of a dominant market position and can therefore be seen as form of monopoly rent.

9.2 Money Does Not Work, Nor Produce Offspring

We have already discussed the fact that money has no intrinsic value, does not perform any work, is not a production factor (capital). These facts are particularly crucial for understanding the moral-ethical aspect of our monetary system. The exponentially growing wealth inequality generated by interest is problematic by itself. However, the situation is even worse when we consider the fact that the wealth of interest recipients does not grow through their own effort but through the effort

[3] And this is no small sum. It has been estimated that the interest component of prices ranges between 12 and 77% (Kennedy, 1990: 25 ff., 2011: 29; Creutz, 1993: 95 f., 106 f.), with this share growing in proportion to the steadily increasing share of debt on corporate balance sheets.

[4] Piketty (2014) has demonstrated through his analyses of tax records that the capital-to-income ratio β has risen globally from about 2.5 to more than 4 since the end of World War II. According to his forecasts, β will increase to more than 6.5 by the end of the century.

of others, namely those who borrow money. The holder of money may very well have earned that money by his or her own hard work. But as soon as that money starts to increase 'by itself', the surplus value **does not represent any additional effort on the part of the holder of money** but, instead, it is generated by those who take out loans and invest productively. The borrowers have to earn the interest that the holder of money receives without moving a finger.

It could be argued that the same is true when real goods such as cars, tools or machines are lent out. Does it make a difference whether real goods or money is loaned out? In fact, it makes an immense difference, but hardly anyone recognizes this due to the fact that money and capital are mistakenly used as synonyms, so money is thought of as if it was a production factor like a tool or a piece of machinery. The conflation of money and capital prevents a clear understanding of the economic as well as the moral-ethical problems inherent in the payment of interest as rent for borrowing money. In classical economics, capital is a production factor alongside land and human labor (see, e.g., Mankiw & Taylor, 2014). Money, however, is not a production factor, but only a **medium of exchange** that facilitates the exchange of production factors or products. Money is therefore not capital, it does not work (Creutz, 1993) and does not produce any offspring, as was remarked upon as far back as Aristotle.[5] Rather, it is people working with the help of machines who generate the interest. For this very reason, Silvio Gesell described interest from money lending as the "right to the foreign labor yield" (Gesell, 1920: 210) and criticized the division of society into "beasts of burden" and "rentiers" (Fig. 9.1), i.e., into those people who work and those who live off it (Gesell, 1949: 27). 400 years earlier, for Martin Luther interest from **lending money was a form of** robbery (Luther, 1841: 75).

In an economic system where interest did not exist, i.e., where loans were free of interest, the only way to gain money would be to produce or sell real products (with real intrinsic value) or services. In such an economy **only real productive effort would generate profits**. Of course, in such an economic system money can be invested lucratively. In other words: even in an interest-free economic system,

Fig. 9.1 Rentier on hammock

[5] In Greek „interest" and "offspring" are the same word: tókos.

concentration of wealth could still occur, and it would even be possible to invest money such that one could live off the profits without working. For example, the founder of a company could once his company is doing well, appoint a manager to run the company, without the owner having to continue working.

However, the entrepreneur creates products and services that serve people and satisfy fundamental human needs, and often they also create jobs thereby contributing to the well-being of all. Moreover, business owners bear **entrepreneurial risk** even if they are no longer actively involved in operating their businesses. Money lenders, however, do not produce any value themselves (see already Fuders, 2010) and do not even bear any entrepreneurial risk. Instead, it is the borrowers who are the ones who create the value and who bear the entrepreneurial risk, because if their investment fails, they still have to pay interest. This brings us to the next point.

9.3 The Borrower Takes the Entrepreneurial Risk and Pays Twice

The last point is particularly interesting, since in monetary theory interest is justified by the risk of loan default. However, it is actually the borrower who bears the entrepreneurial risk, while the lender secures loans with guarantees, mortgages, or the pledging of other real assets. Section 4.2.5 had already outlined that the interest rate therefore is for most loans (practically all loans except government bonds or venture "capital") no risk premium. The distinguishing momentum between a loan and a real economic investment is precisely the fact that the loan has to be repaid **regardless** of the success of the business that is financed by the loan.

The fact that the money lender does not share the entrepreneurial risk and that interest must be paid regardless of whether the investment is successful or not, is ultimately the reason why, after some decades, the financial system starts to decouple from the growth of the real economy. And here we find the reason for the occurrence of inflation, price bubbles and especially the obligation to grow. Figuratively spoken, the economy must keep pace with the growing money supply and debt. If we were able to perfectly link the interest of loans to the entrepreneurial risk of the business financed with these loans—which means the interest rate would turn negative in the event of mismanagement—then the financial markets would not decouple from the real economy. In that case, however, it would not be a 'loan' but, instead, an investment in the real economy, namely a classic corporate investment. Again, the difference between a loan and an investment in the real economy is that the loan earns interest **independently** of the productive process financed by it.

The borrower not only takes entrepreneurial risk but also pays twice. To understand this point, let us return to the barter example discussed in Sect. 5.6.2: A blacksmith lends his plow to a farmer and receives a sack of potatoes from the harvest in exchange. We discussed in Chap. 5 that the sack of potatoes does NOT represent interest but rather a rental fee for borrowing real capital (the plow). The sack of potatoes is real economic profit. What is even more interesting, today, unlike in times of barter, the farmer would first have to borrow money in order to then lease the plow with the borrowed money. After the harvest, he would then have to return the plow and pay the rental fee as well as interest on the borrowed money. So, the farmer pays twice (rental fee and interest). This problem was recognized as far back as Saint Thomas Aquinas (1596). This brings us to the next point.

9.4 Can Interest Even Be Considered Fraud?

As explained, money is merely a medium of exchange. It may indirectly represent a fraction of the economy's total production (Hartmann, 1992), but it is not a real good with intrinsic value and certainly not a production factor (capital). Keynes, as mentioned in Sect. 3.3 correctly compared money with a theater ticket. According to Keynes, to claim that money has intrinsic value would be the same as confusing a theater ticket with the play itself.

For this very reason, lending money for a fee is "unnatural" (Aristotle, [384 BC] 1995a: 1258b) and thus not justified. It makes sense to lend out goods, but not vouchers representing goods. The latter is not only "complete nonsense" (Steiner, 1979: 50) but may even be considered fraud. At least the emperor in Goethe's Faust II characterizes the creation of money out of nothing invented by Mephistopheles (Satan) as **"sacrilege" and "monstrous fraud"** (author's translation from von Goethe, (1976: 187). Goethe was referring to the monetary system (Binswanger, 1985), which even in Goethe's time functioned similarly to how it does today.

And money cannot even be regarded a voucher that represents specific goods. While a theater ticket, to stay with Keynes' example, at least specifies the good or service that it can be exchanged for (namely the play), a bank note **does not entitle the holder to any specific good or service**. It is nothing but a piece of paper, with no intrinsic value. "Money in itself is worth nothing, since it is incapable of generating any value", as the car manufacturer Henry Ford[6] correctly put it.

The owner of a real good relinquishes the usufruct when he lends it. As long as the good is lent out, the owner cannot derive any benefit from it. Also, the good depreciates over time and needs maintenance. Both circumstances justify a rental fee in exchange for lending it. The lender of money, however, does not renounce any usufruct when lending it; after all, the money has not yet been exchanged for a usufruct-generating good. Again, money has no intrinsic value or usufruct for the owner to relinquish. Furthermore, it does not require maintenance, nor does it depreciate.

One might argue that money depreciates due to inflation, and thus interest is justified as compensation for the reduced purchasing power of money. However, the truth is the other way around. As was explained in Chap. 5 inflation is caused by

[6]Authors translation from German "(...) Geld an sich ist überhaupt nichts wert, da es für sich keinen Wert zu erzeugen vermag" quoted in: Seminar für freiheitliche Ordnung (1984: 35).

interest. As explained above, the effect the interest rate has on the money supply is exactly the opposite of what is taught in traditional monetary theory. Bank **deposits grow by interest and so does money supply** (defined as cash plus deposits); the higher the interest rate, the faster money supply grows. And if the money supply increases at a faster rate than the production of real goods and services, inflation will result. There are many theories that try to explain the existence of inflation, as we discussed in Sect. 4.4, but in the end all these theories boil down to the same thing: inflation originates in the money supply (and so aggregate demand) growing faster than the supply of goods and services. None of the many theories of inflation explains why this constantly occurs. The usual explanation that is offered is that the money supply is an exogenous variable controlled by the central bank and if inflation occurs, this is due to mismanagement by the central bank (Friedman, 1966).

We have already established above that this is wrong. The central bank is not a trend setter but a trend follower. The money supply grows as long as there is a positive interest rate paid on deposits; the higher the interest rate, the faster the money supply grows. This is a mathematical rule, and there is nothing that can change this fact. It is **not the fault of the central Bank**, nor can a central bank prevent this from happening, as long as it does not address the design flaw of money.

Returning to the previously mentioned idea that interest represents compensation for inflation: we do not need interest as compensation for inflation because interest is the cause of inflation. Without interest inflation would not exist. In contrast to renting money and the 'snowball effect' (see Sect. 5.6.4) this causes, renting a machine (or any other good) just causes money to flow from one hand to another. The money supply does not grow as a result. And therefore, there is no unjustified redistribution of wealth (Fuders, 2010, 2017).

9.5 A Reward for the Sacrifice of Consumption, or for Improving Productivity?

Renting real capital assets such as cars or machines can be considered a **service**. The service consists in abstaining from the use of the capital asset as well covering the cost of maintenance and depreciation. The foregone use and maintenance justify the rental income. A similar argument is often made when money is lent. The lender sacrifices consumption and thus interest is compensation for deferring consumption (e.g., already Carver, 1883; Süchting, 1995: 437; Sell et al., 2020). However, the comparison with real goods is not apt because money is not a real good or service with intrinsic value but a voucher that by itself does not directly perform any useful service. A voucher does not generate any income or benefit which is foregone by the owner who lends it out. Only when the voucher is exchanged for real goods, e.g., a car, and if this car then in lent out, is it accurate to speak of abstaining from consumption. People who lend money **neither abstain from, nor renounce, nor sacrifice, nor postpone consumption**, because the owner has not yet bought anything with real intrinsic value the use of which they abstain from, renounce, sacrifice, or postpone (Fuders, 2009, 2010).

Instead of a premium for abstaining from consumption, interest is a liquidity or non-hoarding premium as Keynes (1936) correctly understood and as highlighted above. Recognizing that interest is **neither a premium for abstaining from consumption**, nor a risk premium (see above) is crucial to understanding the moral-ethical dimension that results from interest. Even institutions that address the question of interest in the context of the ethics of economics sometimes fall prey to the misconception that interest is a premium for postponing consumption. One might object that money is not indefinitely storable because of inflation, and therefore we need interest to make up for inflation. However, as said in the previous paragraph, the argumentation should go the other way around since inflation is only a side effect of interest.

According to the "productivity theory" of interest, which goes back to J. B. Clark (1888) and F. H. Knight (1921) interest arises on account of the productivity of capital. This idea is usually stated as follows: labor is more productive when combined with capital than without it, therefore capital should be entitled to share in the additional increment of productivity that is made possible by the combination of labor and capital. This argument seems intuitive and logical. Human labor is of course more efficient when combined with capital (machines and tools). But again, money is NOT capital. Money is a medium of exchange with which capital can be bought. In this erroneous theory, again we find the conflation of money and capital. This theory provides a sound rationale for why leasing a machine or tool should be rewarded, but not for leasing money.

9.6 The Money Owners Should Thank the Borrower

9.6.1 Lending Money Is Not a Service and Certainly Not a Product

Because, as we just discussed, the lender of money does not forgo consumption, **lending money is not a service and certainly not a product**, even if so-called "financial service providers" or banks like to call their loans "services" or "products". It is probably correct to call the mediation of lending a "service", but not the lending of the money itself. Money is not a product, which is why a loan is not a product either. The return from lending money, i.e., interest, must be earned by someone in the real economy, usually the borrower. This is also true for those who borrow real goods. However, in the case of borrowing a real good, the borrower receives the utility (usufruct) of the good in return for the rent he pays. On the contrary the borrower of money gets only a piece of paper, which itself has no value. **If there were no goods that could be exchanged for money, money would be worthless**.

The same logic also applies under a gold standard. If there are no goods that can be exchanged for gold money, it would be worthless,[7] which is why the gold standard solves neither the economic (Fuders & Max-Neef, 2014b), nor the moral-ethical problem. We will come back to the gold standard in the "no alternatives" section (Chap. 13). Those who produce valuables (goods with real, intrinsic value) are those who give money its value, not those who lend money. In fact, money owners should **thank borrowers instead of charging them interest**, because it is through the productive investments and work of borrowers who produce real goods and services that the money obtains its exchange value, i.e., purchasing power.

9.6.2 Interest as Opportunity Cost for Foregone Use

It is often argued that interest is necessary to compensate the money lender for foregoing his/her own investment (see e.g., Voigt, 2012). An example of such an argument goes as follows (Wikipedia, 2022):

> The fact that the lender is allowed to demand interest from the borrower at all is due to the economic consideration that the lender himself foregoes the use of his capital during the term of the loan and therefore cannot generate any profits or income from an alternative financial investment. The interest rate on the loan must therefore be classified in economic terms as an opportunity cost for foregone use (Author's translation).

Interest is (unfortunately) the opportunity cost of any productive investment—and we have seen how the growth imperative results therefrom—but not *vice versa*. The real economy has to grow at least at the pace of the interest rate, otherwise investments would not be viable, as we have analyzed in depth above. Nevertheless, the same rationale **must not** be applied the other way around, arguing that interest is compensation for a hypothetical forgone economic profit that money owners could have otherwise earned by investing the money themselves.

Those who want to make profits can invest money **themselves** in the real economy and **at their own risk**. Then, and only then, are profits justified. We must remember that money is not capital, nor is it a resource, nor does it perform work or beget offspring (Aristotle, [384 BC] 1995a; Aquinas, 1596). Why should the money lender receive a share of the profits that **others generate through their own work and risk taking**? To repeat, the conventional understanding results again from the conflation of money and capital.

In the same vein and stemming from the same conflation of money with capital, the following describes a classic misconception of the role of interest (Starbatty, 1977):

> Because it is clear that interest income is not a benefitless income, but that the surrender of capital makes progress in the economic process possible. ... The renunciation of present

[7] Here, the legend of King Midas, mentioned by Aristotle ([384 BC], 1995a: 19 f.), is very illustrative. Everything that was put before Midas is said to have turned to gold. Despite the legendary wealth measured in gold, Midas starved to death because his food also turned into gold.

goods makes it possible for another to improve productivity, and from this the latter can then pay the interest (Author's translation).

However, the money lender does NOT surrender "capital" but money. Again, money is not capital and does not perform any work. With money we can buy capital goods and those capital goods can then be employed in productive processes. It is the capital (machines and tools), not the money, that contributes to productivity. And in the event that the investment is unsuccessful, the borrower still has to pay interest. Regarding the second sentence in the quotation above, we do not renounce present goods as long as we have not bought them. What we renounce is the medium of exchange (the money), which itself has neither intrinsic value nor utility.

Another frequently heard argument to justify the existence of interest is the time **transformation function** ascribed to interest. The critics of interest (Aristotle, Thomas Aquinas and other scholastics), it is argued, did not take into account that in order to be able to compare investments we need to calculate the present value of future cash flows, and to do so we need interest as a discount rate (see, e.g., Voigt, 2012). However, that interest gives time a value and in doing so distorts economic calculations is precisely the problem. In Sect. 5.7.2 it was already mentioned that interest strongly promotes unsustainable forest and aquaculture management in Chile and elsewhere (see also Fuders, 2020). The well-known motto "the faster the better" and that "time is money" results primarily from the fact that due to interest time is not a neutral factor.

9.7 Interest Can Be Seen as a Monopoly Rent

9.7.1 A Monopoly Rent Is Inherently Unjust

The old Greek concept of justice *suum cuique* which goes back to Plato, Aristotle, and Cicero describes a state of distribution as just, in which everyone gets what they deserve according to their performance. And no one should receive compensation in excess of this. The model of perfect competition comes close to this concept. In this theoretical model, everyone earns exactly the amount that corresponds to the opportunity costs of the production factors they use. With the same factor inputs, every market actor—be it as an entrepreneur or as an employee—could earn exactly the same profit elsewhere. Thus, in the model of perfect competition, everyone earns exactly what corresponds to their input of labor, time, know-how, and physical capital. Those who put forth more effort earn more, those who put forth less effort earn less. The opposite of perfect competition is a monopoly.[8]

[8] The opposite of perfect competition is not communism, as some might think, but a monopoly. However, in communism all production is monopolized, which is one of the two major reasons for why communism is highly inefficient.

The monopolist, unlike the producer in perfect competition, can earn a profit that significantly exceeds the opportunity costs of the production factors employed. This profit is called "monopoly rent". In economics the term "rent" describes an income that is not (or not only) based on the effort of the investor. The monopoly rent arises from exploiting the market power of the monopolist. As the sole supplier, monopolists can charge much higher prices than they could if they faced competition. In other words, the high prices charged by monopolists do not reflect the quality of their goods/services but are rather based on the lack of competition. When it comes to essential goods (inelastic demand), such prices can be very high, even though production may be very cheap. Thus, the price charged by the monopolist, as opposed to the dynamics of perfect competition, bear no relation to the cost of production. In essence, monopoly rent is a profit that exceeds the monopolist's effort or performance. Although the price charged by the monopolist is determined by the market, it is perceived as unfair because it is disproportionate to the monopolist's effort. Monopoly profits thus contradict the old Greek concept of justice *suum cuique*. In Sect. 9.10 I will come back to this concept again.

9.7.2 Interest Is Even Worse Than Regular Monopoly Rent

The owner of money has, as Silvio Gesell correctly described (Gesell, 1949), a sort of monopoly position over those who lack money and need to sell goods that perish rapidly. This monopoly position of money is based in law since it is the law that gives money its position as the **only medium of exchange.**[9] Money is merely a medium of exchange. It should be seen as a public "good" that must circulate in order to facilitate the exchange of goods and services. Those who withhold it from circulation prevent others from using it to exchange goods, which is why people should only be able to hold it for a limited period of time. However, in the current monetary system there is no legal or practical disincentive to holding money for extended periods of time. Hence, those who need money must pay those who hold it a premium for lending it (for parting with liquidity), which we call interest (Keynes, 1936). Consequently, interest is not paid for any service or effort that the money owner provides to the borrower but is rather based on the exploitation of the dominant position that the money holder has over the holder of real goods and services. Interest is the profit of a monopolist: **a monopoly rent**.

However, interest is not only a simple monopoly rent, but is even worse: it is a monopoly rent that results in abusing a monopolistic position of a "good" with a perfectly inelastic demand. From microeconomic theory we know that monopolists can exploit their dominant market position more the less elastic is the demand for their goods/services (see e.g., Frank, 2015). In contrast to money, real goods perish

[9] For example, for Germany see: § 14 Abs. 1 Satz 2 BBankG; for Chile see: Decreto Ley 1123–30.07.1975.

and have extremely limited (and costly) storability. The only way to easily store wealth in the short-term is by holding money. This characteristic of money together with the fact that the law grants it an exclusive position as the only medium of exchange causes the demand for money to be almost completely inelastic: everyone needs money. You cannot live without it. Money is as important as a life-saving medicine. Hence, we can conclude that the money holder not only exercises a "simple" monopoly position over the holder of real goods, but rather a monopoly position of an essential "good" with an inelastic demand.

9.8 Interest Is Comparable to Extortion

We have just discussed the fact that interest can be considered as a form of monopoly rent. However, the monopoly rent obtained by the money lender is even more harmful than the monopoly rent obtained by producers of goods and services who enjoy monopolistic privileges. Manufacturers holding dominant positions at least provide valuable goods and/or services, even though their prices may bear no relation to their costs of production. Money lenders, on the other hand, earn income without providing any valuable goods or services. The recipient of interest thus earns income without providing anything valuable in return. This is made possible by the monopoly position of money as the sole medium of exchange established by law. For this reason, interest is essentially a **form of extortion** (Gesell, 1949: 205, 344). Extortion is unlawful enrichment from the assets of others, basically receiving a profit without providing anything in return (Gabler Wirtschaftslexikon, 2022). This is precisely what happens when interest is charged as a "reward for parting with liquidity" (Keynes, 1936: 167).

To explain this perspective, I would like to use an example that the outstanding monetary theoretician Helmut Creutz once gave at a conference.[10] One can compare the charging of interest with a situation in which someone parks his car on a public road in such a way that prevents others from getting through. Let us further imagine that a delivery company is supposed to deliver a piano to a house that is located at the end of this street, which is a dead end. There is no way to get to the house other than via this road. When asked to move his car, the car owner replies that he would be happy to do so for a fee of 5 €. The truck driver, who urgently needs to deliver the piano, now has the option of foregoing the delivery or paying the 5 €. This example illustrates the basic dynamics of interest. Just as the car owner occupies the public road and thus obstructs the flow of traffic, so the holder of money who hoards it rather than returning it to circulation obstructs the flow of money.

Extortion is forbidden by law. Since taking interest can be regarded as form of extortion, it should also be forbidden. But because we are taught from an early age

[10] Geld-Macht-Sinn congress, Köthen, Germany 2012. Gesell (1949) used a similar historical example from when bridge tolls were levied.

that interest is a "price", just like the price of a good, the fact that it is a form of extortion is obscured (see next chapter). Most people believe that receiving interest is a "right" and some court decisions have even confirmed this wrong view (e.g., German Federal Supreme Court: BGH, 2010).

9.9 Market Prices Are Not Always Fair

It is sometimes argued that since interest is a price which is determined by the market, i.e., the relationship between supply and demand, it cannot be considered unfair (Reichmuth & Reichmuth, 2001). In reply to this, it must first be said that money, as discussed above, cannot have a price, since it is not a commodity and has no intrinsic value. As we have already established, money is a voucher that represents goods (with real intrinsic value), but such a voucher has no intrinsic value of its own.[11] Money is an accounting unit by which value is measured, but it does not have value of its own. The rent or "price" for borrowing money, i.e., interest, is determined by the market. But this fact alone does not automatically mean that interest is fair. **Not every market price is automatically fair**. Let us stay with the aforementioned example of the car that obstructs a public street. There is no way to get to the apartment to deliver the piano other than via this street. Since the delivery man cannot carry the piano to the house on foot, he agrees to pay the price that is asked. Let us now change the example a bit. The delivery man does not immediately agree to pay but starts negotiating with the car owner. After a few tough minutes of negotiation, they finally agree on a price of 3 € (instead of 5 €). This amount of 3 € is then a market price that balances supply and demand. Nevertheless, hardly anyone would consider this price as fair. After all, this is a clear case of extortion.

The situation is quite comparable to a money owner who charges interest for returning money into circulation. As outlined above, interest can be seen as extortion. Those who desire to exchange goods or services are not interested in money itself but need money in order to complete transactions. Silvio Gesell spoke of the unnatural, special position of money vis-à-vis real goods which cause people to hold money for bad times or to "press" interest for the lending of money (Gesell, 1949: 205, 344). The reason why this can be compared to extortion is because money is not a good but merely a means of exchange, a voucher intended to simplify the exchange of goods. The fact that money is not a good, but a voucher, is critical for understanding that interest represents an unjust enrichment of the money owner at the expense of those working in the real (productive) economy. The price being extorted may be a market price, but that doesn't mean it's not extortion.

Money can and should be thought of as public property that is supposed to serve all by facilitating the exchange of goods and services. As such it must circulate and

[11] Vouchers that represent a specific good, e.g., a theatre ticket, can of course have a "price". However, this price does not represent the value of the voucher but rather that of the good that it represents (the play). The voucher itself is worthless.

should not be held for prolonged periods under the mattress but should rather be spent or lent. In line with this reasoning, banknotes in most countries belong to the central bank and it is illegal to destroy them (e.g., for the US see U.S.C. § 333). So, why should it be permissible to charge a price for leasing what is, in fact, public property?

9.10 Interest Contradicts All Theories of Justice

In philosophy, various models or theories of justice have been developed, which can be divided into three major schools of thought: (i) egalitarianism, (ii) the old Greek principle of "*suum cuique*" going back to Plato, Aristotle, and Cicero, and (iii) utilitarianism.[12]

The followers of egalitarianism consider a situation to be just in which goods are distributed equally among all people. Many today advocate not complete egalitarianism but the equality of opportunities. Those in a society who are less capable should not suffer for the inherent inequality of capabilities. The concept of "equality of opportunity" approaches the ideal of justice "*suum cuique*", according to which a certain distribution is to be regarded as just in which everyone gets what they deserve according to their individual performance. According to this idea, someone who performs better than someone else must also receive more. However, he should not earn disproportionately more than that, because then he would enrich himself at the expense of others. According to the utilitarian model of justice, the income and distribution of goods are just if a so-called "social welfare function" can be maximized, howsoever this function might be designed. Accordingly, a state is just and good if it serves the common good (the wellbeing of all). This model of justice is not seldom preferred by economic policy makers because, depending on how the common good (the social welfare function) is defined, it can justify any kind of wealth redistribution. At this point, I will not pursue the question of which of these models is closest to the truth or to the majority's sense of justice. There is an abundance of literature on this subject favoring one model of justice or another with good arguments. Instead, the idea is to show that our current monetary system contradicts **all three** models of justice.

The exponentially growing redistribution of wealth unquestionably contradicts the egalitarianism model. But our current monetary system also contradicts utilitarianism, which ultimately seeks to maximize the common good. We had already mentioned the fact that recurring financial crises are a symptom of the current interest-based system (and Chap. 16 will expand upon this). It is obvious that a system that generates recurring financial and economic crises can hardly serve the

[12] On egalitarianism, see, for example: Anderson (2000), Raz (2000), Walzer (2000), Gosepath (2004), Dworkin (2014), and Rawls (2020). On utilitarianism see: Mill (1863). The justice postulate "*suum cuique*" goes back to: Plato ([400 BC] 2011: 333 f.); Aristotle ([384 BC] 1995b: 100 ff.); Cicero ([44 BC], 2002); on this, see, e.g., Bubner (1995).

common good. In addition to financial crises, there are other detrimental consequences of the 'interest economy', such as constantly growing overall debt, the obligation to grow and the environmental destruction that this causes, unemployment, inflation, etc. No social welfare function whatsoever can be maximized in the long run using a model where money "earns" interest. Finally, interest also violates the third justice concept "*suum cuique*", since the redistribution of wealth it causes is based not on the effort or labor of the interest recipients, but on the effort of the borrowers, and interest recipients, unlike lessors of real goods, do not sacrifice consumption, as explained above.

9.11 Speculation

The previous remarks provide the basis for now assessing speculation. Aristotle criticized not only the business of money lending and charging interest, but also the business of the merchant, since a simple trade, unlike an act of production, predominantly serves to earn money, without the trader improving, creating or manufacturing anything. According to Aristotle, the job of the merchant is less morally reprehensible than that of the money lender (Aristotle, [384 BC] 1995a). This is understandable insofar as merchants earn income without producing the products they sell. However, trade facilitates the distribution of goods, and that could be considered a service. The trader who imports goods from abroad makes them available to consumers at home. He takes the risk that his goods will spoil while in transport or that he will not find buyers for the goods. In the past, the risk of being robbed on the months-long trade routes was high. In this sense, the merchant's profit is remuneration for the service of distributing goods. According to Aristotle, trade which serves the distribution of goods and thus "to supplement and complete the natural self-sufficiency" is not to be blamed (Aristotle, [384 BC] 1995a: 18 f.).

The activity of the speculator, on the other hand, cannot be compared to that of an importer or a wholesaler or retailer. Speculators do not perform any useful service. They do not produce goods, nor do they help make goods available to consumers. While the act of production represents the creation of value, as it increases the quantity of products and services in the economy and thus contributes to the satisfaction of fundamental human needs, the speculator (who buys and sells goods without contributing to their production or distribution), just like the money lender, does not create any value. Similar to money lending, speculation does not increase social welfare in terms of satisfying fundamental human needs. Just like profits from money lending, speculative profits are income without consideration.

Incidentally, speculation and interest are linked in another way. Without interest from money lending, there would probably be little speculation. We have already discussed the fact in Sect. 4.4.4 that speculative bubbles arise as asset price inflation because an ever-growing money supply seeks forms of investment; consequently, price bubbles too are a symptom of our monetary system. Without an ever-growing money supply, in the long run the probability of profiting from pure speculative transactions would be lower, thus making speculation less profitable and less attractive.

If one understands that no value is created by pure speculative transactions, i.e., the speculator makes no contribution to the welfare of society, it is hard to understand why in our society stock market speculators, who earn large financial fortunes from speculation, are so admired. Meanwhile workers, craftsmen, and engineers who have lower incomes but who make contributions to the creation of wealth often receive less recognition. Shouldn't it be the other way around?

9.12 The Prohibition of Interest in the Holy Scripture

9.12.1 Interest Contradicts the Biblical Commandment to Love One another

Having understood the moral-ethical problems inherent in our monetary system, especially the fact that interest allows holders of money to appropriate someone else's effort, and that charging interest can even be regarded as extortion (Gesell, 1949), robbery (Luther, 1841) and fraud (von Goethe, [1832] 1976), interest clearly contradicts the biblical commandment to love one another. For Christians, to love one another is the highest commandment next to the commandment to love God (Bible, Leviticus 19:18; Deuteronomy 6:5; Matthew 22:37–40; Galatians 5:14).[13] We find the commandment to love one another in other religions too (for Islam see Koran, surah 42:23; for Judaism see Torah, Leviticus 19:18). Love for one's neighbor translates into treating others as we would like to be treated; not to do unto our neighbor what we would not want to be done to us (Bible, Matthew 7:12). One could also say that loving our neighbor is the perfect realization of Immanuel Kant's categorical imperative (Kant, 1868).

9.12.2 The Bible, Torah and Koran All Explicitly Prohibit Interest

Charging interest does not only contradict the commandment to love one's neighbor. All three books explicitly prohibit interest in numerous verses.[14] Pope Benedict XIV wrote in the encyclical *Vix pervenit* (1745):

---

[13] Both commandments are equally important (Mathew 22:40), which is consistent with the statement that "God is love" (Bible-NIV, 1. John 4:8).

[14] Bible: Exodus 22:24; Leviticus 25:36–37; Deuteronomy 23:20–21 and 24:10–13; Ezekiel 18:8; 18:13 and 22:12; Luke 6:35; Nehemiah 5:6–11; Psalm 15:5; Torah: Exodus 22:24; Leviticus

> The sin called usura, which has its proper seat and origin in the loan contract, is based on the fact that someone demands back for himself from the loan more than the other has received from him (...). Any profit exceeding the sum lent is therefore illicit and usurious (Benedicti XIV, 1854: 297, author's translation).

The Bible, Torah and Koran, usually (depending on the translation) use the term "usury" instead of "interest". This is because, unlike in modern usage, (see, e.g., § 291 German Criminal Code; for the US see different State laws that specify the maximum legal interest rate at which loans can be made), the term "usury" used to be a synonym for interest (Kirshner & North, 1995: 430; Heinsius, 1822: 686; Prien, 1992: 98). Only quite recently, perhaps in the past 150 years, has "usury" been reinterpreted to mean an immorally high interest rate. The original meaning of the term usury as used in the Bible, is "any profit" that exceeds the amount lent as we can read in the encyclical *Vix pervenit*. The modern redefinition, in which usury is no longer a synonym of interest, paved the way for the lifting of the general prohibition of interest by the Christian churches.

Here it should be noted that there is one verse in the Bible from which it could be concluded that interest would be permissible. In Deuteronomy 23:20 (Bible-NIV, 2015) we read:

> You may charge a foreigner interest, but not a fellow Israelite, so that the LORD your God may bless you in everything you put your hand to in the land you are entering to possess.

However, **all other verses** in the Bible are very clear and do not admit any exceptions. And since anyway we are called to love one another and treat everyone as brothers and sisters, there would be no foreigner to charge interest. Furthermore, it might be the case that this one verse which seems to permit the charging of interest to strangers might actually be due to a faulty translation. This notion is supported by the fact that in the original Latin version of the Bible the word "stranger" does not appear. The same verse in Latin reads:

> Sed alieno fratri autem tuo absque usura id quo indiget, commodabis: ut benedicat tibi Dominus Deus tuus in omni opere tuo in terra, ad quam ingredieris possidendam (Deut. 23:20, Biblia Sacra Vulgata, 2007).

The word "alieno" does not necessarily mean stranger but "to alienate, to give up, to lose possession, to transfer by sale, to estrange" (Olivetti, 2003).

25:36–37; Deuteronomy 23:20–21 and 24:10; Ezekiel 18:8; 18:13 and 22:12. Koran: surah 2:275–279. On the biblical prohibition of interest see e.g., also: Brockhaus-Enzyklopädie, 1994: 561; Petersen, 2005. On the prohibition of interest in the Koran see: Baqir Sadr, 2011.

9.12.3 The Reinterpretation of Usury as Immoral High Interest

It is not surprising that the reinterpretation of the strict prohibition of interest into a prohibition of excessive interest goes hand in hand with the conflation of money and capital in economics, a development in which Marxists played a decisive part. Both in Karl Marx's main work as well as in Marxist textbooks, it is explained why money should be considered capital (Marx, 1872; Bogdanoff, 1925). Basically, the expression "capitalism" coined by Marxism is also rooted in this mistaken conflation of money and capital. The conflation of money and capital leads to the above-mentioned misunderstanding, which is pervasive in economics today, whereby every kind of profit is considered "interest", regardless of whether it is derived from productive activity in the real economy or from money lending.

Of course, there is nothing wrong with earning a return from productive labor equal to the opportunity costs of (real) production factors invested. But as long as money is regarded as a production factor, it is difficult to see why it should make a difference if profits are generated from money lending or from real productive activity. Note that in the Koran, Muhammad explicitly condemns the conflation of real economic profits and interest on loans. He admonished that those who said "purchase transactions and interest lending are one and the same thing" would be punished (Koran, Surah 2:275–279).

9.12.4 Is Interest on Loans for Productive Investments Morally Justified?

Unfortunately, neither the Bible nor the Koran provide an explanation for the reason behind the prohibition of interest, which leaves room for interpretations, such as the often-drawn distinction between supposedly morally justified interest on productive loans and immoral interest on consumer loans (for further references see Leutzsch, 2000; Lachmann, 2016). According to this view, the Bible would only forbid charging interest on loans to brethren in distress, i.e., the exploitation of the misfortune of others. Loans that do not exploit an emergency, i.e., especially those for investments in productive activities (productive loans), would therefore not be subject to the prohibition of interest.

This gray area can be attributed to the fact that neither Hebrew nor Greek distinguishes between interest and usury. Therefore, it is only an immoral high rate of interest that would be subject to the prohibition (Müller, 2011; Vieweger & Mockler, 2002). However, if no distinction is made between "interest" and "usury" in either Hebrew or Greek (nor in other languages such as old German), doesn't this support the view that there should be no distinction made between the two? Further support for this view is provided by the fact that even in present day translations of the Bible both terms are used as synonyms (sometimes even in the same verse). For example, in the Lutheran Bible, 2009: Exodus 22:25) reads: "If you lend money to one of my people, to a poor man next to you, you shall not deal with him as a usurer; you shall not take any interest from him".[15]

In view of the fact that even a low interest rate leads, through the effects of compound interest (which is impossible to avoid as explained above, Sect. 4.3), to an exponential increase in credit and debt, which gradually causes the money supply to decouple from the growth of the real economy, and in view of the hereof resulting devastating consequences for the economy and society, any profit from money lending is to be disapproved. Incidentally, the encyclical *Vix pervenit* **explicitly rejects** the aforementioned idea that loans for consumption should be regarded as "bad" while loans for productive investments are considered "good" (Benedicti XIV, [1745] 1854: 297).

The true resolution to this issue is not to distinguish between loans for consumption and loans for production but rather to distinguish between real economic profits and interest, as proposed in Sect. 5.6.5. Any profit from money lending is economically problematic since it creates a need for growth, and thus promotes environmental destruction, and is also morally indefensible since it leads to unfair distribution of income. It is irrelevant whether borrowed money is spent on consumer goods or is invested productively.

9.13 Did Aristotle and Thomas Aquinas Lack a Sound Capital Markets Theory?

In business ethics, it is often argued that the ancient critics of interest, such as Aristotle, St. Thomas Aquinas and other scholastics lacked a sound capital market theory (see, e.g., Voigt, 2012; Lachmann, 2016). In fact, one can argue that it is the other way around. Our present-day capital market theory, in which money is regarded as capital, i.e., as a production factor, is based on the fact that it is apparently not recognized that money is only a voucher representing goods, not a good itself and therefore cannot be considered productive. It seems that the ancient philosophers, religious teachers and fathers of canon law had a much better capital market theory than we do today.[16] At that time money was not yet confused with capital. In those days, probably few would have concluded that money could "work" for us, as many believe today (Consulting, 2021). As mentioned, Aristotle even makes fun of the fact that interest in Greek is a synonym of the word "offspring"

[15] Author's translation. The original words in German are: "*Wenn du Geld verleihst an einen aus meinem Volk, an einen Armen neben dir, so sollst du an ihm nicht wie ein Wucherer handeln; du sollst keinerlei Zinsen von ihm nehmen*".

[16] By "we" I mean the western world. In Sharia-based Islamic banking, money is correctly viewed not as actual capital but as "potential capital" (Iqbal & Mirakhor, 1987: 2; van Gruening & Zamir, 2008: 7).

assuring his readers that money will surely not copulate and have offspring (Aristotle, [384 BC] 1995a; similar Aquinas, 1596).

Readers trained in economics might now be willing to object that modern economic theory very well distinguished between "financial" capital and "physical" or "real" capital (e.g., Frank, 2015: 523 ff.). However, both concepts nevertheless then are usually being mixed up in the very same textbook or model (e.g., Frank, 2015). The confusion is inherent in the expression "financial capital" since it suggests that money represents capital, i.e., a production factor. The same is true for the expression "capital market" (Copeland et al., 2008: 35 ff.), which today is understood not as a market for means of production (machines, etc.) but a market where money can be borrowed.

Since today money and capital are conflated, no distinction is made between economic profits drawn from the real economy and interest. Hereof then results a popular but erroneous view according to which the prohibition of interest in Holy Scripture (and the criticism of interest brought forth by Aristotle, Thomas Aquinas or other scholastics) were to be considered as inconsequent. According to this erroneous view, also in former times interest was sometimes permitted. For example, the lender hands over a sum of money to a craftsman and becomes a partner, sharing profits **but also eventual losses**, he or she therefore may lawfully claim a part of the profit. This is what Thomas Aquinas termed a "*societas*" (Aquinas, 1596: II-II, L.77, C.3).

Some believe that such a *societas* is nothing else but a tricky form to circumvent the biblical prohibition of interest (see, e.g., Voigt, 2012). However, it is not. Thomas Aquinas correctly differentiates between the lending of money, which requires the payment of interest **independently** of the success of the business financed by it, and an investment in the real productive economy where both profits and losses are shared. This form of investment in Aquina's *societas* is not a loan, and the shared profit hence is not interest but the shared profit of the common business (*societas*). Members of the *societas* participate in the creation of **real values** as well as in the **risk** and **eventual profits**.

A capital theory in which money is considered a production factor is not only worthless, but even misleading. It has the effect that profits from productive activities and from lending money are not distinguished. As a result, it fails to recognize the fact that interest represents enrichment based on the labor of others as Silvio Gesell (1949) correctly put it.

9.14 Conclusions

In this chapter we have discussed two crucial points. First, the recipient of interest does not work productively for the interest received, but instead it is the borrower who **works** and **creates value** and **takes risk**. Second, money is only a medium to facilitate the exchange of goods and services but has no intrinsic value itself. Money is therefore not a resource and not a production factor, which is why the lending of money, unlike the lending of real goods, does not involve a sacrifice of consumption, but rather only a sacrifice of liquidity.

Bearing this in mind, we can better understand why John Maynard Keynes saw in interest the cause of the many "objectionable features of capitalism" (Keynes, 1936: 221). Aristotle (1995a [384 BC]: 1258b) used to differentiate three forms of income: (i) productive work (the creation of real values), (ii) trade and (iii) earning interest. Earning interest is in his opinion the form of earning that should most be reproved. Because money does not work, those who receive enough interest that they can live off it, essentially live at the expense of others; they exploit society. One might compare them to people who live off social welfare even though they could actually work (Fuders, 2009; Fuders & Max-Neef, 2014a). Hence, any measures intended to increase social justice which do not address the monetary system are by definition inadequate.

In a monetary system in which money circulates without the need for interest—one with true monetary neutrality—the distribution of wealth would automatically be more equitable. All earnings would then correspond to the creation of real value and thus be a contribution to the common good. Profits could then only be earned by performing some kind of productive work. For example:

- products
- services
- lending real goods (with intrinsic value)

Those who earn money in such a system would have contributed in some way to society. Even the profits of a monopolist seem less harmful than income derived from money lending. Although the monopolist's profit exceeds the opportunity costs of the production factors employed (Frank, 2015; Mankiw & Taylor, 2014) and can be considered unjust to the extent of the excess returns, the monopolist, unlike the interest recipient, at least produces a product or service that contributes to the common good. In the absence of interest, economic success would depend, at least more so than today, on the "efficiency, strength, love and wisdom of people and not by holding money or privileges" (Gesell, 1938: VIII, author's translation from German).

With a monetary reform along the lines proposed by Silvio Gesell and which will be elaborated upon in the following chapters, we would approach the goal of a fair economic system that socialist systems aim to achieve via an artificially planned economy, but which cannot be achieved via socialism.

References

Anderson, E. (2000). Warum eigentlich Gleichheit? In A. Krebs (Ed.), *Gleichheit oder Gerechtigkeit* (pp. 117–171). Frankfurt.

Aquinas, T. (1596). *Summae Teologicae*. Apud Iuntas.

Aristotle. (1995a). Politik 1st Book. In *Aristoteles Philosophische Schriften in sechs Bänden. Band 4*. Felix Meiner Verlag.
Aristotle. (1995b). Nikomachische Ethik. In *Aristoteles Philosophische Schriften in sechs Bänden. Band 3*. Felix Meiner Verlag.
Baqir Sadr, S. M. (2011). *Unsere Wirtschaft Iqtisaduna*. M-haditec.
Benedicti XIV. (1854). *Enzyklika "Vix pervenit" ad italiae episcopos de usura 1 Nov 1745*. Stahel.
BGH (Bundesgerichtshof). (2010). *Urteil 242/2010 – Bundesgerichtshof entwickelt Grundsätze zur Berechnung laufender Zinsen in Prämiensparverträgen bei unwirksamer Zinsänderungsklausel fort*. [Online] Available at: https://www.bundesgerichtshof.de/SharedDocs/Pressemitteilungen/DE/2010/2010242.html. Accessed 9 Dec 2022.
Bible-Luther. (2009). *Die Bibel Übersetzung nach Martin Luther – Altes und Neues Testament – Text der Lutherbibel in der revidierten Fassung von 1984*. Dt. Bibelges.
Bible-NIV. (2015). *NIV Zondervan study bible: New international version*. Zondervan.
Bible-NRS. (2011). *New revised standard version bible* (Anglicised ed). HarperCollins.
Biblia Sacra Vulgata. (2007). *Biblia sacra iuxta Vulgatam versionem*. Deutsche Bibelgesellschaft.
Binswanger, H. C. (1985). *Geld und Magie – Deutung und Kritik der modernen Wirtschaft anhand Goethes Faust*. Edition Weitbrecht.
Bogdanoff, A. (1925). *A short course of economic science* (2nd ed.). Communist Party of Great Britain.
Brockhaus-Enzyklopädie. (1994). *Bd. 24* (19th ed.). Bibliographisches Institut & F. A. Brockhaus.
Bubner, R. (1995). Gerechtigkeit herrscht, wo jeder das Seinige tut. In P. Fischer (Ed.), *Freiheit oder Gerechtigkeit – Perspektiven politischer Philosophie* (pp. 176–193). Reclam.
Carver, T. (1883). The place of abstinence in the theory of interest. *The Quarterly Journal of Economics, 8*(1), 40–61.
Cicero, M. T. (2002). De officiis. In C. Horn (Ed.), *Philosophie der Gerechtigkeit. Texte von der Antike bis zur Gegenwart* (pp. 79–87). Suhrkamp.
Clark, J. B. (1888). Capital and its earnings. *Publications of the American Economic Association, 3*(2), 8–69.
Consulting. (2021). *8 efficient, guaranteed ways to make your money work for you*. [Online] Available at: https://www.consulting.com/how-to-make-your-money-work-for-you. Accessed 15 Oct 2021.
Copeland, T., Weston, J., & Shastri, K. (2008). *Finanzirungstheorie und Unternehmenspolitik. Konzepte der Kapitalmarkorientierten Unternehmensfinanzierung* (4a ed.). Pearson.
Creutz, H. (1993). *Das Geldsyndrom. Wege zu einer krisenfreien Marktwirtschaft*. Wirtschaftsverlag Langen Müller/Herbig.
Creutz, H. (2018). *Das Geldsyndrom* (4th ed.). Kubo.
Dworkin, R. (2014). *Was ist Gleichheit?* Suhrkamp.
Frank, R. (2015). *Micoreconomics and behaviour* (9th ed.). McGraw-Hill.
Friedman, M. (1966). *Capitalismo y Libertad*. Ediciones RIALP.
Fuders, F. (2009). Die natürliche Wirtschaftsordnung als Option nach dem Zusammenbruch. *Aufklärung & Kritik, 16*(2), 128–145.
Fuders, F. (2010). Warum der Zins auch moralisch nicht zu rechtfertigen ist. *Humane Wirtschaft, 2*, 26–29.
Fuders, F. (2017). Neues Geld fÜr eine neue Ökonomie – Die Reform des Geldwesens als Voraussetzung fÜr eine Marktwirtschaft, die den Menschen dient. In *Finanzwirtschaft in ethischer Verantwortung – Erfolgskonzepte fÜr Social Banking und Social Finance* (pp. 121–183). Springer-Gabler.
Fuders, F. (2020). *The 'tragedy of the commons' and the role of the money interest rate*. Springer.
Fuders, F., & Max-Neef, M. (2014a). Local money as solution to a capitalistic global financial crisis. In *From capitalistic to humanistic bussiness*. Palgrave Macmillan.
Fuders, F., & Max-Neef, M. (2014b). Dinero, deuda y crisis financieras. Propuestas teórico-prácticas en pos de la sostenibilidad del sistema financiero internacional. In *Economía Internacional. Claves teórica-prácticas sobre la inserción de Latinoamérica en el mundo*. LATIn.

Gabler Wirtschaftslexikon. (2022). *Erpressung.* [Online] Available at: https://wirtschaftslexikon.gabler.de/definition/erpressung-36203. Accessed 21 June 2022.

Gesell, S. (1920) (Reprint 1988). *Die natürliche Wirtschaftsordnung durch Freiland und Freigeld* (4th ed.). Verlag für Spzialökonomie.

Gesell, S. (1938). *Die natürliche Wirtschaftsordnung* (8th ed.). Genossenschaft Freiwirtschaftlicher Schriften.

Gesell, S. (1949). *Die natürliche Wirtschaftsordnung durch Freiland und Freigeld* (9th ed.). Rudolf Zitzmann.

Gosepath, S. (2004). *Gleiche Gerechtigkeit – Grundlagen eines liberalen Egalitarismus.* Suhrkamp.

Hartmann, G. (1992). *Grundlagen der allgemeinen Volkswirtschaft* (26th ed.). Merkur.

Heinsius, T. (1822). *Volksthümliches Wörterbuch der deutschen Sprache – mit Bezeichnung der Ausprache und Betonung fÜr die Geschäfts-und Lesewelt.* Hahn.

Iqbal, Z., & Mirakhor, A. (1987). *Islamic banking.* International Monetary Fund.

Kant, I. (1868). Metaphysik der Sitten – Zweiter Theil. Metaphysische Anfangsgründe der Tugendlehre. In *Immanuel Kant's sämmtliche Werke – In chronologischer Reihenfolge, Bd. VII* (pp. 175–304). Leopold Voss.

Kennedy, M. (1990). *Geld ohne Zinsen und Inflation – Ein Tauschmittel, das jedem dient.* Permakultur-Verlag.

Kennedy, M. (2011). *Occupy Money – Damit wir zukünftig alle die Gewinner sind.* Bielefeld: J. Kamphausen.

Keynes, J. M. (1936). *General theory of employment, interest and money.* Hartcourt.

Kirshner, J., & North, M. (1995). *Von Aktie bis Zoll. Ein historisches Lexikon des Geldes.* C.H. Beck.

Knight, F. H. (1921) (Reprint 1965). *Risk, uncertainty & profit.* Harper & Row.

Kremer, J. (2009). Eine andere unsichtbare Hand des Marktes: Von den blinden Flecken der Volkswirtschaftslehre. *Humane Wirtschaft, 1*, 2–12.

Lachmann, W. (2016). *Ist das Zinsverbot mit der Sozialen Marktwirtschaft vereinbar?* [Online] Available at: http://www.wirtschaftundethik.de/Grundfragen/Ist_das_Zinsverbot_mit_der_Sozialen_Marktwirtschaft_vereinbar.html. Accessed 12 Apr 2016.

Leutzsch, M. (2000). Das biblische Zinsverbot. In R. Kessler & E. Loos (Eds.), *Eigentum: Freiheit und Fluch. Ökonomische und biblische Entwürfe* (pp. 107–144). Kaiser.

Luther, M. (1841). Ecclesiates. In G. von Otto (Ed.), *Martin Luthers reformatorische Schriften – Vollständige Auswahl alles Wichtigen.* Eichler.

Mankiw, G., & Taylor, M. (2014). *Economics.* Cengage Learning.

Marx, K. (1872). *Das Kapital – Kritik der politischen Oekonomie. Erster Band, Buch I: Der Produktionsprozess des Kapitals* (4th ed.). Meissner.

Mill, J. S. (1863). *Utilitarianism.* Parker.

Müller, M. (2011). *Altersvorsorge Eine biblisch-theologische Analyse und die Bedeutung für heute.* Werkstatt für Gemeindeaufbau.

Olivetti, E. (2003). *Latin dictionary.* [Online] Available at: https://www.online-latin-dictionary.com/latin-english-dictionary.php?parola=alieno. Accessed 6 Apr 2022.

Petersen, C. (2005, Mai/Juni). Wenn Dein Kind Dich morgen fragt.... *Humanwirtschaft*, pp. 1–4.

Piketty, T. (2014). *Capital in the 21st century.* Harvard University Press.

Plato. (2011). Politeia. In U. Wolf (Ed.), *Platon – Sämtliche Werke Band 2* (pp. 211–537). Rowohlt.

Prien, H.-J. (1992). *Luthers Wirtschaftsethik.* Vandenhoeck und Ruprecht.

Rawls, J. (2020). *A theory of justice.* Harvard University Press.

Raz, J. (2000). Strenger und rhetorischer Egalitarismus. In A. Krebs (Ed.), *Gleichheit oder Gerechtigkeit* (pp. 50–80). Frankfurt.

Reichmuth, K., & Reichmuth, R. (2001). *Der RealUnit – Zur Quelle der Geldwertstabilität.* Ott.

Reuters. (2014). *Ifo-Chef – Niedrigzins kostet Deutsche 300 Milliarden Euro.* [Online] Available at: https://www.reuters.com/article/deutschand-ezb-ifo-idDEKCN0JJ0JD20141205. Accessed 7 Dec 2022.

Schieritz, M., & Uchatius, W. (2014, November 13). Mark Schieritz/Wolfgang Uchatius, Der stärkste Trieb der deutschen – Ausgespart! *Die Zeit,* pp. 13–15.

Sell, F., Sauer, B., & Engelkamp, P. (2020). *Einführung in die Volkswirtschaftslehre*. Springer Gabler.

Seminar für freiheitliche Ordnung. (1984). Zins in der Geschichte. *Fragen der Freiheit, 167*, 30–36.

Starbatty, J. (1977). Eine kritische Würdigung der Geldordnung in Silvio Gesells utopischem Barataria (Billigland). *Fragen der Freiheit, 129*, 5–31.

Steiner, R. (1979). *Die soziale Grundforderung unserer Zeit – In geänderter Zeitlage, Zwölf Vorträge, gehalten in Dornach und Bern v. 29. 11. bis 21.12.1918*. Rudolf-Steiner-Verlag.

Süchting, J. (1995). *Finanzmanagement, Theorie und Politik der Unternehmensfinanzierung* (6th ed.). Gabler.

Suhr, D., & Godschalk, H. (1986). *Optimale Liquidität – Eine liquiditätstheoretische Analyse und ein kreditwirtschaftliches Wettbewerbskonzept*. Knapp.

van Gruening, H., & Zamir, I. (2008). *Risk analysis for Islamic banks*. The World Bank.

Vieweger, H.-J., & Mockler, M. (2002). *Hans-Joachim Vieweger und Marcus Mockler, Kann denn Börse Sünde sein. Geld mit gutem Gewissen anlegen – geht das?* Brunnen Verlag.

Voigt, S. (2012). *Telepolis – Geld und Zinsen.* [Online] Available at: https://www.heise.de/tp/features/Geld-und-Zinsen-3396545.html. Accessed 29 Dec 2021.

von Goethe, J. W. (1976). *Faust – Kommentiert von Erich Trunz*. C.H. Beck.

Walzer, M. (2000). Komplexe Gleichheit. In A. Krebs (Ed.), *Gleichheit oder Gerechtigkeit* (pp. 172–214). Frankfurt.

Wikipedia. (2022). *Entry "Zinsverbot"*. [Online] Available at: https://de.wikipedia.org/wiki/Zinsverbot. Accessed 2 Jan 2022.

Part IV
Natural Money as Solution

Chapter 10
Gesell's Solution

Thaler, thaler, you must wander
from one hand to the other....

(German nursery rhyme)

10.1 Money That Circulates Naturally

10.1.1 Interest as 'Necessary Evil'?

We had already discussed (Chap. 2) the fact that money is the counterpart of debt and that there is no way to avoid this. It was further explained that this fact is not inherently problematic, as some would believe (e.g., "positive money" movement). Rather, we concluded that many of the problems the world faces today are directly connected to the fact that both deposits and debt grow because of interest. Thus, both the level of debt and the money supply grow steadily. The question arises though, if interest is not a sort of 'necessary evil' for money to circulate, to be made available for loans and thus investment? I have been asked this questions a couple of times. The answer is that, yes, as long as money is designed as it is today, we need interest to make money circulate. Interest is the drug that makes money, the blood of our economic organism, to circulate. However, money is a human construct and thus could be changed in such a way as to make it flow by itself. Then, interest would no longer be necessary in order to make money circulate.

In Sect. 3.5 it was explained that interest derives from the **unnatural property** of money that, in contrast to real goods, can be easily and almost indefinitely stored.[1] This is the root of our psychological penchant to hoard surplus money as much as

[1] Yes, inflation causes the purchasing power of money to erode over time. However, as was outlined above, the causality goes the other way around. Inflation is a secondary symptom of the interest induced growth of the money supply.

© The Author(s), under exclusive license to Springer Nature Switzerland AG 2023

F. Fuders, *How to Fulfil the UN Sustainability Goals*,
https://doi.org/10.1007/978-3-031-37768-6_10

possible, which Keynes (Keynes, 1936: 168) described as the *liquidity-preference*. However, money is supposed to circulate in the economy in order to perform its function as a medium to facilitate the exchange of goods and services. Our natural, instinctual impulse to hoard is perverted by the fact that our money has this "monopolistic-like" position (Gesell, 1949: 237) compared to real goods due to the fact that it does not perish and can always and everywhere be exchanged for goods and services. While everything in nature is subject to the cycle of growth and decay, only money is exempt from the transience of all earthly things (Gesell, 1949). The unnatural, special position of money incentivizes us to hoard money for bad times or to charge interest for returning it to circulation. Accordingly, interest is a reward for not hoarding money at home; it is a "*reward for parting with liquidity*" as Keynes (1936: 167) accurately defined it.[2] Hence, the existence of interest derives from money's unnatural storability combined with our instinctive urge to save.

But if households can only be prevented from hoarding money 'under the mattress' by offering interest as an incentive to lend it (to a person or a bank), this leads to all the negative consequences previously described, namely the exponential increase of deposits and debt, inflation, asset price bubbles, and the growth imperative of the (real) economy. People must therefore be prevented from hoarding, but this must be accomplished without the existence of interest. This is what Silvio Gesell's solution—discussed in the next section—is designed to achieve.

10.1.2 Gesell's Solution: Natural Money

How can we make money flow without interest being necessary as a reward for parting with liquidity? This is the question that occurred to the German-Argentine merchant Silvio Gesell 100 years ago. After emigrating from Germany to Argentina he experienced firsthand two consecutive financial crises and arrived at a diagnosis attributing the crises to the fact that money does not circulate as it should. He published a series of papers and later combined them into his masterpiece "*Die natürliche Wirtschaftsordnung*" (The Natural Economic Order), the last and nineth edition published in 1949.[3]

As a solution, he proposed what he called "*Freigeld*" (free money). It is "free money" because it is free of interest and because it sets people free (Gesell, 1949). Personally, I prefer to call it "natural money", since it does not, as our current form of money does, deviate from nature. Conventional money deviates from nature due to its unnatural (artificial) storability. Natural money, on the contrary, perishes just

[2] That interest is a reward for parting with liquidity was emphasized by Keynes, but Keynes did not invent this idea. For example, John Locke as far back as 1883 had noted that without the interest, people prefer to hoard money at home (Anderegg, 2007).

[3] Even though written form his domicile of choice, Buenos Aires, the work has never been published in Spanish. An English translation of the sixth edition (by Phillip Pye) sold out quickly and has never been republished.

as real goods do. To achieve this, tax stamps must be placed on the money bills at set intervals e.g., every three months in order for the bills to be officially accepted. The tax stamps are not supposed to be very expensive; just 1% or 2% of the value of each bill. This hoarding fee—sometimes also called "*demurrage fee*" (Lietaer et al., 2012: 6) or "*carrying costs*" (Keynes, 1936: 357)—, although it is not high, provides a powerful incentive to either spend money, invest it productively or lend it out. But, unlike our current money, this money will not be lent out for interest. Instead, the money owner can be grateful to the borrower, since the borrower promises to pay back the bills with the corresponding stamps affixed. Since the money perishes in the same manner as real goods do, it cannot be hoarded indefinitely, and thus circulates without the necessity for interest as a reward for parting with liquidity. Such money would lose its special position compared to real goods, and money holders could no longer "blackmail" those in need of money and "press" interest (Gesell, 1949: 205, 344), to "exploit the scarcity-value of capital" as John Maynard Keynes put it (Keynes, 1936: 376).

10.1.3 Blood Must Flow Without the Need for Drugs

The foregoing analysis suggests the conclusion that if we want to formulate a new model of a market economy that is not perverted by the need to grow and by constantly increasing income inequality, we need a form of natural money that is perishable just like real goods are. To come back to the metaphor of money as the blood of the economy, **banks are then the heart of the economy** that make money flow. Unfortunately, since money, as it is designed today, has this design flaw, it is 'sticky' and does not flow smoothly. Banks (the heart of the economy) cannot effectively perform their function of making money circulate through the economy and therefore interest is needed as the drug that causes money to circulate. And as with any drug that is applied for a prolonged period, harmful side effects result. But Silvio Gesell gifted us the brilliant idea of natural money that doesn't need a drug to make it circulate, money that flows by itself. Such money only serves to facilitate the exchange of goods and services and not to accumulate "virtual wealth" as Frederick Soddy (1933a, b; 1934: 24 ff.) called momentary units. Such money would serve the economy and not *vice versa*.

10.1.4 Keynes and Fisher Praised Gesell's Proposal

It is noteworthy that the most famous economists of the last century, John Maynard Keynes and Irving Fisher, considered Gesell's proposal a possible solution for the Great Depression of the 1930s (Keynes, 1936; Fisher, 1932; Fisher & Cohrssen, 1933). Keynes was even convinced that "the future will learn more from the spirit of Gesell than from that of Marx" (Keynes, 1936: 355). Unfortunately, thus far this prediction has not proven to be true. Everyone knows Marx and only very few have ever heard of Silvio Gesell. At the Bretton Woods conference where the World Bank and the International Monetary Fund were founded and the US dollar was established as currency for international trade, Keynes had proposed a plan to establish an international currency based on the idea of Gesell's natural money. He called this demurrage-fee equipped currency "*Bancor*" (Keynes, [1941] 1980: 42 ff.; on this see Betz, 2010). Unfortunately, Keynes failed in convincing the international community of the advantages of such a currency.

Irving Fisher considered himself a "modest apostle" of Silvio Gesell (Fisher, 1948: 6) and even wrote a book of his own on the concept of free money called "Stamp Scrip" (Fisher & Cohrssen, 1933). However, even though both famous economists praised Gesell's proposal as a potential solution to the Great Depression both considered them either incomplete or not sufficiently sound. I will discuss their objections in Sect. 10.3 and will argue that it is actually Keynes and Fisher who did not fully understand Gesell's analysis.

10.1.5 How to Practically Implement Natural Money Today

Many who hear Gesell's proposal for the first time, object that stamped money sounds cumbersome and impractical, especially in today's world where most transactions are electronic. Of course, modern technology makes possible more convenient ways of assessing a hoarding fee to money. Indeed, the fact that today most payments are electronic makes the implementation of Gesell's ideas much easier. For example, as an alternative to the proposal to use stamps and paper banknotes, today we could use electronic banknotes that devalue over time (Berger, 2009) or electronic pre-paid cash cards. In fact, today there are many local and regional so-called complementary currencies that are based on Gesell's idea. Many of them use bills with expiration date (see Fig. 10.1). After expiration the bill is worthless but can be exchanged at the local currency bank for a new one paying the hoarding fee (Fuders, 2009a, 2010, 2016a: Fuders & Max-Neef; 2014a, b).

Since Gesell's tax stamps can be viewed as a negative interest rate (Lietaer et al., 2012), another way to practically employ Gesell's proposal would be through a

Fig. 10.1 Regional money bill "Lutzetaler" that circulated in Köthen, Germany during the time of the "Macht-Geld-Sinn" Congress 2012, front and backside. Backside with expiration date. (Source: own photography)

negative central Banks interest rate policy, however, not of the variety currently (until March 2022) implemented by the ECB but one in which negative interest rates would not only apply to deposits of commercial banks at the Central Bank, but **also to cash** (Fuders, 2010, 2015, 2016b; Fuders & Max-Neef, 2014b; Fuders & Bockting, 2019). Different ideas have been proposed to achieve this (Buiter, 2005; Seltmann, 2010; Agarwal & Kimball, 2015; Assenmacher & Krogstrup, 2018). In fact, any negative interest rate policy that does not apply to cash will never effectively break through the so-called "zero-lower bound", since the negative rate can then be evaded by holding cash, i.e., by hoarding cash under the mattress (Fuders, 2016b; Fuders & Löhr, 2016; Fuders & Bockting, 2019).

It should be noted, however, that next to cash, only demand deposits but not time deposits should be subject to this hoarding fee (the negative interest rate) if we want to ensure that funds are lent out. After all, people are supposed to spend their money, invest it or lend it out interest-free. In Gesell's natural money system, banks will lend customers' money only if customers expressly request that they do so in order to avoid the hoarding fee that would otherwise be incurred. Customers would then agree not to withdraw the money for a fixed period (months or years) in exchange for not being charged negative interest.

Ideas for how to implement natural money in today's world will be discussed in Chap. 12. Chapter 13 will be dedicated to discussing many proposed alternatives to Gesell's idea, e.g., "100% money", "positive money", "full sovereign money", "MMT", "indexed money", "gold standard", "time banks". We will conclude that there is no true alternative to natural money.

10.2 The Natural Money Banking System

10.2.1 The Role of Banks: And Why Banks Would Profit

It is important to highlight the fact that banks would probably profit from a natural money system since the probability of financial crises would be greatly reduced. In fact, major crises might not exist at all. We have already discussed the fact that our current system is comparable to a *Ponzi* scheme and produces **major crises at regular intervals**. Chapter 16 will analyze this point in greater depth. If the monetary system is less prone to crises, this is good for all stakeholders, especially banks.

In addition, negative interest rates that also apply to cash, as proposed by Gesell, would reduce the risk of bank failures for another reason. Let me explain. With the ECB's current (March 2022) policy (negative interest rates on holdings of commercial banks at the Central Bank, but no negative interest on cash), it is mainly small banks that pass on negative interest to their customers, while large banks can absorb the costs without passing them on to customers. Therefore, customers of smaller banks are likely to move their funds to larger institutions that do not charge negative interest on deposits. This increases the probability of bank failures. If, on the other hand, a sufficiently high negative interest rate was also applied to cash, banks would automatically have to charge negative interest on demand deposits since they would be liable for the costs of the tax stamps. This would have the advantage that negative interest would **apply uniformly throughout the banking sector**, thus avoiding market distortions.

And banks would **not even necessarily earn less**. Instead of an interest margin they would earn commissions for brokering credit. As in the current system, a Gesellian natural money system would need intermediation between those needing to borrow money and those with excess money to lend. In contrast to the current system, banks would not automatically loan out customer money but only do so for those who expressly request that they do so. The incentive that would motivate depositors to lend would no longer be the receipt of positive interest but rather the avoidance of negative interest (the hoarding fee). The only way to avoid paying the hording fee would be to lend money.

Banks would naturally charge commissions for intermediating between those seeking loans and those looking to avoid negative interest by lending their excess funds. The profits earned via brokerage fees would not necessarily be less than those earned by banks under the current system. Banks would still perform the crucial function of intermediating between savers and borrowers, thus making money, the blood of the economy, circulate. And they still profit from this activity, while facing a greatly reduced risk of losses due to financial crises.

10.2.2 Brokerage Commissions Are Not Interest

Some might ask if brokerage commissions are not, in fact, the same as interest? Is it not just a different name for the same thing? However, even though brokerage fees might at first glance look like camouflaged interest they are in fact completely different. A brokerage fee is remuneration for a service. As is the case for any other good or service, when money flows from one person to another the money supply is not altered. On the other hand, interest causes deposits and thus the money supply and total debt to grow, as explained in Sect. 4.3.

And there is another difference between an interest margin and a brokerage commission. The interest margin is paid by the borrower. On the contrary, a brokerage fee could be paid by either the borrower or the lender. Both borrowers and lenders benefit from the conclusion of a loan contract. One needs money and the other wishes to avoid hoarding fees. The contracting parties should therefore naturally share the costs of performing such a transaction.

10.2.3 Insurance Against Loan Defaults

One might ask how firms could be provided with risk "capital" (as discussed above a more accurate term would be "loans for risky investments") in an economic system without (positive) interest on loans. Holders of money would probably not choose to provide interest-free loans for risky investments just to escape the hoarding fee (negative rate). Risk or venture capital refers to those loans which cannot (or not completely) be secured via mortgages or other guarantees. Normally lenders to risky enterprises would require a higher rate of return, a rate in excess of a simple "reward for parting with liquidity" (Keynes, 1936: 167). So, while lenders could probably be found for relatively low risk loans, it might not be so easy to find lenders for riskier ventures.

But insurance companies could offer insurance against loan defaults. Loans could then be interest-free. Banks would earn commissions for intermediating between borrowers and lenders. Insurance companies would assume the risk with the premiums paid by borrowers. From the borrower's perspective, this is the same as paying interest. But unlike in the case of loans, the money supply would not change as a result. The insurance premiums would flow from borrowers to insurers with no money supply growth resulting. No pressure to grow or potential inflation would be generated.

10.2.4 Private Equity Investments Remain Possible

In a Gesellian natural-money system investing in the real economy would of course still be possible. Companies could therefore still obtain investment "capital". Just as loan agreements have specific timeframes, it would be possible to make **equity investments** for specified periods of time in order to give entrepreneurs planning security. The amount repaid would not necessarily be equal to the amount invested but would depend on the companies' business results. Thus, as with any equity investment, investors would naturally share entrepreneurial risk. Any positive "return" on the sum invested would correspond to the real economic profits of companies. Such investments would have no effect on the money supply. The "interest" of investors in such a scenario would be the prospect of profit. This form of corporate investment is what St. Thomas Aquinas (1596: II-II, L.77, C.3, reply to objection 5) referred to as "*societas*".

In order to give small investors the opportunity to participate in the equity capital of companies and also to enable small companies to find investors, professional associations (e.g., chambers of industry and commerce, etc.) could create funds which aggregate small deposits and invest in the equity of affiliated companies.

10.2.5 Risk-Free Profits from Investments Will No Longer Exist

Equity investments naturally have the disadvantage for the investor that they are not completely risk-free. This means that in an interest-free monetary system, savers would not be able to earn a risk-free return on their money simply by depositing money in a bank. The beauty of a savings deposit, namely its risk-free nature, is precisely the problem.[4] Interest makes savings deposits (and debt) grow independently of the development of the real productive economy, which is why they decouple from the real economy. Money must therefore not earn interest independently of the development of the real economy.

A gradual deterioration of savings, the "aging" of money, as Rudolf Steiner (1979: 179)[5] called it, is natural. It corresponds to the aging of real goods. In a natural money system, savers could, however, avoid erosion of their savings by lending them, as explained above. Banks would probably charge a commission to facilitate such lending. But this fee would not necessarily be borne solely by the saver (the money lender) but could be charged to the borrower, or it could be shared by both the saver and the borrower. From the saver's point of view, paying such a fee would be worthwhile, as long as it was less than the hoarding fees which would be avoided by lending. To protect the money of the poor from devaluation, we might consider applying the hoarding fee only to sums above a certain amount.

10.2.6 The Money Supply Automatically Adjusts to GDP Growth

If (positive) interest is no longer necessary to make money circulate in the economy, then the money supply will probably adjust automatically—as if controlled by an invisible hand—to the growth of the economy. This is because a growing economy requires additional investment which will lead to an increase in loan volume and thus an increase in the money supply via the money multiplier (see Sect. 5.5). In an economic downturn, on the other hand, the loan volume would fall, and the money supply would decrease. The latter is inconceivable in our current monetary system, since as outlined above (Sect. 5.6.4) interest forces a constant expansion of credit volume.

In our current system, should there be a decline in the credit volume, banks are likely to get in trouble and even go bankrupt. Money will then not be saved banks anymore (or will be saved in smaller amounts) and thus there will be an insufficiency of money circulating in the economy. This leads to deflation, such as we

[4] In Johann Wolfgang von Goethe's ([1832] 1976) drama "Faust II" the monetary system was invented by Mephistopheles (Satan).

[5] On Steiner's view of natural money see Suhr (1988: 34 ff.; 58 ff.; 85 ff.).

experienced in the Great Depression (and will likely again experience in the near future). Such a situation is what Keynesian economics refer to as the "liquidity trap" (based on Keynes, 1936: 169 ff.). In contrast, in an interest-free monetary system, the money supply would adjust to the rhythm of economic growth via the money multiplier. It is worth recalling that business cycles are amplified by interest (Fuders, 2009a, 2010; Fuders et al., 2013). In a monetary system without interest, economic crises would probably not occur, and economic cycles would be far less pronounced than they are today.

10.2.7 The Role of the Central Bank

In a Gesellian money system, inflation and deflation would no longer exist. Changes in the money supply would automatically correspond exactly to the growth of the economy's economic output. When loans are repaid, the money supply would decrease to the same extent as it expanded upon the issuance of those loans. The problem inherent in our current monetary system does not lay in credit expansion per se, (as followers of the Austrian school of economics believe), but in the fact that interest requires constant expansion of debt. In an interest free monetary system, the **money supply will not decouple from the performance of the real economy** since the money supply could only increase through an increase of the 'natural demand' for loans as explained above (Sect. 5.5), that means, through new investments, all the while new investments will bring along an additional increase in production of goods and services.[6]

If the economy entered a boom phase to such an extent that the demand for investment and thus loans could not be met, then the central bank could increase the hoarding fee in a similar manner to how today the central bank adjusts interest rates. The incentive to lend money would thus increase leading to more rapid circulation of money. The danger of deflation caused by hoarding of money would thus be mitigated. An increase in the velocity of money has the same effect as an expansion of the money supply.

If the circulation of money was still insufficient to avoid deflation the central bank could then print (natural) money and lend it out at zero interest to commercial banks which would then lend to its customers free of interest (but still charging a commission). This would increase the money supply, thus allowing the economy to grow. The task of the central bank in a natural money system would be, as it is today, to adjust the money supply to GDP growth. Contrary to the situation today, this would then really work.

---

[6] Of course, the money supply could increase if the central bank prints money.

10.2.8 No Bank Runs

The worst-case scenario in a fractional reserve banking system occurs when bank customers wish to withdraw more money than banks have available. A so-called "bank run" then occurs in which people lose confidence in the creditworthiness of banks and therefore withdraw their money. In such circumstances banks are forced to temporarily close, declaring a "bank holiday", and if the panic has not subsided by the time they reopen they are forced to declare themselves insolvent. The insolvency of one bank has a negative impact on confidence in other banks, and the resulting panic can cause a chain reaction of bank failures. This problem is related to the fractional reserve system and would not occur if banks were required to keep 100% of their demand deposits with the central bank, an idea Irving Fisher (1936a, b) proposed almost 90 years ago.

Such a 100% reserve system would prevent bank runs and thus stabilize the financial system to a certain extent. However, not all financial crises would be avoided. Interest would still exist. Time deposits would still exist. Banks would still be forced to constantly increase loan volume to be able to pay interest on time deposits.

In a Gesellian natural money system, demand deposits would not be used by banks unless depositors requested that they do so. Hence, a natural money system would automatically be a 100% reserve banking system (but not *vice versa*), and bank runs would no longer occur.

10.2.9 No Speculation Against Free Money

In the current world financial system, any currency can become an object of speculation (Salazar Gomez et al., 2021). Switzerland has been struggling with this problem for about a decade. The Swiss franc has become so expensive versus the euro and the US dollar that the Swiss central bank had to periodically intervene in the markets by buying euros and USD. If the national currency appreciates too much against foreign currencies than this hurts the export-oriented industry.

Currency speculation would be much less likely to occur with a currency that is subject to a hoarding fee. In addition, such a currency would not serve as a reserve currency for foreign central banks. The only reason why someone would acquire such a currency on the foreign exchange market would be in order to purchase goods & services from the issuing country. The adoption of natural money could represent a great relief for export-oriented industries in countries like Switzerland.

10.3 Discussion

10.3.1 Critique of Keynes and Fisher

Even though Keynes viewed Gesell's natural money as a tool to fight economic depressions, he was not convinced that it represented a general solution to the problem of unemployment. He believed Gesell's analysis of the phenomenon of interest to be imprecise. According to Keynes, Gesell had not fully understood the concept of *liquidity-preference* (Keynes, 1936). However, this is not true. Rather, as discussed above, liquidity preference arises due to the special position of money in comparison to real goods: (i) that money can easily be exchanged for goods and (ii) that money is easy to hoard. I would even dare to go so far as to say liquidity preference is, in fact, the starting point of Gesell's analysis, even though Gesell did not use the same words as Keynes. It seems that it was Keynes who did not fully understand the underlying reasons for the existence of people's preference for liquidity. It does not result from a psychological penchant to save as Keynes (1936) believed but the other way around. It is the special, unnatural and monopoly-like position of money vis-a-vis real goods that causes people to prefer to hold money. To eliminate this special advantage that money has over real goods and to make the hoarding of money as unattractive as the hoarding of goods was Gesell's key innovation.

Irving Fisher adopted Gesell's ideas on money since he saw money hoarding as a cause of deflation. However, like Keynes he saw in stamp-scrip money merely a way to fight deflation during deep economic depressions, but not as a solution to inflation, financial crises, unemployment and income inequality. The fact that interest is a reward for parting with liquidity and is caused by money's unnatural hoardability is not explicitly discussed in his text "Stamp Scrip" (Fisher & Cohrssen, 1933). He even explicitly rejected this explanation in a later publication (Fisher, 1937). Rather, in his opinion the main cause of deflation is debt (Fisher, 1932, 1933). He did not recognize that interest is a major cause of ever-growing debt.

10.3.2 Carrot or Stick: Does It Make a Difference?

Gesell's hoarding fee—which serves to make money as perishable as real goods, and thereby causing it to circulate more consistently—could, as previously discussed, be thought of as equivalent to a negative interest rate. Whereas in the current system positive interest incentivizes us to lend money, in the Gesellian natural money system negative interest on demand deposits and cash would incentivize people not to hoard it (and to lend it free of interest).[7] But does it make a difference whether we create a positive incentive to lend or a punishment for not lending money?

In fact, it makes an enormous difference whether we apply a positive or negative incentive, not only from an economic point of view, but also **from a moral point of view**. In order to prevent people from harming others, should they be punished for doing so, or should society reward people for refraining from doing harm? This is an important question in ecological and environmental economics. For example, should we subsidize a company for applying cleaner production methods or prohibit harming the environment, and in case someone infringes the law they would be fined? According to the famous Coase Theorem (after Nobel Prize winner Ronald Coase) the outcome should be the same (on this see also Daly & Farley, 2011).

But even if the outcome regarding pollution is the same, can it be considered morally equivalent? Is rewarding for not polluting not comparable to the mafia charging a "protection fee" to restaurant owners who they would otherwise harm? Such an attitude could be deemed not only unjust but also **unnatural**. Naturally we should not harm others and expect that others will not harm us. This is the basic law in Immanuel Kant's Metaphysics of Morals (Kant, 1868) termed the "categorical imperative".[8] Naturally, most people would therefore intuitively want laws to make us to behave according to this principle. In Philosophy of Law, this is the basic argument for the existence of laws: to regulate the conditions in which we live together in order to promote peace and harmony and thus increase freedom. However, laws only result in freedom, peace and justice if they apply Kant's categorial imperative (see e.g., Schachtschneider, 1994, 2006, 2007). On the contrary, despotic laws do not necessarily result in freedom, peace or justice.

What I try to point out in this section is that positive incentives are not always preferable to negative ones, as many psychologists believe. The same reasoning can be applied to positive interest. Money can be thought of as a **public good**. It is designed to facilitate the exchange of goods and services. To fulfil this function, it must circulate. Hoarding huge amounts of money under the mattress chokes the economy and thus harms society. People should therefore be forced to allow money

---

[7] Indeed, we need either a positive or a negative incentive to not hoard money at home but to lend it out (or to bring it to the bank). A zero-rate policy cannot work. If we neither change the design of our monetary system, i.e., apply a negative rate on hoarding (Gesell), nor apply a positive rate for lending (current system), money will not adequately circulate. This is, in fact, what was happening until March 2022 in many parts of the industrialized world, especially in Japan, Europe and the US where interest rates for bank deposits were zero or close to zero. With a zero-interest rate and without reforming the design of our money such that it is not storable physically at home, money does not flow adequately. This is the reason why central banks have been forced to print money—nowadays referred to as *Quantitative Easing*—to prevent their economies from plunging into deflation. Instead of printing ever more money of the type that does not adequately flow, which then again ends up failing to circulate, central banks could introduce "natural" money that circulates naturally, i.e., they could attach a negative rate to money, including cash.

[8] In Christianity we call this the love-thy-neighbor principle, which according to Christ is the most important commandment of all (Bible, Mathew 22:40).

to circulate via a fine (negative interest rate or hoarding fee) rather than through a reward for not hoarding, for not doing harm.

10.3.3 *Would People Circumvent Natural Money Using Gold or Bitcoin?*

Regarding Gesell's proposal, Keynes posed the question of whether people would try to evade the Gesellian hoarding fee by holding gold or other valuables, which are more suitable for hoarding than depreciating currency. However, if we invest in gold or Bitcoin the circulation of money is not affected, and the money supply will neither increase nor decrease as a result. For example, if Fritz sells on ounce of gold to Hans for 100 Gesellian natural money units, these money units move from Hans to Fritz. The natural money that Hans no longer possesses are now held by Fritz. The **total money supply has not changed**. People can buy as much gold, Bitcoin or other assets as they wish. The proper functioning of the natural money system would not be affected. Gold or Bitcoin only pose a problem for the monetary system if people start using them not as assets but as money, as means to facilitate the exchange of goods and services.

But there is no reason to believe that gold would establish itself as currency. People would still be able to accumulate reserves in gold, just as they can accumulate any other goods. After all, people buy gold coins or Bitcoin as stores of wealth but still pay their bills with the national currency. In Schwanenkirchen and Wörgl, two cities where Gesellian natural money systems were successfully implemented in the 1930s (see Sect. 14.2), people did not attempt to circumvent the Gesell money using gold.

10.3.4 *Would Natural Money Promote Growth or Increase Inflation?*

A common misconception regarding Gesell's proposal is that negative interest rates would stimulate the economy and cause inflation. And since economic growth has a negative impact on the natural environment some are of the opinion that Gesell's proposal should be rejected (see, e.g., Riedel, 2014). It is indeed gratifying to see that more and more economists are coming to understand that there can be no such thing as infinite economic growth. However, as discussed in Sect. 5.7, it is positive interest, not negative interest, that forces the economy to grow. Only immediately following a reduction of interest rates will the money multiplier expand the money supply and foster growth. As we have seen, empirically interest rates and economic growth are positively correlated in the medium and long run. Notwithstanding significant short-term fluctuations, in the long run, when interest rates are high, economic growth rates usually are too, and *vice versa*.[9] In Sect. 5.1 we discussed why this is the case. The only way to achieve an economy with stable money and no growth imperative without increasing unemployment is to abolish positive interest and to keep money circulating by means of a negative interest rate (hoarding fee).

10.3.5 The Hoarding Fee Will Not Be Higher Than Today's Inflation

Maintaining the purchasing power of money over time is not possible in any monetary system. Money is a medium of exchange. It cannot be a store of value at the same time. As long as money is easily and indefinitely storable, interest will exist as a reward for parting with liquidity; and as long there is interest, currency cannot be stable, i.e., free of inflation and deflation. The most stable money in terms of exchange value is probably one with a hoarding fee (demurrage fee, carrying costs, Gesell tax, negative interest rate), although the hoarding fee itself also causes a steady loss of value. However, the hoarding fee is unlikely to be higher than today's inflation rate. So, we have to make a choice. Either we adopt a monetary system with a hoarding fee, or there will be inflation and financial (debt) crises. A monetary system with a hoarding fee is definitely preferable.

10.3.6 Are Negative Rates Unnatural?

As outlined above many believe interest to be natural, consequently they consider a negative rate (and a hoarding fee can be interpreted as a negative rate) to be unnatural. One possible reason for this is probably again the basic misunderstanding in economics today, which can be seen, among other things, in the fact that the medium of exchange money is regularly confused with the production factor capital. Ultimately, the misunderstandings about the effect of interest on the money supply mentioned above also stem from this basic misunderstanding.

Exemplary is the statements of a German economist. He found that negative interest rates would naturally not be possible and would even run counter to natural human behavior. The "natural" interest rate, he explained, were a compensation for a difference of the value of goods over time. This "natural interest" could not become negative; for that would mean, in his view, that an apple that is available in 100 years would be preferred to an apple that is available today (Investment, 2014).

[9] This is, by the way, the opposite of what economic theory teaches. According to orthodox economic theory, economic growth and the interest rate are supposed to be inversely correlated: According to this model, the lower the interest rate, the higher is economic growth, see Stiglitz (1998); Larroulet and Mochón (2003); Dornbusch et al. (2009).

Similar statements can be found frequently. The idea of a "natural" interest rate (sometimes also "real equilibrium rate") probably goes back to Knut Wicksell and has since then found its way into the textbooks of monetary theory (Wicksell, 1936; Woodford, 2003).

However, to stay with the example, money does not represent an apple, nor any other real good with intrinsic and subjective value, and certainly not capital. As mentioned, a couple of times throughout this book, in classical economics capital refers to the third factor of production besides human labor and land (Frank, 2015): it refers to machines or other means of production. Money is a medium of exchange that can be used to exchange capital or other goods (such as apples). Money itself, however, unlike machines or apples, has no independent value, does not work and does not get any offspring, as already Aristotle ([384 BC] 1995) correctly pointed out, who therefore correctly considered the interest from money lending as unnatural. Not negative, but **positive interest rates are unnatural**.

10.3.7 Natural Money Is Neutral Money

Silvio Gesell's idea is a powerful tool for avoiding excessive money hoarding and in doing so to avoid the existence of the interest. Again, interest, which ancient philosopher Aristotle considered unnatural (Aristotle, [384 BC] 1995), is a secondary phenomenon of the unnatural design of our current financial system that makes money infinitely storable. In contrast, Silvio Gesell's money is **natural money**, since it is as perishable as real goods. Creditors would lend such money freely since it enables them to pass on the cost of the tax stamps to borrowers and they are therefore better off than if they simply held their cash (Gesell, 1949). In such a system, the money supply and the real productive output of the economy would **no longer diverge**. Such interest free currency would serve solely as a means to facilitate the exchange of goods and services and not as a means to store wealth. It would also be a powerful tool to combat the 'economy of accumulation'.

In the current system, the possibility of hoarding money not only gives rise to interest but also creates an unnatural **imbalance in the bargaining positions** between owners of goods and owners of money. While owners of goods must sell their goods as soon as possible due to their perishability, owners of money can wait, postponing their purchases (Gesell, 1949). With Gesell's natural money this imbalance in bargaining positions would disappear. Consequently, it would embody the concept that conventional economic theory refers to as "monetary neutrality"[10] (Stiglitz, 1998: 187; Mankiw & Taylor, 2014: 587), a condition that does not hold true under today's financial system (Suhr, 1989; Humphrey, 1991, Sener, 2016).

[10] Monetary neutrality is a concept that states that changes in the money supply only affect nominal variables and not real variables (see. e.g., Stiglitz, 1998).

10.3.8 Gesell Unites Neoclassical and Keynesian Interest Theories

Gesell resolves the contradiction between the neoclassical and Keynesian theories of interest. According to the neoclassical view, interest is the price of lending money that results of money supply and demand for money (demand for credit). According to Keynes, it is a liquidity premium. We can think of it this way: Gesell's hoarding fee neutralizes the liquidity advantage (compared with real goods) of money. Money becomes as perishable as physical goods. People would therefore be more inclined to spend it. In other words, the velocity of circulation of money grows, which has the same effect as if money supply would grow (see Irving Fisher's quantity equation (2)). An increased money supply means, *ceteris paribus*, that the price of money (more precisely, the price of lending money) falls.

10.4 We Can Finally Approach "Perfect Competition"

The non-neutrality of money is probably the **main reason for market failure**, a fact which is unfortunately not widely understood. If we reform our money along the lines proposed by Silvio Gesell and also fight classic market imperfections, we could thereby establish a **sustainable** and **fair** economic order, a market economy that serves people and not vice versa. This would allow us to approach the ideal of so-called "perfect competition".

10.4.1 Small Is Beautiful

"Perfect competition" describes a theoretically ideal situation in which an infinite number of small and medium-sized producers do business with an infinite number of consumers. Producers compete with each other trying to **best satisfy the needs of consumers**, and no producer has the power to unilaterally increase prices without offering an improved product in the eyes of the consumer. This leads to a situation (when also guaranteeing monetary neutrality) where **everyone gains according to their efforts** when satisfying consumer needs. Although no individual producers or consumers consciously try to promote social welfare, the competitive equilibrium automatically produces—through an "invisible hand" to cite Adam Smith's ([1776] 1952: 194) famous expression—the greatest possible social welfare for two reasons:

First reason*:* Due to competition prices fall to the lowest possible level: price is equal to the lowest point on the Long-term Average Total Cost curve (LATC in Fig. 8.3.1). Hence, there is no less costly (more efficient) way to produce. The last unit of production is worth exactly the same to the buyer as the sum of the costs of the resources necessary to produce it (price equals marginal costs). Although companies would prefer to sell at higher prices and consumers to buy at lower prices, neither party has an ability to transact at any price other than the equilibrium price. Perfect competition thus results in **efficiency of resource allocation**: all possibilities of mutual gains through exchange are perfectly realized. In other words, the profits of **all market participants** (consumers and producers) are maximized (see, e.g., Frank, 2015).

Second reason: the competitive equilibrium contributes to **social justice** by producing a balance of income between producers and workers, since all producers receive incomes equal to the opportunity costs of the resources and time invested in their production. In other words, business owners do not necessarily earn more than their employees; if they earn more, this is because they have either taken greater risks, invested more capital (not money), more know-how or more time than the workers. The difference in earnings between owners and employees is therefore fair because it corresponds to the different amounts of resources invested.

In simpler words, perfect competition requires the production of the best products at the best prices (what in microeconomics we term **allocative efficiency**) while everyone gains according to their efforts (**social justice**). Since consumers can choose between many producers and products and producers can enter and leave the market without barriers, this in turn increases **liberty**, in fact, perfect competition would represent the maximization of economic freedom (Fuders, 2009b, d, 2011).

10.4.2 Perfect Competition: No Longer a Theoretical Utopia

Some would say that perfect competition is a utopian model, that in its pure form does not exist in the real world. That is true, until now. A frequent criticism of the concept of perfect competition is that there is no stable point of equilibrium between supply and demand. However, this is mainly due to our current monetary system which generates a constant growth of money supply and therefore inflation. In addition, our current monetary system generates ever-increasing income inequality that many erroneously attribute to the market economy.

However, classic market imperfections also cause inefficiencies and income inequality. For example, the profit of the **monopolist** is out of proportion to the opportunity costs of the production factors invested. He gains an outsized income that is not proportional to his effort but is instead due to the exploitation of a dominant market position. In the end, it is the consumer who pay the inflated prices, which is why the monopolist gains an extraordinary profit at the expense of the consumer. **Price fixing** (cartels) between manufacturers have a similar effect to exploiting a dominant position: competition is eliminated, and the consumer must pay an excessive price. Finally, **unfair competition**, such as advertising fraud, can also generate profits in excess of the normal remuneration of the factors of production. To fight these three classic market imperfections (the abuse of monopolistic market positions, price fixing and unfair competition) are the three "pillars" of competition law in most countries (for the UE see, e.g., Mestmäcker & Schweitzer, 2004; Fuders, 2009b; for the Mercosur see Fuders, 2008).

In economics we use the term "workable competition" to refer to the state that is as close as we can realistically get to the theoretical ideal of 'perfect' competition (Clark, 1940; Kantzenbach, 1967). Workable competition has the power to reduce class differences (Friedman, 1966; Böhm, 1961, 2010; Herdzina, 1999), because the more competition there is, even if it is not perfect, the lower prices will be, which is advantageous for households, but not for producers, who lose monopolistic profits compared to situations with less competition (see, e.g., Frank, 2015).

Hence, competition, even if it is not perfect, leads to a more even distribution of income between households and companies. Franz Böhm, one of the most renowned members of the Freiburg school of economics once called competition the "most marvelous and genius instrument to disempower in history" (Böhm, 1961: 22).[11] Competition can thus help to prevent the establishment of dominant market positions, which also threaten freedom (Emmerich, 2001; similarly: Kantzenbach, 1967). The more competitive markets are, the better for consumers. The distribution function refers not only to the relationship between households and companies, but also to the relationship between competitors. Since in the model of perfect competition each market participant earns just as much as they could earn elsewhere with the same investment of production factors, also workable competition ensures that competitors are remunerated in proportion to their performance (Herdzina, 1999; Tolksdorf, 1994; Frank, 2015) while at the same time maximizing economic freedom (Fuders, 2011).

These dynamics apply, even only imperfectly, under the current monetary system. This is why it is of the utmost importance to fight classic market imperfections even if we choose not to reform our monetary system.

10.4.3 The Role of the Government

Economic history shows that markets do not cure the classic market imperfections by themselves. The task of promoting workable competition is therefore also viewed as a responsibility of government, with the ultimate aim of maximizing the welfare and freedom of market participants (Mestmäcker, 1984; Schachtschneider, 2007). In addition to fighting classic market imperfections, we also need to reform our monetary system. Neutral and natural money is the foundation upon which the three "pillars" of perfect competition must be built, as Fig. 10.2 illustrates.

[11] Author's translation, in German "großartigste und genialste Entmachtungsinstrument der Geschichte".

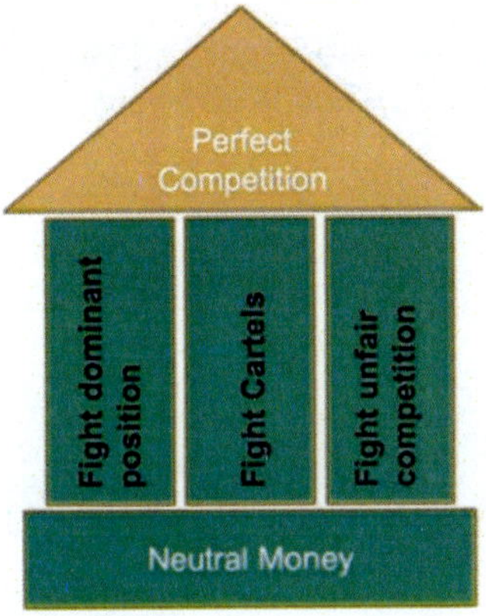

Fig. 10.2 Perfect competition and neutral money. (Source: own work)

In a study conducted at Universidad Austral de Chile (Namatama et al., 2018) analyzing market concentration and the degree of development (measured via the UN Human Development Index) in 100 countries, it was found that there is i) no clear positive correlation between GDP and development, and ii) on average developed countries have significantly lower degrees of market concentration than developing countries, i.e., they have more competition. The study highlights the positive effects of competition, even though we are still far from the ideal of perfect competition.

In many developing countries such as Chile, for example, governments seem to have misinterpreted the notion of free markets as markets free of any regulation (Fuders et al., 2019). Chile, like most other developing countries are far from the ideals of 'perfect competition' and so-called 'free markets'. Markets are highly concentrated (monopolized) and unfair competition as well as price agreements are not consequently prosecuted. But how can we break up monopolies?

Among the possible solutions is the provision of (financial and administrative) **help for start-ups**, so that as many small and medium sized enterprises (SMBs) as possible exist in every market. **Education** is also important for enhancing competition (Fuders & Michaelis, 2016). In the long run, good education that produces capable, professional graduates leads to an increased number of start-ups and hence more SMBs and thus to more competition. The introduction of so-called 'dual training', as is common in Germany, Austria and Switzerland could be an option for other countries too. It lasts only 2–3 years and is less academic and more practical than university studies. Trainees work in companies and are taught basic theoretical knowledge in a vocational school 2 days a week or in afternoon classes. The training ends with an examination before the respective professional body (e.g., the Chamber of Industry or the Chamber of Commerce).

It can be surmised that the high level of prosperity and the relatively low degree of market concentration (monopolization) in German speaking countries is due, among other things, to this dual training system. Through this system, knowledge about production techniques, processing methods, agricultural cultivation methods, etc., is passed on to future generations. The reason why there is a relatively large proportion of SMBs in Germany can too be attributed to this system of dual training (Fuders & Michaelis, 2016). It is repeatedly emphasized that it is the SMB sector that secures Germany's prosperity (BME, 2014). This is probably true not only because, as is often argued, most people in Germany are employed in the SMBs, but also because the many SMBs have to compete with each other, reducing the likelihood that markets will become too concentrated.

10.4.4 Perfect Competition Translates into 'Perfect Solidarity'

We have discussed the idea that it is possible to come close to perfect competition if we fix the design flaw in our money and thus get rid of the most important (and least recognized) reason for market failure. We have also established that workable competition forces producers to offer the best possible products at the best possible prices. In this way, competition (if it works and is not biased by the non-neutrality of money) **causes us to behave as if we loved one another**.

To further clarify, consider the following example. The Ibis hotel group promises all customers that if a problem is not resolved within 15 min, the customer will be given a night's stay for free. One might think that such an offer is a realization of neighborly love. But, most probably, Ibis is compelled to make this offer by the pressure of heavy competition in the hotel market. Here, for the consumer at least, competition is wonderful. It forces the producer to behave in a way that could be interpreted as neighborly love.

Of course, it would be better if we actually behaved like this out of love for our neighbors, i.e., out of desire to contribute to the common good and not because competition obliges us do so. If we all recognized that making money is not an end in itself and that the purpose of life is to use our talents to contribute to the common good (which, in fact, could be seen as the realization of the neighborly-love principle), then we would offer best products at best prices without the pressure of competition being necessary. We would do so because we wanted to make the buyer of our product or service happy, which in turn would make us happy too. Then we could transition from a market economy to a true economy of neighborly love (Fuders, 2017; Fuders & Nowak, 2019), in which we achieve allocative efficiency and happiness by serving each other through the use of our talents. Since we have not yet come to this deeper meaning of life, we need competition to obligate us to behave as if we loved one another.

10.4.5 Competition Is Not Incompatible with Cooperation

Markets and the concept 'competition' are being increasingly challenged. One plausible driver of this dissatisfaction might be the disparity in wealth observed in many countries (e.g., Bárcena et al., 2018). Also, the destruction of the environment, excessive egoism, individualism, and the so-called 'elbow-society' in which everyone seems to think only of themselves are associated with competition. A frequent criticism of the competitive economy is that instead of competition, one should rely on cooperation (e.g., Latouche, 2003, 2009; Felber, 2009; Traverso, 2015). Yet, the destruction of the environment and growing inequality are not caused by the competitive economy, but as explained throughout this book by our monetary system.

Free market competition does not impede cooperation. On the contrary, **a company is a form of cooperation** between different stakeholders to produce certain products or services. Also, companies can **cooperate with each other**, to a certain extent. However, there are some restrictions since not all forms of cooperation are good for society. For example, competition law prohibits price fixing between companies, for a good reason. Not only does price-fixing between competitors limit the freedom of third parties (legal-philosophical justification of competition law, on this see e.g. Emmerich, 2001; Mestmäcker & Schweitzer, 2004; Fuders, 2011) and decrease efficiency (economic justification of competition law, on this see e.g. Fuders, 2009b, 2011); such agreements (called collusion or cartels) also contradict neighborly love because they serve to concentrate market power with the aim of selling a product at a higher price. Price agreements between competitors thus serve to enrich the company owners unjustifiably at the expense of the consumer.

For the same reason, competition law in many countries also prohibits agreements on advertising or research budgets as well as the division of markets (see e.g., Fuders, 2009b). To sum up: competition does not mean that market participants cannot cooperate. And not all forms of cooperation are automatically desirable.

10.4.6 Competition Limits Selfishness

Many associate selfishness and individualism with the competition-based market. Some even argue that this is what Adam Smith's "invisible hand" would be about (e.g., Frank, 2015). Adam Smith wrote:

> (…) and by directing that industry in such a manner as its produce may be of the greatest value, he intends only his own gain; and he is in this, as in many other cases, led by an invisible hand to promote an end which was no part of his intention. Nor is it always the worse for the society that it was no part of it. By pursuing his own interest, he frequently promotes that of the society more effectually than when he really intends to promote it. (Smith, [1776] 1952: 19).

Often, this statement is misinterpreted as asserting that the egoistic, self-interested behavior of market participants, controlled by an "invisible hand" causes efficient allocation of resources (e.g., Frank, 2015; Common & Stagl, 2008). In fact, the opposite is true. It is **despite our selfishness** that competition compels us to behave as if we loved one another: to offer the best products at the best prices, what in economics we term *allocative efficiency* (in the long-term competitive equilibrium, the price is equal to the minimum point of the long-term total average cost curve, see

Fig. 8.3.1), which means there is no less expensive, (i.e., more efficient) way to produce. It is not selfishness that drives competition. Rather, **competition limits our selfishness**.

10.4.7 Neighborly Love Would Render Competition Unnecessary

Selfishness is not a virtue, but the opposite of neighborly love. As outlined above, competition promotes allocative efficiency **even though** we are egoists. Whenever the mechanism that compels us to produce the best possible products at the best prices (*allocative efficiency*) does not work, we speak of *market failures*. A monopoly, for example, is considered a structural market failure and the opposite of so-called perfect competition (e.g., Frank, 2015). While in perfect competition each market player earns exactly the amount that equals the opportunity cost of the production factors invested (i.e., the amount he could have earned elsewhere with the same investment of capital, time, diligence, and knowledge), the monopolist can demand prices that are out of all proportion to the cost of production permitting him to gain a so-called 'monopoly rent'. Achieving monopoly rents is selfish behavior. Someone who loves their neighbor will not exploit market power even if they hold a dominant market position. Instead, they would set a reasonable price relative to production costs. Unfortunately, selfishness is not regarded as a negative characteristic in economics today, which might, in part, be due to a questionable interpretation of Adam Smith's "invisible hand".

In fact, all market failures can be seen as a failure of keeping our selfishness in check. For example, those who love their neighbor will not collude with competitors to artificially create market power with the objective of imposing prices that the customer would not pay in a state of 'workable competition'. Also, unfair competition, such as misleading advertising would not exist. The three pillars of competition law (the prohibition of the abuse of dominant position, the prohibition of price cartels and unfair competition) would not be necessary if everyone respected and practiced the commandment to love one another.

Also, our monetary system, identified as the main (but possibly least recognized) reason for market failure is, in fact, the result of this lack of love for one's neighbor. To "press" interest as Gesell (1949: 205, 344) put it, or "to exploit the scarcity-value of capital" as Keynes (1936: 376) described it in some smoother words, is a selfish behavior. If we were not selfish and everyone practiced neighborly love, we would not need any competition law and maybe not even monetary reform in order to incentivize us to produce the best products at the best prices (on this see Fuders, 2017). At the end of this book, in Sect. 17.3, I will come back to this utopian perspective and outline a model of an economy based on the love-your-neighbor principle.[12]

10.5 A Natural Land Property Order

10.5.1 Introduction

The proposed monetary reform should be accompanied by land reform as proposed by Silvio Gesell and, before him, Henry George and John Stuart Mill (Gesell, 1949; George, 1881; Mill, 1885). Land is not subject to decay and has a similar superiority over other real goods as money does in our present economic system (Suhr, 1988; similar Keynes, 1936). A monetary reform as suggested by Gesell without corresponding land reform would lead to people investing in land as a store of value. After some time, most of the land would end up in the hands of a few, just as today the largest share of financial assets is held by a relatively small percentage of the population.[13]

In addition, land is not a product but a gift from nature which is a powerful argument that from a moral-ethical point of view land should belong to everybody. But it is not only about land. The reform should also include the regulation of the exploitation of other natural resources. These are actually to be regarded as the property of all mankind (Gesell, 1949), or at least as the property of the people in whose territory they are located. In many countries natural resources are considered public property (e.g., in Chile according to the current constitution).[14]

The following paragraphs will discuss the idea that in the economics of property rights we should distinguish between man-made products and (pure nature-made) natural resources (see also Fuders, 2019; Fuders & Pastén, 2020). In conventional economics, private property rights are considered indispensable to allocative efficiency, basically **regardless** of the nature of the goods. Also, according to conventional economics, one way to avoid the overexploitation of freely accessible resources (*open access regimes*) is by privatization. This argument is based on the idea that only market goods (excludable and subject to competition) can be efficiently allocated by the market. The sale of significant parts of Chilean forests to private companies in the 1980s was based on this economic rationale. However, as we will see, powerful arguments exist that allocative efficiency can be achieved if private property rights exist for man-made products, but not necessarily for natural resources, which ideally should remain freely accessible. We will also argue that

[12] The idea of an economy based on neighborly love instead of competition has already been laid down in Fuders, 2017 and Fuders and Nowak, 2019.

[13] For an exhaustive analysis of the possibility of gaining economic rents from land ownership see the publications of Löhr (2012, 2013, 2018).

[14] "Disposición Tercera Transitoria", Constitution from 1980 in conjunction with "Disposición 17ª Transitoria", Constitution from 1925.

this is NOT a Marxist idea but, on the contrary, a way to avoid monopoly rents and **approach to the ideal of perfect competition**.

However, we will also argue that (i) open access regimes are not possible anymore given existing exploitation technologies and that (ii) land reform is less important than monetary reform. And since the concept can be easily confused with Marxism, we will conclude that it is better to **stay with the current land ownership system**, but to tax away monopolistic rents resulting from the unequal distribution of land, as was proposed by Henry George 150 years ago.

10.5.2 The So-Called "Natural Dividend" Is a Monopoly Rent

Contrary to conventional economic thinking, assigning property rights to natural resources does **not necessarily** prevent nature from being excessively exploited (Fuders, 2020). Instead, we can argue that the lack of sustainability in the management of natural resources has its origins in our monetary system wherein interest (as the opportunity cost of any productive investment) exerts constant pressure on businesses to achieve profits at least as high, which is why the economy is forced to grow at least at the same pace. In other words, **regardless** of whether or not natural resources are freely accessible goods, they will be over-exploited in the long term due to the logic of constant economic growth. Chile provides a case in point. Although the coastal areas are divided into exclusive management zones (*áreas de manejo*) and the forests are predominantly privately owned, neither aquaculture nor forestry are sustainably operated (regarding aquaculture see Sommer, 2009 and Allsopp et al., 2008; regarding forestry: see Pastén et al., 2020; Manuschevich, 2020; Donoso & Romero, 2020). Here, unsustainable resource management cannot be explained by free access to resources.

In addition to the fact that the privatization of natural resources does not necessarily avoid over exploitation, it also allows for a so-called "natural dividend" (see, e.g., Daly & Farley, 2011: 226) or "natural resources income" (Neher, 1990: 51; Conrad & Clark, 1995: 89, 94; Acquatella et al., 2013: 1). The term "natural dividend" expresses the idea that the earnings originate mainly from the reproductive force of nature, not from the effort of man (Daly & Farley, 2011). In other words, such income can be considered unjust, i.e., not based on the owner's effort. Such earnings in excess of normal compensation for the effort of the investor, i.e., the opportunity costs of the production factors invested, do not exist in perfect competition, and arise from market inefficiencies. In fact, to earn a "natural dividend" is only possible if exploitation is limited to a **single owner**, whether private or public.

Given all of this, it can be asserted that the "natural dividend" is in fact a **monopoly rent**, and accordingly can be viewed as an **unjust** profit (for a more detailed discussion see Fuders, 2020). Any profit that exceeds the recuperation of total costs—which include opportunity costs, that is, the profit that one could have earned by expending the same effort on the best alternative opportunity—is income that is not based on the producer's effort. In perfect competition, such earnings do not exist.

10.5.3 With Traditional Harvesting Methods Free Access Would Lead to Allocative Efficiency

If we think of a natural dividend as monopolistic income, then we could argue that free access regimes are actually desirable, since their result corresponds to that produced by perfect competition (Fuders, 2020). Let us examine this idea further using the oceans as example. When there are no artificial restrictions on access, more competitors (fishermen) enter the industry. This increases the fish supply on markets and pushes prices to the lowest possible level, that is, to the point at which they only cover the total costs. **This is a desired situation**. In this situation, price equals marginal income and marginal costs, thus allowing for "efficient allocation of resources" or "allocative efficiency" (Frank, 2015: 330; Common & Stagl, 2008: 310).

This is precisely the situation that existed for centuries: an undefined number of small fishermen (suppliers) fishing without any restrictions generating a large supply of fish in relation to the demand and, consequently, low prices. The price of fish used to be so cheap that even until the present day it is a custom for Christians to eat fish on Fridays and during the Holy Week. Originally, Christians were supposed to fast on these days, but many decided to eat fish instead of fasting because fish was considered such a low value food that eating it was similar to fasting. This is an example of perfect competition that **functioned for thousands of years, without significant over-exploitation of the ocean.** That natural resources should not be privatized was emphasized by John Stuart Mill (1885), Henry George (1881) and Silvio Gesell (1949), based on the same rationale of avoiding monopoly profits for individuals with exclusive exploitation rights.

Unfortunately, what has been said is only true for traditional harvesting methods and does not necessarily apply today. The problem is that harvesting techniques changed, and this in turn has to do with the growth imperative, as explained in Chap. 4. Gigantism promoted by the growth imperative has impacted the fishing and forestry industries. As we have discussed above, the world has become relatively small in relation to exploitation technologies (as illustrated in Fig. 10.3). Free access combined with today's technologies endanger all natural resources with exhaustion, even our oceans.

Some might object that technologies do not develop solely due to the growth imperative. For example, the so-called "induced innovation" theory (Hicks, 1973) also provides an explanation of how technological innovation arises in certain areas in the face of increasing population growth and resource scarcity. However, the trend towards ever-increasing efficiency is certainly driven, at least in part, by the compulsion to either grow or to save costs. How else about 99% of all inventions have been introduced in the last 500 years? Our fishing boats, for example, have

Fig. 10.3 Fisher boats ever growing

basically transformed into industrial barges during the last hundred years. For many thousands of years prior, they did not change at all or only quite insignificantly. The growth imperative can also be recognized in the technical evolution. We could ask, why did this exponential technical evolution not start three or four thousand years ago, but only relatively recently? The answer is that our current monetary system has only existed for about 700 years, and it is during this period that the pace of technological evolution accelerated so much.

10.5.4 Distinction Between Man-Made Products and the Produce of Nature

Since the excessive exploitation of nature does not result from free access to resources, but from the growth imperative that interest has exerted for the past centuries, it can be argued that natural resources, especially land and soil, mineral resources, water and air (emissions trading can be considered as a form of privatization of the air, see Lohmann, 2006), should remain freely accessible (Fuders, 2020;

Fuders & Pasten, 2020). From the perspective of allocation theory, free access in the absence of the growth imperative (and gigantism) is the most efficient approach.

No doubt, **we need private ownership of man-made goods**, otherwise no one would produce them. For example, imagine a coffee loving country, whose government declared drinking coffee as a human right and therefore prohibited private ownership of coffee machines. Would coffee machines be produced or sold in this country? Probably not. Anyone who produced a coffee machine would not be able to sell it, since as soon as it was produced, the producer would lose ownership of it. It also seems fair that someone who produces goods is rewarded for the creation of the added value, i.e., for his labor. Private property has always existed, even though it was not always explicitly anchored in laws. The biblical commandment "Thou shalt not steal", for example, would be superfluous if private property rights were not thought to be necessary to assure a living together in justice, peace and harmony.

However, one might ask whether it is right and just to consider natural resources as private property. Here, private ownership is not necessary in order to incentivize their production. It makes sense, therefore, to distinguish between goods made by man and those produced by nature and assign **private property only to man-made goods**. Yet, this would not automatically result in natural resources being freely accessible to all humans and animals.

10.5.5 Free Access for Everyone to Natural Resources?

Free access to natural resources is not only required in order to achieve allocative efficiency but is also beneficial from the point of view of justice. After all, nature was created for everyone. Holy Scriptures of Christians, Jews and Moslems recognize that the earth and the heavens are a gift from God (Bible, Genesis 15:18–21, Deuteronomy 10:14 and Psalm 115:16; Tanakh, Genesis 15:18–21 and Deuteronomy 10:14; Koran, Surah 3:26, 4:131 and 17:100–104). For those who do not believe in divine creation we can say that land as well as all natural resources are a gift from nature. This argument is taken up by Henry George (1881). He uses an interesting metaphor. If someone is the first to take a seat in a railroad car, can that person then prevent other passengers from taking other seats simply because he was there first? Of course not.

Regarding land ownership, why should a person who first occupied a piece of land be entitled to exclude others from using it simply because he came first? Tribes of aborigines all over the world do not recognize property rights as applied to nature, but they do respect property rights for anything which is the produce of man having applied labor to transform natural resources (Du Plessis, 2011; Boyd, 2021; Atleo & Boron, 2022; Smithonian Institution, 2023). "Chief Seattle", one of the last chiefs of the U.S. indigenous population, is said to have said in a speech to the Governor of Washington Territory (today Washington State) that the earth cannot be sold; for man belongs to the earth, not the earth to man.[15]

In Europe private property in land, as we know it today, is a relatively recent development. Until the time of the French Revolution, private property of land was a privilege of the nobility. Land, as well as natural resources, belonged to the king and his network of nobles.[16] The commandment "*Thou shalt not steal*" (Bible-21st Century King James, 1994, Exodus 20:15; see also Tanakh, Exodus 20:13; Koran: 2:188) therefore refers first of all to goods produced by man (Fuders, 2019). Now, it is worth asking whether the simple effort of harvesting justifies claiming property rights over those resources. It might be argued that harvesting a natural resource is indeed sufficient effort to justify ownership of the harvested resources. Timber or fish could thereby belong to whoever felled the tree or set the net. But the **access** to forests and fishing grounds and other natural resources has to be **available to everyone**. This is a powerful argument that pure natural resources (where man did not plant or water) should not be privatized.

10.5.6 Gesell's Freiland: Equal Opportunities and Access to Plant and Harvest

What has been said so far does not apply to agricultural and forestry products that have been planted by man himself. These should not remain freely accessible since the effort of planting must be rewarded. The value added by virtue of the effort of planting and watering justifies private property (see already Fuders, 2019, 2020). Interestingly, this aligns with the epistle to the Corinthians: "The one who plants and the one who waters have one purpose, and they will each be rewarded according to their own labor" (Bible-NIV, 2015: 1 Cor 3:8). Today, however, this is not always the case as the owner of land often earns more than the opportunity costs of the production factors invested. This additional remuneration has been described as a "*natural dividend*" but actually constitutes a monopoly rent and results from the appropriation of soil, mineral resources or water by private parties, as discussed above.

So, on one hand, access to land needs to be restricted so that those who plant can also harvest. On the other hand, property rights that enable private parties to extract monopoly rents should be avoided. And from a moral point of view: shouldn't all people have equal opportunities and access to land and natural resources? How can

[15]The original speech is lost. There are many versions circulating but with uncertain authenticity. One of the most academic versions was published by Kaiser (1987). By the same logic, native populations do not own land either.

[16]The nobles made up less than 3% of the population. The king needed the nobles to enforce his power and to govern different regions. The nobles were given some authority to create laws and rules for the regions they governed but they had to obey the king. Peasants could possess land (and even inherit it), but it was not full ownership as we understand it today (which includes the unrestricted right to sell or deny access to others). For German regions see: Posch (1963); Tremel (1963).

we grant access to land and other natural resources while at the same time (i) assuring sustainable use and (ii) allowing those who plant, water and harvest to receive just compensation for their efforts? The answer is, while private property of pure natural resources like land and water should not exist, objects which are produced by the use of land and water as well as any improvements to land such as crops or buildings should be private property. This is the idea of Silvio Gesell's natural land property order. He termed it "free land" (*Freiland*).

10.5.7 Not to Be Confused with Marxism

Gesell's Freiland proposal must not be confused with Marxism. The argument is that **only** commodities which are pure products of nature (land and natural resources) should be common property. Gesell did not believe in communal ownership of the means of production, as proposed by Karl Marx (1872). Gesell's land order corresponds to a **natural** practice that has existed for thousands of years. It is, as stated above, to be advocated from the point of view of allocation theory. And appears fair since less monopoly profits through "natural dividends" can be earned. We previously observed that prior to the French Revolution private ownership of land, water and air did not exist in Europe. Only the king and his nobles owned land (a power structure with parallels to current federal governments). Everyone else was a tenant (peasant), paying rent, or 'tributes' drawn from the goods produced. The more land used for agricultural production the more tribute was paid. This system has been said to be the origin of the obligation to pay taxes (Pfeifer, 1997).

Only later was this governmental structure weakened when populations came to view the situation as unfair. The divine origins of government were questioned, and social class struggles emerged. These led in the aftermath of the French Revolution to land reforms in many European countries. Even though such reforms may at first sound just, this is not necessarily the case as it made possible the monopolization of land by a relative few (now private owners). In the Americas from the very beginning of European colonization land has been privately owned and very unequally distributed. The European kings promised land titles to incentivize emigration to the new colonies. A small group of landlords owned major parts of the new colonies, and this extremely unequal distribution of land exist especially in South America until the present day.

10.5.8 The Natural Economic Order (Money and Land Reform Together)

A reform of the monetary system in the way Gesell proposed with money that is not easily storable would incentivize people to invest in land, even more than it is already the case. The problem of people 'hoarding' land was recognized by Silvio

Gesell. This is why Silvio Gesell in his masterpiece "The Natural Economic Order" proposed land reform alongside monetary reform.

It would then not be possible anymore that a few families own significant parts of a country, as it is the case, e.g., in many Latin American countries today. In the proposed system anyone who wants to hold land for whatever reason—may it be for edifications or for agriculture—pays a lease or rent to the State, while the heirs have priority in terms of being able to continue leasing (Gesell, 1949). With such a reform we can comply with the above defined rule that pure nature-made resources (here land and water) are to be hold by the public hand while the products that humans produce making use of the natural resources should be privately owned. The ground on which self-used houses are built could be exempted from the leasing rent (Fuders & Pastén, 2020). This solution would require the State to buy back all land properties from current private owners (George, 1881; Gesell, 1949).

The more land one wishes to hold, e.g., for agriculture, the more earnings can be generated and therefore the more rent the holder must pay. The rent is thus proportional to the income generated through land use. In fact, even though this sounds like a sea change from the current system of private land ownership, it is not so different. Under the current system we pay real estate taxes in most countries. Does it make a difference if we call that payment "rent" or "tax"?

While in socialistic systems all production factors are publicly owned and in capitalism all production factors privately owned, in the Natural Economic Order as proposed by Silvio Gesell private property exists for man-made goods (regardless whether it is a final product or factor of production) while natural resources, especially land, are publicly owned. While in socialism as well as capitalism the current monetary system is not questioned, in the Natural Economic Order money is perishable as real goods and thus serves only as a means of exchange, not as a means of accumulation. Figure 10.4 gives a schematic overview of the three systems.

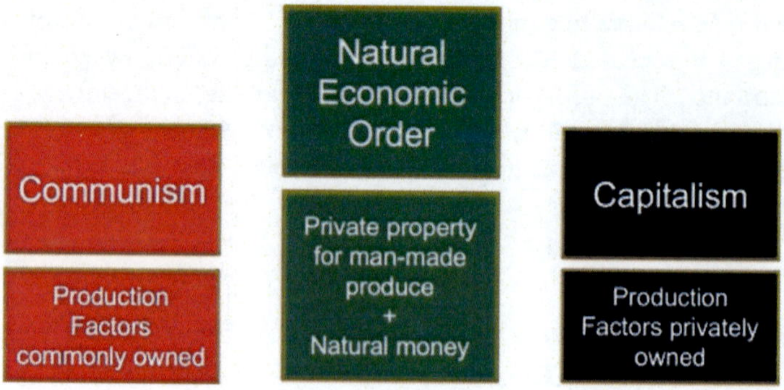

Fig. 10.4 Natural economic order vs. communism and capitalism. (Source: own work)

10.5.9 *Money Reform Is Far More Important Than Land Reform*

Silvio Gesell first published a series of papers and small books in German and Spanish on the problems inherent in the design of money and only later discussed the land question, probably influenced by the very popular writings of Henry George. However, in his masterpiece "The Natural Economic Order" Gesell discussed the land question before the money question, and by choosing this order he seems to give priority to the land question. I do not subscribe to this point of view and believe, as did Keynes (1936), that monetary reform is far more important than land reform, for the following three reasons:

First reason: In Chap. 5 we examined why the money supply in the current system always grows, be it through interest money creation or, when the monetary system is close to collapse and interest rates are close to zero, because the central bank prints money (nowadays called *Quantitative Easing*) in order to keep money circulating in the economy. It is this ever-growing money supply that causes price bubbles in investment markets, especially the real estate market. With neutral money these price bubbles would be of much smaller magnitude. Without price bubbles in real estate the issue of unequal land distribution would be much less of a problem.

Second reason: Price bubbles in real estate would not be such a big problem if land was more equally distributed. The problem could be addressed by finding a solution to the unequal distribution of land ownership. Land does not necessarily have to be publicly owned. For example, the State could acquire land from the current owners and resell it to others. In fact, the problem can be understood as a case of abusing a dominant market position, which is prohibited in competition law in most countries in the world (for the European Union see: Fuders, 2009b; for the Mercosur see Fuders, 2008). Generally, this prohibition of the abuse of dominant market position is applied to monopolistic practices by companies, but it can be applied to the case of land ownership too. For example, leaving large areas of land fallow in anticipation of rising prices could be seen as an abuse of a dominant position.[17]

Third reason: To lease land and to lease money are two completely different things from a philosophical point of view. Land is a real good. Money is not. We receive real value and real utility when we "borrow" land. The borrower of money receives no real value but only a medium of exchange (which can then be used to buy goods and services with real value). To once again cite Rudolf Steiner (1979:

[17] Here, the so-called essential facility doctrine (see, e.g., Mestmäcker & Schweitzer, 2004; Fuders, 2009b) could be applied to the monopolistic aspects of land ownership too. For example, if someone holds a significant portion (dominant position) of all land in a region and leaves it fallow, the owner could be forced to rent the land at market price to farmers, in the same way that today companies with dominant positions in the telecommunications and energy sectors are required to rent their infrastructure (essential facility) to smaller competitors.

50), it is completely "insane" to borrow a voucher (money) that represents a real good instead of the real good itself. I am convinced that everything that deviates from nature is harmful in the long run. Our monetary system contradicts the laws of nature, especially physics (for an in-depth analysis see Soddy, 1933a, b, 1934) and it therefore makes sense that significant negative consequences result

Since the problems inherent in land ownership are less relevant than the money question, and since it would be very problematic to return all privately owned land back to public ownership, it would be preferable to address the problem via tax reform instead of land reform. This topic has been extensively discussed by John Start Mill, Henry George, Silvio Gesell and recently Dirk Löhr.

10.5.10 Tax Reform Instead of Land-Property-Order Reform

As explained in the previous section the unequal distribution of land can be viewed as abuse of monopolistic power. Hence, one way to address the problem is to continue to **allow private ownership** of land and other natural resources but to tax monopoly rents.

Today in most countries in the world real estate is already subject to taxation, however, in a very inadequate and even unjust way: monopoly rents are not efficiently taxed away (for in-depth analysis of the argument see Löhr, 2012, 2013, 2018; Kriese et al., 2019). Real estate prices are especially high in regions with good infrastructure, but this value is primarily created by the community, not the landowner (George, 1881; Daly & Farley 2011; Löhr, 2013, 2018). The system of taxation in most countries does not distinguish between land and improvements to land, such as buildings. Landowners are disincentivized to invest in expensive, high quality and sustainable construction, since this would result in higher taxes.

On the contrary, in most countries fallow land incurs little or no tax liability. This creates a powerful incentive to leave plots idle, at least while real estate prices are rising for speculative reasons. We see this in Berlin as well as in small Latin American cities such as Valdivia. The best land often remains fallow (Fig. 10.5). Increasingly, investors buy prime real estate for speculative reasons only. Local residents who need land for homes are therefore unable to afford to buy in such areas. The current tax system works hand in hand with the monetary system. The monetary system generates an ever-growing money supply that then finds its way into real estate.

150 years ago, Henry George (1881—first edition form 1879) proposed a real estate tax where the tax applies only to the value of the raw land, i.e., excluding the value of buildings or crops. A tax based only on raw land value disincentivizes land speculation and in so doing eliminates speculative rents (George, 1881), while it would not disincentivize to build expensive high-quality constructions. That is to say, landowners would not be punished for building long lasting, sustainable and architecturally beautiful buildings, as is currently the case.

Fig. 10.5 Photo of an idle plot in the center of Valdivia, Chile. (Photo: Felix Fuders)

The proposed form of taxation respects the distinction made above between land as a (pure nature-made) natural resource that belongs to all, and the work performed on this land by man. The rise of the value of raw land (excluding the value of buildings and crops) can be regarded as a monopolistic gain for the owner, since it results from the assignation of property rights and not from his effort. It therefore seems **fair and just** to tax this increase of value which can then be redistributed by the State to benefit everyone (George, 1881).

10.5.11 A Ray of Hope: Real Estate Tax Reform in Germany

The proposal of Henry George, even though he could be considered a best-selling author of his time, unfortunately has, until now, not been given much attention by politicians. Such a plan is contrary to the current land tax systems in Germany, Chile and most other countries in the world where the basis for the assessment of real estate taxes is the value of the property as a whole, including both raw land value and improvements. However, there seems to be a ray of hope. Thanks to the persistent and fruitful works of Dirk Löhr & colleagues who have been promoting real estate tax reform in Germany based on the insights of Henry George and Silvio

Gesell for almost two decades now (Wikipedia, 2022), in 2020—2 years after the German Constitutional Court criticized the current tax system as unjust—one of the 16 federal states of the Republic of Germany reformed its real estate tax system along these lines (Bockting, 2020). There is hope that other German federal states will follow.

Note that such tax reform would not necessarily result in an increase in overall taxes since governments could use the additional income to lower other taxes. Henry George was even convinced that with real estate taxes **alone** all public spendings could be financed (George, 1881). In fact, as mentioned above, the origins of taxes have roots in the tributes paid by citizens to the King as a sort of leasing fee for the use of land. If the raw land value tax is sufficiently high it would not be necessary to tax production, commerce or labor as most countries do today. Hence, a taxation system with a single tax on raw land would not disincentivize wealth creation (George, 1881). Indeed, the **international competitiveness** of a country's economy could be enhanced by such a system.

10.5.12 Summary

In the classic liberal view all production factors, including natural resources, are privately owned. By contrast, Marxism argues that all production factors should be publicly owned. In this chapter we have advanced a third view. Private property should exist for products made by man, while pure natural resources such as land, water, air, mineral resources, native forests, oceanic fishing grounds, where man has not added any value private property should not exist. Such a system would not only be fairer but in conjunction with a Gesellian monetary reform it can theoretically improve allocative efficiency.

However, as outlined in Sect. 5.7.5 open access together with the growth imperative and modern exploitation technologies is the worst-case scenario in terms of sustainability. Even if we manage to change our monetary system to a Gesellian natural money system and get rid of interest and the obligation to grow, the advance of technology cannot be reversed. Thus, even in absence of a growth imperative today any open access regime risks overexploitation. And regarding the psychological effect and the risk that many would intuitively associate such a land property reform with Marxism, we concluded that is preferable to **maintain private ownership** on land while combining it with a tax reform to levy eventual monopolistic gains. In addition, a tax reform is much easier to implement than a land reform since the public sector will not have to buy back the currently privately owned land.

10.6 Is the Natural Economic Order a Christian Economic Order?

10.6.1 Gesell's Natural Economic Order Is Truly Natural

In the previous paragraphs we have seen that both our current monetary system as well as our land use system are completely unnatural. That our monetary system, and especially interest, is an unnatural human construct was, as mentioned several times throughout the text, recognized by both Aristotle ([384 BC] 1995) and St. Thomas Aquinas (1596). Nobel prize winner Frederick Soddy explored the unnatural character of our monetary system in three books (Soddy, 1933a, b, 1934). He demonstrates that interest contradicts the laws of physics resulting in significant negative consequences for society. On the contrary, the slow "rusting" of savings (Gesell, 1949: 238), the slow "aging" of money, as Rudolf Steiner called it (Steiner, 1979: 179), is natural. It corresponds to the natural decay of real goods (on this see also: Suhr, 1988: 34 ff.; 58 ff.; 85 ff.). Regarding the system of land tenure, native populations all over the world seem to recognize private property of goods made with the assistance of nature but not of nature itself (see Sect. 10.5.5), even though these rules are not, or not always expressly fixed in written laws.

10.6.2 The Natural Economic Order and Holy Scripture

Such a Natural Economic Order is probably the closest we can get here on earth to what could be considered a "Divine Economic Order".[18] In other words, the Natural Economic Order is a Christian Economic Order, a Jewish Economic Order as well as an Islamic Economic Order. We previously discussed that the Bible (Holy Scripture of Christians), Tanakh (Holy Scripture of Jews) and Koran (Holy Scripture of Moslems) all prohibit interest.[19]

In the same way as Gesell's Natural Economic Order does, Holy Scriptures permit private property for man-made goods but not necessarily for land or natural resources to which man has added no value. The commandment "you shall not steal" (Bible, Exodus 20:15; Tanakh, Exodus 20:13; Koran, 2:188) would be

[18] A true and complete "Divine Economic Order" most probably cannot be achieved until heaven and earth have fused into the New Jerusalem (New Kingdom or New Paradise) since today here on earth we are not free of the influence of the evil one (Bible, 1. John, 5:19).

[19] This is no coincidence. All three religions are related. The Jewish Tanakh is contained in the Christian Bible as part of the so called "Old Testament". Next to the Old Testament the Bible consists of the "New Testament" which contains biographic data and teachings of Jesus (the Gospels), as well as some other books. The Koran is the Bible amended by Muhammed, which is why the Koran says to consult those who read the book that came before the Koran and specifically refers to the Torah (which is an essential part of the Tanakh) and Gospels in case of doubts (Koran, Surah 10:94; see also Surah 7:157).

senseless if private property was not foreseen. Yet, according to the Holy Scriptures of the three religions land belongs to God (Bible, Genesis 15:18-21, Deuteronomy 10:14 and Psalm 115:16; Tanakh, Genesis 15:18-21, Deuteronomy 10:14; Koran, Surah 3:26, 4:131 and 17:100-104). However, here some reader trained in theological studies might now object that the Bible (or Tanakh) would surely permit private property on land.[20] For example, in Leviticus 25.14 (Bible-NIV, 2015) we read:

> If you sell land to any of your own people or buy land from them, do not take advantage of each other.

In Numbers 27:6-7 and 36:2 we learn that land holdings could also be passed down to descendants. However, in the following verses it is made clear that the price paid is **not for the land but for the crops**. In Leviticus 25:15-16 (Bible-NIV, 2015) we read:

> You are to buy from your own people on the basis of the number of years since the Jubilee. And they are to sell to you on the basis of the number of years left for harvesting crops. When the years are many, you are to increase the price, and when the years are few, you are to decrease the price, because **what is really being sold to you is the number of crops**. (Highlighted by the author).

Leviticus 25:23-24 (Bible-NIV, 2015) is even clearer and states that the land is owned by God, and we are only "foreigners and strangers" who may hold land in "possession" but not in ownership:

> The land must not be sold permanently, because the **land is mine** and you reside in my land as **foreigners and strangers**. Throughout the land that you hold as a **possession**, you must provide for the **redemption of the land**. (Highlighted by the author).

"Possession" is something different from "ownership". The possessor uses a good without being the owner. And what is meant by "redemption of the land"? This has to do with the Jubilee (Leviticus 25:10-11, 13, 15), another interesting concept worth discussing. In the "Jubilee" every fiftieth year all debts are to be canceled. All land possessions go back to their original possessor. The jubilee year was (and in fact still is) necessary since people got (and get) indebted and sometimes had (and have) to sell themselves as slaves. The jubilee year made them free again. In Leviticus 25:10-11 and 13 (Bible-NIV, 2015) we read:

> "Consecrate the fiftieth year and proclaim liberty throughout the land to all its inhabitants. It shall be a jubilee for you; each of you is to return to your family property and to your own clan." (…) In this Year of Jubilee everyone is to return to their own property.

This means land that has been "sold" goes back to the original owner. In other words, it has not really been sold **but only leased** and the price paid was a leasing fee. So, according to Holy Scripture, private property was allowed, yet it was not possible to sell land but only to lease it for a maximum of 49 years. The Jubilee was necessary since people did not adhere to the rule not to charge interest and not to sell land so in the absence of the Jubilee inequality would grow constantly. God knew this, so He created the jubilee-year-rule, which meant that any "selling" of land

[20] In the following I will refer to Bible and Tanakh. For a detailed analysis of the Islamic Economic Order see Baqir Sadr (2011).

would only be temporarily, for a maximum of 49 years.[21] Certainly, it would be better to not charge interest and observe that land belongs to God and can only be administered (possessed) by man. The immense inequality in possessions would be avoided from the beginning.

10.6.3 Gesell's Solution Instead of the Jubilee

Unfortunately, the Jubilee was only respected for a certain period. In fact, as early as the times of governor Nehemias the Jubilee was no longer observed. The Bible and Tanakh describe the effect this had: the same effect as we see today. Greed and interest ran rampant, and inequality grew to such an extent that people had to sell themselves, their land and vineyards. Nehemiah urged creditors to cancel their debts (Bible and Tanakh, Nehemias 5:1-13). Gesell's Natural Economic Order would probably make such a Jubilee unnecessary or, at least, significantly less urgent.

Some might argue that today we have financial crises, which serve as a sort of "natural" Jubilee. Indeed, the deep financial crises, which occur every 50–80 years cause a reset of money supply and debt. However, this is only a monetary reset. Land does not revert to its original owners. Inequality is therefore not completely reset by financial crises.

Interestingly, for Christians we find the same commandment to cancel or "forgive" debts in the Latin version of the most important prayer given by Jesus to mankind, the "our-father" prayer (Bible, Mathew 6:9-13). In its Latin version in the Gospel according to Matthew we ask God to forgive us our debts as we would forgive the debts of others (in Latin: "*dimitte nobis débita nostra*"),[22] not "trespasses" or "sins" as we pray in the modern version of the same prayer in English. Debt is a wider term than "trespass" or "sin"; it includes monetary debts.[23]

[21] Again, here we see that there must have been private, temporary owners of land, namely the family after the land division under Moses and Joshua. Christians, Jews and Moslems assume that the land distribution was justly made or inspired by God at that time. But what about today where many people are not religious, and how far back can we trace the history to find out who was the rightful owner? Probably also the people expelled or enslaved during the Israelite land seizure after the Exodus would see the matter differently. Interesting in this context is the social system given by God at that time, which the land-owning Israelites had to adhere to (see Exodus 23:10,11; Leviticus 23:22; Deuteronomy 15:4,5 and book Ruth, chapter 2 ff.), apart from the fact that mankind should be aware that everything on earth belongs to God anyway (Psalm 24:1).

[22] Biblia Sacra Vulgata, 2007: Mathew 6:9-13. The original Gospel of Matthew was not in Latin but in the language of the Hebrews (see preface to the Gospel of Matthew in: Lord, 2019). This original writing was then translated into Greek and later into Latin. The Greek word was "*opheilema*", a debt that required reparation, which was also recoverable, i.e., the root word also has something to do with inability to pay.

[23] Some conclude that this is the root of Jesus Christ's dispute with the Pharisees who initially approved of His teachings until they realized that he proposed to prohibit the charging of interest to anyone, including strangers, and that the Lord asks us to forgive debts, which would have meant a reduction of their income (Hudson, 2018).

10.6.4 Holy Scripture Asks Us to Be Good Administrators of Nature

Nowadays it seems quite *in vogue* to criticize the Judeo-Christian culture and even to claim that the over-exploitation of our natural resources and our insatiable and unsustainable lifestyle is rooted in the formulations of the Tanakh and the Old Testament of the Bible, which invite us to dominate the earth and subdue nature in order to place it at the service of human beings.[24] In Genesis 1:28 we read (Bible-21st century King James Version, 1994):

> And God blessed them, and God said unto them, 'Be fruitful and multiply, and replenish the earth, and subdue it; and have dominion over the fish of the sea, and over the fowl of the air, and over every living thing that moveth upon the earth.'

However, although in our modern language the word to "dominate" has taken on a somewhat negative connotation, to have dominion does not mean that we are allowed to or even should destroy the earth. On the contrary, the word "dominion" stems from Latin "*dominus*" which means "lord or master" (Etymology Dictionary, 2022).[25] Hence, we are called to be good lords or masters of the things that were given to us, like a good father (or mother) who is supposed to manage the resources of the household in responsible and efficient ways. That is, dominion is associated with a great deal of responsibility, for which one must give account (Bible, Revelation 11:18). Throughout the Bible (and Tanakh), it is repeatedly evident that dominion is accompanied by a duty to be good stewards of the earth and to care for what has been entrusted to us (e.g., Bible/Tanakh, Genesis 2:15; Jeremiah 2:7; Bible, 1 Peter 4:10).

It seems that the cultural roots most directly responsible for the modern ecological and social crisis have more to do with the Enlightenment and other ideals of western modernity; that is with those rationalist, utilitarian and scientific traditions that declared themselves enemies of religion (Riechmann, 2010; Azkarraga & Fuders, 2019), and which have given a perfect pretext to not confront the problems inherent in our unnatural monetary system and land tenure system. In fact, all major religions have elements that go well in line with modern ecologist ideals (Azkarraga & Fuders, 2019). In the roman catholic tradition we can see it in the integral ecology defended by several Saints and Popes, especially noteworthy is St. Francis of Assisi who can be seen as one of the first representatives of what Arne Naes (1973) would later call "deep ecology" (Azkarraga & Fuders, 2019). Regarding the Popes, the first important references to the ecological problems are found in Paul VI's (1971)

---

[24] Unfortunately, my dear colleague Manfred Max-Neef, who was involved in the design of the Human-Scale-Development approach (Max-Neef et al., 1991) to which I will refer in Chap. 15, also fell for this interpretation (Max-Neef, 1992).

[25] Note that this part of the Bible was not originally written in Latin, but in ancient Hebrew. And here the expression to "rule" ("*radah*") refers to the usage or to make arable ("*kabasch*"), in the sense of extracting honey, to press or produce wine, to care or provide for a flock, etc., not to destruction (see preface in Lord, 2019).

apostolical letter *Octogesima adveniens* followed by Pope Benedict XVI who in a couple of writings instructed us to be good and responsible stewards of what we were given by God, especially his encyclical *Caritas in veritate* (Benedict XVI, 2009). And last but surely not least Benedict XVI's administrative follower Jorge Bergoglio, who in *Laudato si'*, makes frequent reference to the encyclical *Caritas in veritate*, calls on us to care for our "common house", the earth (Bergoglio, 2015).

10.6.5 Some Further Reflections

It is in fact obvious that a Natural Economic Order must be, for those who believe in God, a "Divine Economic Order" for a simple reason: if nature was created by God, then everything or every circumstance that contradicts nature automatically contradicts the divine order. And if I do not believe in God then I probably believe in evolution and the big-bang theory. But who created the circumstances that made evolution and the big bang possible? Who is the creator of gravity, of physics and natural laws, who made the elements, etc.? Neither evolution theory nor big-bang theory give adequate answers to these questions. If we understand that all elements and all natural laws are of divine origin, then we can only conclude that nature in all its variety and beauty is made by God. A natural order is then indeed a divine order.[26]

References

Acquatella, J., Altomonte, H., Arroyo, A., & Lardé, J. (2013). *Rentas de Recursos naturales no renovables en América Latina y el Caribe*. Cepal.

Agarwal, R., & Kimball, M. (2015). *Breaking through the zero lower bound – IMF working paper WP/15/224*. IMF.

Allsopp, M., Johnston, P., & Santillo, D. (2008). *Challenging the aquaculture industry on sustainability*. Greenpeace/University of Exeter.

Anderegg, R. (2007). *Grundzüge der Geldtheorie und Geldpolitik*. Oldenbourg.

Aquinas, T. (1596). *Summae Teologicae*. Apud Iuntas.

Aristotle. (1995). Politik 1st book. In *Aristoteles Philosophische Schriften in sechs Bänden. Band 4*. Felix Meiner Verlag.

Assenmacher, K., & Krogstrup, S. (2018). *Monetary policy with negative interest rates: Decoupling cash from electronic money – IMF working paper wp/18/191*. IMF.

Atleo, C. G., & Boron, J. (2022). Land is life: Indigenous relationships to territory and navigating settler colonial property regimes in Canada. *Land, 11*(5). https://doi.org/10.3390/land11050609

[26] In this context it is important to highlight that Darwin's theroy of evolution does not in fact explain the existence of the different species inhabiting the earth but only the phenotypical adaptation of these species to different geographic circumstances. The theory of evolution can only explain the formation of races within a given pre-existing species. And again, also the formation of races is divine creation if we argue that the coincidences (on which mutations and recombinations of genes depend) come from God.

Azkarraga, J., & Fuders, F. (2019). La economía ecológica, paradigma de futuro. In C. A. Cioce Sampaio et al. (Eds.), *Ecossocioeconomias – Promovendo Territórios Sustentavéis* (pp. 35–58). edifurb.

Bárcena, A., Cimoli, M., García-Buchaca, R., & Pérez, R. (2018). *La ineficiencia de la desigualdad.* Santiago de Chile: Cepal.

Baqir Sadr, S. M. (2011). *Unsere Wirtschaft Iqtisaduna*. M-haditec.

Benedict XVI, P. (2009). *Caritas in veritate*. Libreria Editrice Vaticana.

Berger, W. (2009). Die Finanzmarktkrise – Ergebnis einer Fehlkonstruktion, die wir korrigieren können. *Hintergrund, 1/2009*, 13–16.

Bergoglio, J. M. (2015). *Laudato si'*. Libreria Editrice Vaticana.

Betz, T. (2010). Keynes' Bancor-Plan reloaded Eine moderne Idee kommt endlich in Mode. *Zeitschrift für Sozialökonomie, 47*(164/165), 38–49.

Bible-21st century King James Version. (1994). *The Holy Bible*. Deuel Enterprises.

Bible-NIV. (2015). *NIV Zondervan study Bible: New International Version*. Zondervan.

Biblia Sacra Vulgata. (2007). *Biblia sacra iuxta Vulgatam versionem*. Deutsche Bibelgesellschaft.

BME, B. f. W. u. E. (2014). *German Mittelstand: Motor der deutschen Wirtschaft – Zahlen udn Fakten zu den deutschen mittelständischen Unternehmen*. BME.

Bockting, B. (2020). Baden-Württemberg führt Bodenwertsteuer ein! *Fairconomy, 3*, 3.

Böhm, F. (1961). *Demokratie und ökonomische Macht – Institut für ausländisches und internationales Wirtschaftsrecht, der J.-W.-Goethe-Universität, Frankfurt – Kartelle und Monopole im modernen Recht*. Müller.

Böhm, F. (2010, Reprint from 1933). *Wettbewerb und Monopolkampf: eine Untersuchung zur Frage des wirtschaftlichen Kampfrechts und zur Frage der rechtlichen Struktur der geltenden Wirtschaftsordnung*. Nomos.

Boyd, T. (2021). *Native perspectives: Land ownership.* [Online] Available at: https://www.grandcanyontrust.org/blog/native-perspectives-land-ownership. Accessed 1.10.2022.

Buiter, W. (2005). Overcoming the zero bound on nominal interest rates: Gesell's currency carry tax vs. Eisler's parallel virtual currency. *IEEP, 2*, 189–200.

Clark, J. (1940). Toward a concept of workable competition. *The American Economic Review, 30*, 241–256.

Common, M., & Stagl, S. (2008). *Introducción a la Economía Ecológica*. Reverté.

Conrad, J., & Clark, C. (1995). *Natural resource economics – Notes and problems* (5th ed.). Cambridge University Press.

Daly, H., & Farley, J. (2011). *Ecological economics – Principles and applications* (2nd ed.). Island Press.

Donoso, P. J., & Romero, J. E. (2020). Towards a new forest model for Chile: Managing forest ecosystems to increase their social, ecological and economic benefits. In F. Fuders & P. Donoso (Eds.), *Ecological economic and socio ecological strategies for forest conservation* (pp. 159–170). Cham: Springer Nature.

Dornbusch, R., Fischer, S., & Startz, R. (2009). *Macroeconomía* (10th ed.). McGraw-Hill.

Du Plessis, W. (2011). African indigenous land rights in a private ownership paradigm. *Potchefstroom Electronic Law Journal, 14*(7), 45–261.

Emmerich, V. (2001). *Kartellrecht*. Munich: Beck.

Etymology Dictionary. (2022). *Online Etymology Dictionary.* [Online] Available at: https://www.etymonline.com/word/dominion#:~:text=mid%2D15c.%2C%20%22lordship,%22house%2C%20household%22. Accessed 24.1.2022.

Felber, C. (2009). *Kooperation statt Konkurrenz – 10 Schritte aus der Krise*. Wien: Deuticke / Zsolnay Verlag.

Fisher, I. (1932). *Booms and depressions – Some first principles*. Adelphi.

Fisher, I. (1933). The debt deflation theory of great depressions. *Econometrica, 1*(4), 339–357.

Fisher, I. (1936a). *100% Money* (2nd ed.). Adelphi.

Fisher, I. (1936b). 100% Money and the public debt. *Economic Forum*, 406–420.

Fisher, I. (1937). *Feste Währung – Zur Entwicklungsgeschichte der Idee*. Lautenbach.

Fisher, I. (1948). *Feste Währung – Illusion und Wirlichkeit*. Freiheit-Verlag.

Fisher, I., & Cohrssen, H. (1933). *Stamp scrip*. Adelphi.

Frank, R. (2015). *Microeconomics and behaviour*. McGraw-Hill/Irwin.

Friedman, M. (1966). *Capitalismo y Libertad*. Ediciones RIALP.

Fuders, F. (2008). *Die Wirtschaftsverfassung des Mercosur*. Duncker & Humblot.

Fuders, F. (2009a). Die natürliche Wirtschaftsordnung als Option nach dem Zusammenbruch. *Aufklärung & Kritik, 16*(2), 128–145.

Fuders, F. (2009b). *EG-Wettbewerbsrecht – Ein Kurzlehrbuch nach der Systematik der Prüfungsschemata aufgebaut und mit Fällen*. SVH-Verlag.

Fuders, F. (2009d). Zum so genannten freien Wettbewerb – Versuch einer Begriffsdefinition. *Aufklärung & Kritik, 16*(2), 146–150.

Fuders, F. (2010). Alternative concepts for a global financial system – An answer to the present world financial crisis. *Estudios Internacionales, 166*, 45–56.

Fuders, F. (2011). Zum so genannten freien Wettbewerb und dem vermeintlichen Verbot vertikaler Kartelle im EU-Wettbewerbsrecht. *Zeitschrift für Wirtschaftspolitik, 60*(3), 271–284.

Fuders, F. (2015). Auch das Bargeld braucht einen Negativzins. *Fairconomy, 11*(Special Edition: Wirtschaft Natürlich), 10.

Fuders, F. (2016a). Smarter money for smarter cities: How regional currencies can help to promote a decentralised and sustainable regional development. In *Decentralisation and regional development – Experiences and lessons from four continents over three decades* (pp. 155–185). Springer.

Fuders, F. (2016b). Negativzinsen ja – aber richtig. Issue 4/2016, pp. 6–11.

Fuders, F. (2017). Neues Geld für eine neue Ökonomie – Die Reform des Geldwesens als Voraussetzung für eine Marktwirtschaft, die den Menschen dient. In *Finanzwirtschaft in ethischer Verantwortung – Erfolgskonzepte für Social Banking und Social Finance* (pp. 121–183). Springer-Gabler.

Fuders, F. (2019). Boden, Wasser und Luft: Warum Privateigentum mit der Abschöpfung von Monopolrenten einhergehen sollte. In U. Kriese, D. Löhr, & H. Wilke (Eds.), *Grundsteuer: Zeitgemäß! – Der Reader zum Aufruf* (pp. 229–246). Kubo.

Fuders, F. (2020). *The 'tragedy of the commons' and the role of the money interest rate*. Springer.

Fuders, F., & Bockting, B. (2019). Negativzinsen werden noch nicht richtig verstanden. *Fairconomy, 15*(1/2019), 12–16.

Fuders, F., & Löhr, D. (2016). Die EZB und Silvio Gesell – Eine neue Partnerschaft? Eine Einschätzung zwei Jahre nach der bemerkenswerten Rede Benoît Cœurés. *Humane Wirtschaft, 6*(2016), 4–9.

Fuders, F., & Max-Neef, M. (2014a). Local money as solution to a capitalistic global financial crisis. In *From capitalistic to humanistc business*. Palgrave Macmillan.

Fuders, F., & Max-Neef, M. (2014b). Dinero, deuda y crisis financieras. Propuestas teórico-prácticas en pos de la sostenibilidad del sistema financiero internacional. In *Economía Internacional. Claves teórica-prácticas sobre la inserción de Latinoamérica en el mundo*. LATIn.

Fuders, F., & Michaelis, N. (2016). Marktkonzentration: Ein großes Hindernis für eine Nachhaltige Ökonomie. In *Jahrbuch Nachhaltige Ökonomie* (pp. 271–294). Metropolis.

Fuders, F., & Nowak, V. (2019). The economics of love: How a meaningful and mindful life can promote allocative efficiency and happiness. In C. Steinebach & A. Langer (Eds.), *The economics of love: How a meaningful and mindful life can enhancing resilience in youth—Mindfulness-interventions in positive environments, Springer, Cham 2019, ISBN 978* (pp. 259–277). Cham: Springer.

Fuders, F., & Pasten, R. (2020). Allocative efficiency and property rights in ecological economics: Why we need to distinguish between man-made capital and natural resources. In *Ecological economic and socio ecological strategies for forest conservation – A transdisciplinary approach focused on Chile and Brazil* (pp. 43–56). Springer.

Fuders, F., Mondaca, C., & Azungah Haruna, M. (2013). The central' banks dilemma, the inflation-deflation paradox and a new interpretation of the Kondratieff waves. *Economía (U. de los Andes), 36*, 33–66.

Fuders, F., Jacques, V., & Arellano, P. (2019). Efectos de la competencia en la educación superior en Chile – una mirada transdisciplinaria desde la economía y el derecho. In *Vances y Perspectivas para el Desarrollo Territorial* (pp. 155–171). Ediciones UACh.

George, H. (1881). *Progress and poverty – An inquiry into the cause of industrial depressions and of increase of want with increase of wealth – The Remedy* (5th ed.). Appleton.

Gesell, S. (1949). *Die natürliche Wirtschaftsordnung durch Freiland und Freigeld* (9th ed.). Rudolf Zitzmann.

Goethe, J. W. V. (1976). *Faust – Kommentiert von Erich Trunz*. C.H. Beck.

Herdzina, K. (1999). *Wettbewerbspolitik*. Lucius & Lucius.

Hicks, J. (1973). *The theory of wages* (2nd ed.). Macmillan.

Hudson, M. (2018). *... and forgive them their debts: Lending, foreclosure and redemption from bronze age finance to the jubilee year.* ISLET.

Humphrey, T. M. (1991). *Nonneutrality of money in classical monetary thought Thomas M. Humphrey*. Federal Reserve Bank of Richmond.

Investment, D. (2014). *Verstehen die Kunden, was geschieht, kommt es zum Bank Run – Interview mit Thorsten Polleit.* [Online] Available at: http://www.dasinvestment.com/negativzinsen-teil-3-verstehen-die-kunden-was-geschieht-kommt-es-zum-bank-run/. Accessed 7.1.2022.

Kaiser, R. (1987). Chief Seattle's speech(es): American origins and European reception – Almost a detective story. In B. Swann & A. Krupat (Eds.), *Recovering the word: Essays on native American literature* (pp. 497–536). University of California Press.

Kant, I. (1868). Metaphysik der Sitten – Zweiter Theil. Metaphysische Anfangsgründe der Tugendlehre. In *Immanuel Kant's sämmtliche Werke – In chronologischer Reihenfolge, Bd. VII* (pp. 175–304). Leopold Voss.

Kantzenbach, E. (1967). *Die Funktionsfähigkeit des Wettbewerbs* (2nd ed.). Vandenhoeck u. Ruprecht.

Keynes, J. M. (1936). *General theory of employment, interest and money*. Hartcourt.

Keynes, J. M. (1980). Proposals for an international currency union. In D. Moggridge (Ed.), *The collected writings of John Maynard Keynes, volume XXV, activities 1940-1944* (pp. 42–66). Cambridge University Press.

Kriese, U., Löhr, D., & Wilke, H. (2019). *Grundsteuer Zeitgemäß – Der Reader zum Aufruf*. Kubo.

Latouche, S. (2003). *Would the West actually be happier with less? The world downscaled*. Le Monde Diplomatique, p. 1.

Latouche, S. (2009). *Farewell to Growth*. Cambridge: Polity Press.

Larroulet, C., & Mochón, F. (2003). *Economía*. McGraw-Hill.

Lietaer, B., Arnsperger, C., Goerner, S., & Brunnhuber, S. (2012). *Money and sustainability: The missing link – A report from the Club of Rome – EU Chapter to the Finance Watch and the World Business Academy*. Triarchy Press.

Lohmann, L. (2006). Carbon trading – A critical conversation on climate change, privatisation and power. *Dev. Dialogue, 48*, 1–159.

Löhr, D. (2012). The euthanasia of the rentier – A way toward a steady-state economy? *Ecological Economics, 84*, 232–239.

Löhr, D. (2013). *Prinzip Rentenökonomie – Wenn Eigentum zu Diebstahl wird*. Metropolis.

Löhr, D. (2018). Boden – die verkannte Umverteilungsmaschine. *Zeitschrift für Sozialökonomie (ZfSÖ)*, 3–19.

Lord, T. (2019). *Elberfelder Studienbibel*. SCM R. Brockhaus.

Mankiw, G., & Taylor, M. (2014). *Economics*. Cengage Learning.

Manuschevich, D. (2020). Land Use as a socio-ecological system: developing a transdisciplinary approach to studies of land use change in South-Central Chile. In F. Fuders & P. Donoso (Eds.), *Ecological economic and socio ecological strategies for forest conservation* (pp. 79–97). Springer.

Marx, K. (1872). *Das Kapital – Kritik der politischen Oekonomie. Erster Band, Buch I: Der Produktionsprozess des Kapitals* (4th ed.). Meissner.

Max-Neef, M. (1992). *From the outside looking in – Experiences in Barefood economics* (2nd ed.). Zed.

Max-Neef, M., Elizalde, A., & Hopenhayn, M. (1991). *Human scale development – Conception, application and further reflections*. Apex.

Mestmäcker, E. J. (1984). *Der verwaltete Wettbewerb – Eine vergleichende Untersuchung über den Schutz von Freiheit und Lauterkeit im Wettbewerbsrecht*. Mohr.

Mestmäcker, E. J., & Schweitzer, H. (2004). *Europäisches Wettbewerbsrecht*. Beck.

Mill, J. S. (1885). *Principles of political economy*. Appleton.

Naess, A. (1973). The shallow and the deep, long-range ecology movement. A summary. *Inquiry, 16*(1–4), 95–100.

Namatama, N., Fuders, F., & Monti, G. (2018). *An empirical examination of the relationship between GDP growth rate and development in 100 countries and a city*. Universidad Austral de Chile.

Neher, P. (1990). *Natural resource economics – Conservation and exploitation*. Cambridge University Press.

Pastén, R., Nazal, N., & Fuders, F. (2020). Subsidizing green deserts in Southern Chile: Between fast growth and sustainability of forest management. In F. Fuders & P. Donoso (Eds.), *Ecological economic and socio ecological strategies for forest conservation* (pp. 59–78). Springer.

Paul VI, P. (1971). *Octogesima adveniens – APOSTOLIC LETTER*. Libreria Editrice Vaticana.

Pfeiffer, W. (1997). *Etymologisches Wörterbuch des Deutschen*. Dtv.

Posch, F. (1963). Bauer und Grundherrschaft. In F. Posch (Ed.), *Das Bauerntum in der Steiermark. Ein geschichtlicher Überblick* (pp. 11–16). Historischer Verein für Steiermark.

Riechmann, J. (2010). *Entre la cantera y el jardín*. La oveja roja.

Riedel, F. (2014). Rettet die Sparer! Negative Zinsen sind das Warnsignal einer kranken Wirtschaft. Es ist Zeit für radikale Lösungen. *F.A.Z.*, Issue 25.12.2014.

Salazar Gomez, L. L., et al. (2021). Long memory estimation in a non-Gaussian bivariate process. *Applied Mathematics and Computation, 420*, 126871. https://doi.org/10.1016/j.amc.2021.126871

Schachtschneider, K. A. (1994). *Res publica res populi – Grundlegung einer Allgemeinen Republiklehre. Ein Beitrag zur Freiheits-, Rechts- und Staatslehre*. Duncker & Humblot.

Schachtschneider, K. A. (2006). *Prinzipien des Rechtsstaates*. Duncker & Humblot.

Schachtschneider, K. A. (2007). *Freiheit in der Republik*. Duncker & Humblot.

Seltmann, T. (2010). Umlaufsicherung von Banknoten – Wie sich die Einführung der Liquiditätsgebühr bei Geldscheinen systemkonform realisieren lässt. *Humane Wirtschaft, 2*(2010), 10–16.

Sener, U. (2016). *Die Neutralitätstheorie des Geldes Ein kritischer Überblick* (2nd ed.). Universitätsverlag Potsdam.

Smith, A. (1952). *An inquiry into the nature and causes of the weallth of nations*. W. Benton.

Smithonian Institution. (2023. *Native knowledge 360 – The different views of land*. [Online] Available at: https://americanindian.si.edu/nk360/manhattan/different-views-land/different-views-land.cshtml. Accessed 10.4.2023.

Soddy, F. (1933a, Reprint 1961). *Wealth, virtual wealth and debt – The solution of the economic paradox*. Briton.

Soddy, F. (1933b). *Money versus man*. Dutton.

Soddy, F. (1934). *The role of money – What it should be contrasted with what it has become*. Routledge.

Sommer, M. (2009). Acuicultura Insostenible en Chile. *Revista electrónica de Veterinaria, 10*(3), 1–23.

Steiner, R., 1979. *Die soziale Grundforderung unserer Zeit – In geänderter Zeitlage, Zwölf Vorträge, gehalten in Dornach und Bern v. 29. 11. bis 21.12.1918*. Dornach: Rudolf-Steiner-Verlag.

Stiglitz, J. E. (1998). *Macroeconomía* (2nd ed.). Ariel.

Suhr, D. (1988). *Alterndes Geld – Das Konzept Rudolf Steiners aus geldtheoretischer Sicht*. Novalis Verlag.

Suhr, D. (1989). *The capitalistic cost-benefit structure of money – An analysis of money's structural nonneutrality and its effects on the economy*. Springer.

Tolksdorf, M. (1994). *Dynamischer Wettbewerb*. Springer Gabler.

Traverso, M. (2015). *Economía Consciente – La transformación espritual de la economía que comienza por uno mimso*. Buenos Aires: Kier.

Tremel, F. (1963). Die bäuerlichen Besitzrechte und Besitzverhältnisse. In F. Posch (Ed.), *Das Bauerntum in der Steiermark. Ein geschichtlicher Überblick* (pp. 18–24). Historischer Verein für Steiermark.

Wicksell, K. (1936). *Interest and prices – A study of the causes regulating the value of money*. Macmilllan.

Wikipedia. (2022). *Entry "Grundsteuer: Zeitgemäß!"* [Online] Available at: https://de.wikipedia.org/wiki/Grundsteuer:_Zeitgemäß!. Accessed 22.1.2022.

Woodford, M. (2003). *Interest and prices: Foundations of a theory of monetary policy*. Princeton University Press.

Chapter 11
Gesell's Solution to Achieve the SDGs

11.1 SDG 1 and 2 (No Poverty and Zero Hunger)

The main reason for poverty and hunger is the unequal distribution of wealth. Our planet provides plenty resources for all people to live with dignity and without hunger. Even if world population doubled, our planet would still be capable of providing sufficient food for everyone. In countries where people are dying of hunger, we find very rich people who live extravagantly. Acknowledging that there are other reasons that contribute to income inequality, such as market concentration, this book stresses that the main cause of the increasingly unequal distribution of wealth is the gap between the exponentially growing financial assets and debts, as Fig. 8.1 clearly depicts. While an ever-smaller number of people receive ever more interest, an ever-larger number of people are increasingly indebted. The only way to sustainably guarantee a fair distribution of wealth is to eliminate the root of inequality, that is to change the unnatural design of money. We can observe this wealth gap not only within countries, but also between countries, since some countries are creditors of others (Creutz, 2018). Thus, our financial system also (exponentially) increases inequality between rich countries and poor ones.

11.2 SDGs 3, 4 and 6 (Good Health and Wellbeing, Quality Education, Clean Water and Sanitation)

Many ideas exist to improve access to healthcare, to make cities heathier and increase wellbeing, using for example a "systems approach" (Gatzweiler, 2020; Gatzweiler et al., 2020) or simply through smart development planning (see, e.g., Dick et al., 2016). However, sanitary and healthcare facilities are expensive. Wellbeing requires access to basic goods, a clean environment, good infrastructure

© The Author(s), under exclusive license to Springer Nature Switzerland AG 2023
F. Fuders, *How to Fulfil the UN Sustainability Goals*,
https://doi.org/10.1007/978-3-031-37768-6_11

and intelligent development planning, all of which cost money. The same is true for high-quality education as well as clean water and a decent sanitary infrastructure (Bárcena et al., 2018). Hence, good health, wellbeing and education are strongly connected to the fair distribution of wealth.

11.3 SDG 5 and 10 (Gender Equality and Reduced Inequalities)

11.3.1 Existing Inequalities Are Being Exacerbated

The underlying cause of most types of inequality is the inequality of wealth distribution. The main reason for unequal wealth distribution is our unnatural monetary system as it has been functioning for about the past 700 years. Income inequalities that exist for other reasons are exacerbated by the current monetary system. Viewed from this perspective, although gender inequality does not appear to be primarily caused by the monetary system, it may be exacerbated by it.

11.3.2 Gender Inequality

There may even be a direct link between the monetary system and gender inequality. Consider the following example: In 2006 I participated in a conference at UN ECLAC (Economic Commission for Latin America and the Caribbean) co-organized by the German foundations Konrad-Adenauer-Stiftung y Friedrich-Ebert-Stiftung. Under the title "The social market economy: new challenges—new solutions"[1] the conference analyzed how Latin America can learn from Europe, especially Germany. The tenor of the conference was that in Latin America growth and prosperity could be enhanced by more women entering the labor markets. Women, I learned at the conference, are production factors. Women working as housewives and thus not available to the labor market could be considered a misallocation of production factors. I must admit that I was shocked. A woman (or man) serving their family at home is, according to this logic, an idle production factor, as if childcare and managing the household are of no value to society.[2]

Maybe we can reinterpret the mantra-like demand for gender equality in the light of the pressure and the need to grow economically emanating from our current monetary system. Also, the greater the number of people entering the labor market,

[1] "La economía social de mercado: nuevos desafíos – nuevas soluciones", ECLAC, Santiago de Chile, 7–8.11.2006.

[2] That managing a household is precisely where the word "economics" (from Greek *oikonomia*) stems from, seems to have been forgotten.

the lower is the price of labor, *ceteris paribus*, which enhances the corporate profits and economic growth.

11.3.3 Gesell's Mütterrente: Another Extraordinary and Modern Idea

Silvio Gesell (1949) proposed to use part of the revenue the government receives from land rent to pay a mother's stipend (Mütterrente). This very modern idea was intended to economically compensate the work that women performed at home, particularly childcare and education. This work is extremely valuable for society and should not be viewed as a misallocation of production factors (Azkarraga et al., 2011). Gesell's proposal of a mother's stipend[3] would not only make women (or men) less dependent on their partners but would also provide an incentive to not abandon the important and valuable work of childcare in order to engage in economically remunerative work.

11.4 SDG 7 (Affordable and Clean Energy)

That clean energy is affordable is not only key for protecting the environment but also from a socio-economic point of view. In this context, some speak of "energy democracy" (Stephens, 2019). Technologies for clean and almost free energy exist; today they are much cheaper than in the past. For those who live far from the cities, off grid solar, wind or hydropower can even be cheaper than connecting to the conventional power grid (Debus et al., 2016). Even though these technologies are less expensive now they may still be too expensive for the poor. Again, the fair distribution of wealth can be considered crucial for broad use of clean energy.

11.5 SDG 8 (Decent Work and Economic Growth)

11.5.1 Economic Growth Cannot Be Sustainable

As discussed in Chap. 7, what this goal tries to address is unemployment and precarious working conditions that are usually associated with high unemployment rates. Steady economic growth is incompatible with the idea of sustainability since in the long run it **must** result in increased use of natural resources, a fact that Aristotle pointed out 2300 years ago. To reach sustainability through economic growth is like trying to square a circle (Azkarraga & Fuders, 2019). Constant

---

[3] Today we could of course find a different, gender-neutral term, e.g., "childcare stipend".

economic growth is the main reason for unsustainability in *stricto sensu*. The idea that economic growth is necessary to decrease unemployment is based on the empirical observation that unemployment and GDP growth are negatively correlated (so-called Okun's law). However, the dichotomy whereby we must increase output if we want to prevent unemployment from rising is rooted in the growth imperative inherent in our monetary system. As explained above, if a company does not manage to grow, it will try to reduce costs, and since labor is usually the most important cost category, there is constant pressure to either grow or to replace humans by machines, i.e., to improve productivity. Since the laws of physics tell us that 'sustainable growth' cannot exist, the only way to avoid unemployment and precarious working conditions is to reform the monetary system so that we can break the linkage between growth and unemployment.

11.5.2 Universal Basic Income

Next to reforming the monetary system, another interesting measure would be a so-called universal basic income or basic income guarantee. The universal basic income (UBI) is a concept that was proposed in the USA as early as the 1960s (Theobald, 1967; Pechman & Timpane, 1975) and has been the subject of controversial debate in many countries around the world for many years now. It could be implemented parallelly to a Gesellian money reform. A UBI would be paid to **every citizen** regardless of their economic circumstances. Without going into the question of whether and how such a UBI could be financed, there are advantages of such an income guarantee that usually are not associated with it.

(a) Happiness and allocative efficiency

A UBI could substantially improve allocative efficiency. This is because if we were no longer forced to work in order to survive, we would be able to prioritize happiness and fulfillment over monetary gain when seeing employment. In other words, we could even manage to make our hobby our profession. Our profession would then be our vocation. If people work with joy, they will probably be more efficient than if they work only for the sake earning their daily bread. For example, a chef who loves to cook and cooks with joy will cook better and could therefore be even more economically successful. This is because an activity gives us joy when we can make use of our individual talents, which are different for each person. When we use our talents and gifts, we feel joy in our work, because we feel that our time is being used wisely—in microeconomics we would say "efficiently"—, we feel that we are serving the common good (Fuders & Nowak, 2019).

If all people used their talents, this would not only serve themselves since they would perceive inner satisfaction, happiness or "self-actualization" as Maslow (2005) called it. Rather, it would also serve the realization of neighborly love since when everyone works at what they do best, everyone benefits.

In economics, this would be called *allocative efficiency*, the best form of resource allocation. On the contrary, it is arguable that we harm our neighbors if we perform a job which we are not suited to and thus prevent someone who would be better suited to it from performing that employment. Unfortunately, the use of talents or "creative and synergic potential" as Manfred Max-Neef et al. (1991: 38) call it, is ignored in most economic models. Economic models only speak of production factors, and the only reason to use these is it to obtain a profit measurable in monetary units. In Sect. 17.3 I will reflect a bit more on what could be called an 'economy of neighborly love'.

(b) Raising the wage level

A UBI would also be conducive to a fair and humane economic order because the wage level of lower income groups would rise. This is because without the pressure of having to work and to accept any job whatsoever to survive, people would not have to offer themselves on the labor market at ever lower real wages, as is the case for the middle and lower wage groups not only in Chile, but also in Germany and in many other countries around the world. This is a significant difference between UBI and classic unemployment assistance. Classic social aid legislation does not take pressure off the unemployed to find jobs. So, even though this group receives social assistance, their labor is still offered on the market, putting downward pressure on wages. On the contrary, if we had a UBI, many would decide to apply only for jobs they feel especially suited for. As a result, the supply of labor would fall, shifting the labor supply curve to the left, which would cause wages to rise. So, wages would rise, especially in the low-wage segment of the labor market. The following graph depicts this dynamic.

The labor supply curve S in Fig. 11.1 corresponds to the labor supply in the low-wage sector without a UBI. A low real wage level (W/P) is established at the intersection with the labor demand curve D. The supply curve has an almost exponential shape because people with no or poor education crowd the labor market and at a low wage level. This is especially true in countries where no social welfare or unemployment aid is being provided. But it is also true in countries such as Germany where abundant social aid programs exist, but beneficiaries are obliged to offer themselves on the labor market to receive the aid (if the wages do not reach the level of social assistance, the difference is topped up by the social aid program). In other words, if people are obliged to offer themselves on the labor market in order to receive the aid, wages (W/P in Fig. 11.1) will not rise. A UBI would presumably cause the labor supply curve to follow a less exponential course, resulting in a higher wage (W′/P in Fig. 11.1), because many would not enter the labor market at all and might instead dedicate themselves to charity or take on honorary positions. For example, I know a person who after retiring started working voluntarily at a mama cancer self-help group. For years she dedicated at least the same number of hours weekly to charity work as she did to her job. This is because she loved what she was doing.

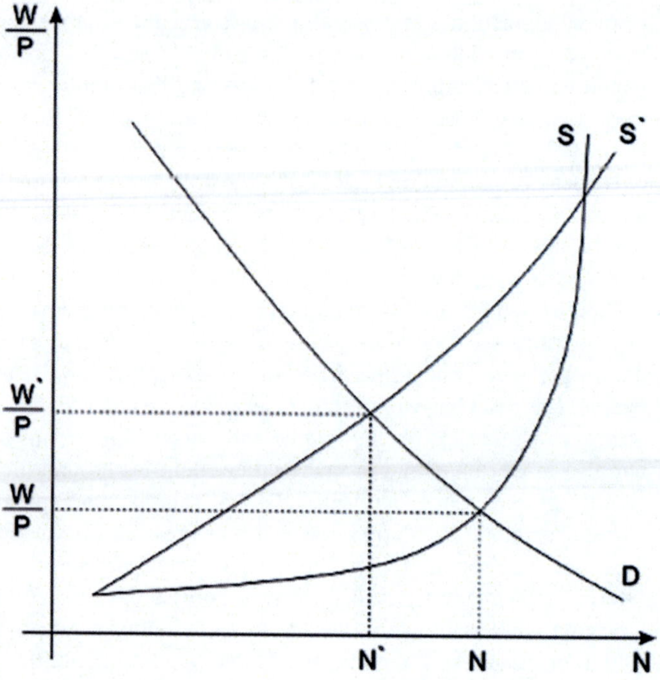

Fig. 11.1 Real wages on the labor market in the low-wage sector, without (S) and with (S′) UBI. (Source: own work)

(c) Aspects of freedom and justice

By raising the wage level, especially in the low-wage sector, a UBI would also contribute to more freedom for people. Since the time of Montesquieu, Rousseau and Kant we understand that freedom is inseparable from equality (see, e.g., Schachtschneider, 1994, 2007). However, equality does not mean that everyone should have the same amount of wealth, but that no one should be so rich that they can buy someone, nor so poor that they have to sell themselves, as already Rousseau, [1762] 2000) claimed.

The discussion about UBI also offers the opportunity to bring in Plato's ideas about justice. According to Plato, justice prevails if all persons dedicate themselves to the tasks which are appropriate for them (Plato, [427 BC] 2011). According to Plato, the government of a State is responsible for ensuring that each person performs their profession according to their abilities and does not interfere with other people's responsibilities. In this sense, a UBI could also promote justice since it could incentivize us to pursue a profession which allows us make use of our talents.

11.5.3 Provision of Land for Small Scale Agriculture

Countries such as Chile with abundant natural resources may be able to finance a basic income from the exploitation of those resources. Chile's revenues from copper mining amount to about 300 USD per capita per month (González & Fuders, 2013). It is arguable that this money rightly belongs to all. This does not mean that the natural resources should necessarily be exploited by publicly owned companies. Another alternative would be exploitation by private companies on behalf of the State via concessions.[4] Chile could also sell energy. Chile is a true renewable energy paradise, and energy could one day be the primary export of Chile (Fuders, 2022). So-called "green hydrogen" technology makes it possible to transform electrical energy into hydrogen, which as liquid gas can be exported to other countries. This could further enable the financing of a UBI.

Countries that do not have resources to finance a UBI could do something else. For example, many countries have much space and fertile land in public ownership that could serve for agriculture but is not seldom left idle, especially in Latin America. These countries could offer each unemployed person a few hectares of land free of rent, so that they could cultivate crops. This would give people a livelihood and enable them to provide themselves with food. And since fewer people would offer themselves on the labor market, wages would increase, at least in the lower-wage segment of the labor market.

In addition, this would also serve to augment the country's food supply security, an issue of ever-increasing importance (see, e.g., Pérez Garcés et al., 2016). Since the new farmers would presumably offer surplus production on the local markets, this would have the further positive effect that food prices would fall due to the increased supply, which would in turn be good for all, especially since this food would probably be healthier and more organic than the food produced today in industrial agriculture.

11.6 SDG 9 (Industry, Innovation and Infrastructure)

Innovations, digitalization and improving infrastructure (industry 4.0) could substantially increase efficiency and therefore decrease the input and throughput of physical material and energy required to produce a certain product or service. Nevertheless, as outlined above, the laws of thermodynamics tell us that if we keep on growing economically (increasing real GDP) we will in the long run use up all of our resources since we cannot produce something out of nothing. Even renewable

[4] Today, however, it is exploited *de facto* on behalf of the companies who sell the copper and take the revenues and pay only a relatively small royalty, which can be seen as a *de facto* privatization of Chilean copper.

resources will end up being overexploited, because if growth continues indefinitely at some point we will exploit the resource at a faster pace than it replenishes.

In addition, any physical product will end up as waste (2nd law of thermodynamics). But even if we managed to recycle everything we produce in a sort of perfect circular economy, we would still destroy and overuse our environment due to the steady increase of material and energy throughput if we keep on expanding production. However, while maintaining our current monetary system any degrowth strategies will cause unemployment and misery. Innovations must be implemented along with the introduction of natural (and neutral) money.

11.7 SDG 11 (Sustainable Cities and Communities)

We may improve sustainability in cities and communities through intelligent development planning (e.g., Dick et al., 2016) and even manage to make cities healthier (e.g., Gatzweiler et al., 2020; Gatzweiler, 2020). However, in the long run steady economic growth cannot be sustainable and will interfere with this goal. If we really aspire to sustainable cities and communities, we will first need to get rid of the obligation to grow (Fuders, 2016).

11.8 SDG 12 (Responsible Consumption and Production)

It is a widespread and modern belief that responsible consumption is the key to achieving greater sustainability (see, e.g., Latouche, 2003, 2009; Pérez, 2016; Becker, 2020; Higgs, 2021). We should recycle more and consume less. However, if we reduce consumption while leaving the current monetary system in place unemployment would increase. As explained in Sect. 4.5, the interest rate for investments in financial assets is the benchmark for the assessment of the profitability of any productive investment (e.g., Copeland et al., 2008), even in ecological economics texts (Common & Stagl, 2008). For most people gaining money is the main (or only) reason for investing.[5] Any business that does not yield a return at least as high as the interest rate will fail (see Suhr, 1988, 1989). Therefore, as we discussed in Sect. 4.5 businesses will always try to achieve a return on investment that is at least as high as what they could earn by investing in financial assets. This is why

[5] However, entrepreneurs could, of course, also start their business because they wished to make use of their talents in order to produce products that contribute to the common good, i.e., to the quality of life of people, and not only for financial gains. If all entrepreneurs understood this, we could come close to what might be called an "economy of neighborly love" (Fuders, 2017; Fuders & Nowak, 2019). In that case, entrepreneurs would care less about interest as an opportunity cost, and it might be sufficient for them to earn enough to live off it. Section 17.3 dedicates some thoughts on this idea.

businesses engage in intelligent and tricky marketing strategies to steadily increase consumption. As long as they can grow at least as fast as the interest rate, unemployment will not rise *ceteris paribus*.

If we really want more responsible consumption without increasing unemployment, we must first change our monetary system. Any regulation aiming for responsible consumption and production will, in the long run, only address the symptoms rather than the root cause as long as we are forced to constantly increase production and thus consumption. To decrease production and consumption (degrowth) while maintaining the current monetary system is not possible unless we accept high unemployment. It is surely more convenient for society to change the monetary system. In Chap. 12 we will see that this is not such a difficult task.

11.9 SDG 13, 14 and 15 (Climate Action, Life Below Water and Life On Land)

11.9.1 We Must Stop Growing

If climate change (SDG 13) originates from an increased use of fossil fuels,[6] then there are, two ways to protect our climate. We can either change to energy resources that do not use fossil fuels, such as solar, wind, hydropower, geothermal or tidal energy. The other way is to economically degrow and so consume less energy. But even if we opted for the technological solution and started using solely energy resources that do not burn fossil fuels, in the long run we still have to stop expanding production since, as pointed out in Sect. 6.5.5, for clean energy technologies to be implemented we need physical space, rivers, materials, etc. Since we do not have infinite physical space and resources—and even renewable resources are only renewable as long as we do not exploit the resource at a faster pace than it replenishes—infinite growth means that someday even with the aforementioned methods for generating clean energy we will still not be able to keep up with the demand for energy.

In order to protect the life in the oceans and on land (SDGs 14 and 15) it is crucial to keep our natural environment clean and to not overuse renewable resources. To achieve this, we need to degrow economically or, at least, not to keep expanding

[6] Note that even though this theory is widely accepted and very plausible, causality has never been proved. That we have higher levels of CO_2 in our atmosphere today is not necessarily the cause of climate change. It could simply be coincidence since climate has always been changing (Gray, 2020; Wilson et al., 2022). Even if we declare the increased CO_2 level in the atmosphere to be the main cause of the current climate change, the changed composition of the atmosphere is not necessarily caused by human activity. It may be natural. Examinations of ice boreholes in the Antarctic have shown changing levels of CO_2 in the atmosphere over the last millenniums (Gray, 2020). Volcanic activity, e.g., could also explain changes in the composition of the atmosphere (Marzoli et al., 2020; USGS, 2022).

production. Any regulations regarding the use of oceans and land such as fishing quotas, agricultural regulations or regulation of housing and urban planning will, if we keep on growing, be no more than curing the symptoms.

As explained throughout this text, the only way to get rid of the obligation to grow, i.e., to enter a *steady-state economy* (Daly, 1991, 1993) without causing unemployment and misery is to establish a monetary system that does not force the productive economy to constantly grow and in which the interest rate does not bias the profit calculations of economic agents.

11.9.2 Modern Harvesting Technology Requires Further Regulations

Technological innovations that have already been introduced in the past—possibly triggered by the growth imperative—cannot be reversed. That is to say, even if we implemented natural money and got rid of the growth imperative and the pressure that this growth imperative exerts on technological evolution, we would not return to using axes to cut trees or small artisanal fishing boats to catch fish. Indeed, when viewed through the lens of Hardin's (1968) "*Tragedy of the Commons*" metaphor, technological advancements have made huge swathes of natural resources as accessible for exploitation as the shared pasture in a village (Fuders & Pastén, 2020). In addition, our monetary system has not yet been reformed and we are therefore still obliged to grow.

This begs the question, how can we protect our natural resources while the current system keeps functioning the way it does, i.e., as long as all productive processes must grow at least at the rate of interest? Given our current monetary system and given current natural resource exploitation methods, it is indeed advisable to restrict access to originally freely accessible natural resources **combined with strict regulations** regarding their exploitation in order to achieve sustainable management.

11.9.3 Restricting Access to Publicly Owned Natural Resources

(a) Concession and quotas for 'pure' nature-made natural resources
For resources that are 'purely' nature-made, i.e., resources that are exploited or harvested by man but where man did not have to water or to sow, concessions or quotas are an adequate solution to fight over exploitation (Fuders, 2019; Fuders & Pasten, 2020).

A **concession** is a contract whereby a private firm is granted the right to extract and market resources on **behalf of the State**. The most common form is the State allows the company to sell the resource and to keep the revenue but is required to pay a royalty to the State. This way citizens benefit from natural resource exploitation. In addition, concessions allow the State to impose rules and supervise exploitation ensuring sustainable practices. The challenge with concessions is for the State to have knowledge of production costs so that the royalty can be set fairly. For example, copper mining in Chile is predominantly based on concessions. However, in Chile the companies market the copper *de facto* not on behalf of the State but for themselves, since they are subject to a relatively small royalty fee (Fuders & Pasten, 2020). Until 2007 they were not charged any royalties. The current model for concessions in Chile, allowing private companies to exploit **and sell** the natural resources while paying minimal royalties, could be viewed as a form of camouflaged privatization. In order to assure that a concession is truly a concession, and not a hidden form of privatization, concessions should incorporate three policies:

- They should expire after a set period.
- They should be assigned to as many individuals or companies as is economically viable. These two policies increase the potential **opportunity for everyone** to access and **generate competition** between concessionaires.
- Instead of a royalty, the concessionaires should be required to sell the resource in the name of the State and transfer the **total return** to the State. They would then be paid a customary return corresponding to their cost structure and production factors invested. The concessionary must disclose its books and cost structure to inspectors of the government.

For the exploitation of old-growth forests and oceans quotas can be assigned to individuals or firms. An important difference between quotas and concessions is that quotas can be assigned to a relatively large (but limited) number of individuals thereby reducing the risk of monopoly. Furthermore, in contrast to concessions, quotas do not require complicated contracts between the concessionaire and the respective government. Concessions therefore are particularly suitable for assigning exploitation rights that require high initial investments such as those that are reliant on economies of scale (e.g., mining activities). This is not the case for the exploitation of primary forests or oceans. Here quotas are preferable. Quotas should only be implemented when effective monitoring can be ensured. Strict monitoring is one of the eight design principles for efficient governance of commonly owned resources that Elinor Ostrom proposes (Ostrom, 1990). If the government manages to effectively control quotas (e.g., by fining those who exceed them), this will lead to the dominant game theoretical strategy where all cooperate (Ostrom, 1990), i.e., a situation where all stakeholders win (Fuders & Pasten, 2020).

Regarding quotas some challenging questions arise:

- Should the quotas be sold or assigned freely? If they were to be sold, how would they be priced?
- If they were free, to whom should they be given and who makes that decision?
- Should they be transferable (marketable) or not?

In the following I will try to answer these questions. For more in-depth reflections see Fuders & Pasten (2020). Any price that a government sets is unlikely to be the market equilibrium price, i.e., the *pareto-optimal* price that balances demand and supply. But a government (or any local decision maker over self-governed commonly owned resources) could assign quotas via auction where the quotas are assigned to those individuals who offer the most. In this manner price can function as mechanism to assign resources to those who value them most highly, e.g., those who perceive the highest utility from harvesting the timber or fish. In addition, selling these rights has the additional benefit that even those who do not make winning bids would benefit indirectly since the revenue collected means that the government can charge less taxes elsewhere.

Alternatively, quotas could be assigned freely to all people. People who don't wish to use them could opt to sell the quotas if they were to be transferable. Thereby, those with greater interest in using the natural resources could buy additional quotas. Here, again price would serve the function of assigning resources to those individuals who value them most highly, a desirable outcome from perspective of *allocative efficiency*.

Nevertheless, a strong argument exists for making quotas non-transferable, or at least limiting the number that any one individual or entity can control. Otherwise, those with the deepest pockets could buy up all of the quotas and monopolize what was originally a commonly owned recourse. This is not to be desired neither from the point of view of justice, nor from the point of view of allocative efficiency, since monopolistic positions per definition are not efficient.[7] In other words: allowing one (or few) individuals to obtain majority portions of exploitation rights can be seen as camouflaged privatization of the resource granting the owner(s) a dominant market position.

Regarding the question of who should be in charge of assigning exploitation rights, this should be done as much as possible on a local level; it may not always be possible, e.g., in the case of fishing rights. All four examples of successful sustainable use of commonly used resources that Elinor Ostrom describes in her masterpiece "*Governing the Commons*" have in common that the use of commonly owned resources is locally regulated. On the local level it is possible to (1) clearly define exploitation rights (boundaries), (2) allow local people to participate (collective choice agreements), to (3) monitor the compliance of established rules and (4) apply sanctions when needed. These are four of the eight design principles Ostrom (1990) proposes for a sustainable common usage of resources. Here, a strongly federal governmental structure, such as found in Switzerland or Germany, has an advantage compared to a more centralized governmental system. Nevertheless, it is also possible that strongly centralized States, such as Chile, could decentralize the allocation of quotas by the regional dependencies of the central government (*regiones*).

---

[7] In microeconomic theory a monopoly cannot be efficient since it does not fulfill the efficiency definition (price equals marginal revenue).

(b) Leasing of land and costal management areas
As an alternative to granting concessions or quotas, the government can also lease land and management areas in coastal zones of the ocean. This model is probably best suited to natural resources that are **not only harvested but also planted by man** such as agricultural and aquicultural produce (Fuders & Pasten, 2020). The pricing of leases could be based either on the amount of land/water leased or on the profits reaped. Since the land rightly belongs to all people, leases should be designed in such a way as to make them remunerative but to not produce monopoly rents. The land on which owner-occupied houses are built could be exempted from the leasing scheme. As in the case of granting concessions, leasing of land allows the State to impose rules. For example, leases might only be made available to those individuals or companies who follow sustainable forestry, agriculture or aquaculture practices.

11.9.4 Restricting Access to Privately Owned Resources

Despite *homo oeconomicus* having a vested interest in preserving future sources of income, privately owned natural resources are nevertheless subject to the risk of being over exploited because of the obligation to grow emanating from our unnatural monetary system (Fuders, 2020). In the case of privately owned natural resources, regulation and subsidies can help prevent unsustainable practices.

(a) Regulations
Although unsustainable practices are common in South American forestry (for instance, planting monocultures of non-native species and the practice of clear cutting), governments could regulate such practices through legislation. For example, in Germany the federal forest law seeks to balance the interests of forests owners with the common good (§ 1, No 3 Bundeswaldgesetz). Forest owners are required to manage their forests in a proper and sustainable manner (§ 11 Bundeswaldgesetz). The respective federal state laws in Germany build on this premise by requiring regional and local governments to establish forest use plans aimed at maintaining the environmental and recreational functions of forests as well as their biodiversity (e.g., Art. 6 Bayerisches Waldgesetz). Similar regulations enacted across Europe have resulted in marked improvements in forest management. For example, clear cutting is now very uncommon in many countries (Paletto et al., 2008; Lundmark et al., 2017).

We could ask the question of what the differences are between concessions for the exploitation of commonly owned resources and the regulation of privately owned resources? Concession contracts contain regulations regarding exploitation. On the other hand, a concession can be thought of as a private right to a public good. In this sense, concessions and quotas can be interpreted as special forms of privatization. Again, independently of the form we use to restrict access to natural resources and oblige resource owners or concessionaires to manage resources in sustainable manners, the true problem is the obligation to grow, which applies to both private and common goods (Fuders, 2020).

(b) Subsidies
An alternative to regulations requiring sustainable practices is offering subsidies to encourage desirable behavior. In Chile, for example, for the past couple years the government has been trying to incentivize sustainable practices by forest owners, in particular the planting of native species and avoidance of monocultures. However, these policies have thus far not been very successful (Manuschevich, 2020; Donoso & Romero, 2020; Pastén et al., 2020). The subsidies are hardly used. People prefer to keep planting exotic species (eucalyptus and pine tree) in monocultures. Apparently, subsidies are not sufficient to incentivize forest owners to change their planting and harvesting methods.

Before I discuss this aspect deeper, I would like to again recall from Sect. 5.7.3 that if we implemented natural (and neutral) money such subsidies would most likely not be necessary. The reason for why forest owners decide to plant fast growing eucalyptus or pine trees in monocultures is precisely because money is not neutral. The positive risk-free interest rate gives time a particularly high value ("time is money" we say) and distorts the profit calculations of investors. In a zero-interest environment it would probably be more profitable to plant slow growing native species since they command higher prices, even considering the fact that they take longer to reach maturity. The only reason why it is currently more lucrative to invest in fast growing species is that time is not neutral. We can counteract this distortion caused by the risk-free interest rate in profit calculations with subsidies to encourage sustainable management of renewable natural resources. Although we should keep in mind that such subsidies only address symptoms, not the root cause of the problem.

An **optimal subsidy** is one that makes the net present value (NPV) of all future revenues from an investment higher than the initial investment. The higher the market interest rate and the longer the period to harvest, the lower is the NPV. In other words: time becomes a more important factor in profit calculations the longer the time to harvest is. Hence, the interest rate is a particularly important factor in long-term investments such as forestry. There is abundant literature on optimal subsidies in general but relatively few regarding optimal subsidies in forestry, agriculture and aquaculture that aim to incentivize sustainable management of those resources. An example of a subsidy that did work relatively well was the Payment for Ecosystem Services (PES schemes), as employed in Brazil (Schmitt Filho & Farley, 2020; Jurema Grimm et al., 2020; Rosa Conceição & Börner, 2020).

Let me conclude this section with a comment regarding subsidies in the European Union (EU). In the EU a major part of its budget (37%) is spent subsidizing industrialized agriculture (European Commission, 2022), when this money could be used to subsidize organic food production, which is still very expensive in the EU compared to subsidized industrial food. If these subsidies were granted only for organic food production, then most agriculture in the EU would be done according to organic food production standards. Likely, this would lead to the production of more organic than non-organic food.

11.10 SDG 16 (Peace and Justice, Strong Institutions)

Wars and financial crisis have often been timely correlated; and this might not just be coincidence (Gesell, 1949; Creutz, 1993). The interest burden of sovereign debt could be paid with captured capital and resources (Creutz, 1993). On the other hand, war results in the destruction of real capital, making new economic growth and payment of interest possible and necessary (Creutz, 1993), all the while increasing the demand for loans. Not only do households need to rebuild destroyed homes and replace destroyed goods, but war also leads the governments of nations at war to increase their levels of national debt. Therefore, wars have the power to delay the collapse of the monetary system by providing an incentive to borrow on the one hand and enabling further economic growth (and thus paying of interest) on the other (Fuders, 2011; Fuders & Max-Neef, 2014).

What applies to the destruction of real capital by war also applies to the destruction that natural disasters wreak. Indeed, this is precisely what could be observed after the second strongest recorded earthquake that took place on February 27th, 2010, in Chile. Contrary to the predictions of Chilean central bank, GDP growth in 2010 was higher than it had been in the previous 16 years before the quake. The demand for all types of loans increased sharply and so did the interest rate as "price" for loans. Section 16.2 explains in more detail how earthquakes are beneficial for the financial sector.

Regarding the demand for "justice" Chap. 9 outlined why our current monetary system not only provokes in exponential growing income inequality but also why this income inequality is highly unjust.

11.11 SDG 17 (Partnerships for the Goals)

This is the only goal not directly connected to our unnatural monetary system. In fact, SDG 17 is not a goal in itself but seeks helping to achieve all other goals. However, even here a reform of our monetary system could help to promote such partnerships. Since money in the Gesellian system is not easily storable and does not yield interest, we have less incentive to work more than can directly be consumed. A monetary system where money cannot be hoarded for a prolonged period has the power to reduce greed, excessive individualism and selfish behaviors in the pursuit of getting rich at the expenses of others.

References

Azkarraga, J., & Fuders, F. (2019). La economía ecológica, paradigma de futuro. In C. A. C. Sampaio et al. (Eds.), *Ecossocioeconomias – Promovendo Territórios Sustentavéis* (pp. 35–58). edifurb.

Azkarraga, J., Max-Neef, M., Fuders, F., & Altuna, L. (2011). *La evolución sostenible II*. Lanki (Mondragon Unibertsitatea).

Bárcena, A., Cimoli, M., García-Buchaca, R., & Pérez, R. (2018). *La ineficiencia de la desigualdad*. Cepal.

Becker, J. (2020). Reasons to escape excess consumerism—And how to do it. *Forbes,* Issue Online: https://www.forbes.com/sites/joshuabecker/2020/04/30/escape-excess-consumerism/?sh=7d1c5fd72fbb

Common, M., & Stagl, S. (2008). *Introducción a la Economía Ecológica*. Reverté.

Copeland, T., Weston, J., & Shastri, K. (2008). *Finanzirungstheorie und Unternehmenspolitik. Konzepte der Kapitalmarkorientierten Unternehmensfinanzierung* (4th ed.). Pearson.

Creutz, H. (1993). *Das Geldsyndrom. Wege zu einer krisenfreien Marktwirtschaft*. Wirtschaftsverlag Langen Müller/Herbig.

Creutz, H. (2018). *Das Geldsyndrom* (4th ed.). Kubo.

Daly, H. (1991). *Steady-state economics* (2nd ed.). Island Press.

Daly, H. (1993). Steady-state economics: A new paradigm. *New Literary History, 24*(4), 811–816.

Debus, V., Fuders, F., Henriquez, C., & Erlwein, A. (2016). *La tecnología apropiada frente al cambio climático*. Universidad Austral de Chile.

Dick, E., Gaesing, K., Inkoom, D., & Kausel, T. (2016). *Decentralisation and regional development—Experiences and lessons from four continents over three decades*. Springer.

Donoso, P. J., & Romero, J. E. (2020). Towards a new forest model for Chile: Managing forest ecosystems to increase their social, ecological and economic benefits. In F. Fuders & P. Donoso (Eds.), *Ecological economic and socio ecological strategies for forest conservation* (pp. 159–170). Springer Nature.

European Commission. (2022). *The common agricultural policy at a glance.* [Online] Available at: https://agriculture.ec.europa.eu/common-agricultural-policy/cap-overview/cap-glance_en. Accessed 22.9.2022.

Fuders, F. (2011). Wie Zerstörung von Sachkapital dem Finanzsystem in die Hände spielt. *Humane Wirtschaft, 2(2011)*, 34–38.

Fuders, F. (2016). Smarter money for smarter cities: How regional currencies can help to promote a decentralised and sustainable regional development. In *Decentralisation and regional development—Experiences and lessons from four continents over three decades* (pp. 155–185). Springer.

Fuders, F. (2017). Neues Geld für eine neue Ökonomie—Die Reform des Geldwesens als Voraussetzung für eine Marktwirtschaft, die den Menschen dient. In *Finanzwirtschaft in ethischer Verantwortung—Erfolgskonzepte für Social Banking und Social Finance* (pp. 121–183). Springer-Gabler.

Fuders, F. (2019). Boden, Wasser und Luft: Warum Privateigentum mit der Abschöpfung von Monopolrenten einhergehen sollte. In U. Kriese, D. Löhr, & H. Wilke (Eds.), *Grundsteuer: Zeitgemäß!—Der Reader zum Aufruf* (pp. 229–246). Kubo.

Fuders, F. (2020). *The 'tragedy of the commons' and the role of the money interest rate*. Springer.

Fuders, F. (2022). Renewable ENERGIES at human scale: Market concentration as obstacle. In *Energías Renovables y Cambio Climático en el G20: Planteamientos desde el sur del mundo para el mundo,* Issue min. 31.30, p. https://www.youtube.com/watch?v=ov5n4bMsr90&t=456s.

Fuders, F., & Max-Neef, M. (2014). Local money as solution to a capitalistic global financial crisis. In *From capitalistic to humanistc business*. Palgrave Macmillan.

Fuders, F., & Nowak, V. (2019). The economics of love: How a meaningful and mindful life can promote allocative efficiency and happiness. In C. Steinebach & A. Langer (Eds.), *The economics of love: How a meaningful and mindful life can enhancing resilience in youth—Mindfulness-interventions in positive environments, Springer, Cham 2019, ISBN 978* (pp. 259–277). Springer.

Fuders, F., & Pasten, R. (2020). Allocative efficiency and property rights in ecological economics: Why we need to distinguish between man-made capital and natural resources. In *Ecological economic and socio ecological strategies for forest conservation—A transdisciplinary approach focused on Chile and Brazil* (pp. 43–56). Springer Nature.

Gatzweiler, F. W. (2020). *Urban health and wellbeing programme: Policy briefs. Volume 1.* Springer.

Gatzweiler, F., et al. (2020). *The little book of the HEALTH OF CITIES—From the pulse of the new urban agenda to the heart of sustainable development.* ImaginationLancaster.

Gesell, S. (1949). *Die natürliche Wirtschaftsordnung durch Freiland und Freigeld* (9th ed.). Rudolf Zitzmann.

González, F., & Fuders, F. (2013). *González Beneficio del cobre—Comparativo de ingresos provenientes del cobre entre fuentes privadas y públicas. Thesis by F. Gonzáles, supervisado by Felix Fuders.* Universidad Austral de Chile.

Gray, R. (2020). Deep Antarctic drilling will reveal climate secrets trapped in 1.5 million-year-old ice. *Horizon,* Issue 20.01.2020, pp. open access: https://ec.europa.eu/research-and-innovation/en/horizon-magazine/deep-antarctic-drilling-will-reveal-climate-secrets-trapped-15-million-year-old-ice

Hardin, G. (1968). *The tragedy of the commons* (Vol. 162, pp. 1243–1248).

Higgs, K. (2021). How the world embraced consumerism. *BBC,* Issue online: https://www.bbc.com/future/article/20210120-how-the-world-became-consumerist

Jurema Grimm, I., Tomaselli Piva, J. H., & Cioce Sampaio, C. A. (2020). Municipal private natural heritage reserves: Uses and attributions of natural protected areas in the City of Curitiba (PR). In F. Fuders & P. Donoso (Eds.), *Ecological economic and socio ecological strategies for forest conservation* (pp. 227–240). Springer Nature.

Latouche, S. (2003). Would the west actually be happier with less? The world downscaled. *Le Monde Diplomatique*, p. 1.

Latouche, S. (2009). *Farewell to growth.* Polity Press.

Lundmark, H., Josefsson, T., & Östlund, L. (2017). The introduction of modern forest management and clear-cutting in Sweden: Ridö State Forest 1832–2014. *European Journal of Forest Research, 136*(2), 269–285.

Manuschevich, D. (2020). Land use as a socio-ecological system: Developing a transdisciplinary approach to studies of land use change in South-Central Chile. In F. Fuders & P. Donoso (Eds.), *Ecological economic and socio ecological strategies for forest conservation* (pp. 79–97). Springer Nature.

Marzoli, A., et al. (2020). Deep CO_2 in the end-Triassic Central Atlantic Magmatic Province. *Nature Communications, 11*, 1670. https://doi.org/10.1038/s41467-020-15325-6

Maslow, A. (2005). *El hombre autorealizado* (16th ed.). Kairós.

Max-Neef, M., Elizalde, A., & Hopenhayn, M. (1991). *Human scale development—Conception, application and further reflections.* Apex.

Ostrom, E. (1990). *Governing the commons: The evolution of institutions for collective action.* Cambridge University Press.

Paletto, A., Sereno, C., & Furuido, H. (2008). Historical evolution of forest management in Europe and in Japan. *Bulletin of theTokyo University Forest, 119*, 25–44.

Pastén, R., Nazal, N., & Fuders, F. (2020). Subsidizing green deserts in southern chile: between fast growth and sustainability of forest management. In F. Fuders & P. Donoso (Eds.), *Ecological economic and socio ecological strategies for forest conservation* (pp. 59–78). Springer Nature.

Pechman, J. A., & Timpane, P. M. (1975). *Work incentives and income guarantees.* The Brookings Institution.

Pérez, D. (2016). Decrecimiento: Necesitar menos para vivir mejor Fuente: Emol.com – https://www.emol.com/noticias/Tendencias/2016/03/18/793800/Decrecimiento-Necesitar-menos-para-vivir-mejor.html. *Emol,* Issue online: https://www.emol.com/noticias/Tendencias/2016/03/18/793800/Decrecimiento-Necesitar-menos-para-vivir-mejor.html

Pérez Garcés, R., Espinosa Ayala, E., & Teran Varela, O. E. (2016). *Seguridad Alimentaria, Actores territoriales y Desarrollo Endógeno, Laberinto Ediciones*. Laberinto Ediciones.

Plato. (2011). Politeia. In U. Wolf (Ed.), *Platon—Sämtliche Werke Band 2* (pp. 211–537). Rowohlt.

Rosa Conceição, H. D., & Börner, J. (2020). Understanding adoption and design of incentive-based forest conservation policies: A case study of the SISA program in acre, Brazil. In F. Fuders & P. Donoso (Eds.), *Ecological economic and socio ecological strategies for forest conservation* (pp. 241–258). Springer Nature.

Rousseau, J.-J. (2000). *Vom Gesellschaftsvertrag oder Prinzipien des Staatsrechts*. Akad.-Verlag.

Schachtschneider, K. A. (1994). *Res publica res populi - Grundlegung einer Allgemeinen Republiklehre. Ein Beitrag zur Freiheits-, Rechts- und Staatslehre*. Duncker & Humblot.

Schachtschneider, K. A. (2007). *Freiheit in der Republik*. Duncker & Humblot.

Schmitt Filho, A., & Farley, J. (2020). Transdisciplinary case study approaches to the ecological restoration of rainforest ecosystems. In F. Fuders & P. Donoso (Eds.), *Ecological economic and socio ecological strategies for forest conservation* (pp. 185–212). Springer Nature.

Stephens, J. C. (2019). Energy democracy: Redistributing power to the people through renewable transformation. *Environment, 61*(2), 4–13.

Suhr, D. (1988). *Alterndes Geld—Das Konzept Rudolf Steiners aus geldtheoretischer Sicht*. Novalis Verlag.

Suhr, D. (1989). *The capitalistic cost-benefit structure of money—An analysis of money's structural nonneutrality and its effects on the economy*. Springer.

Theobald, R. (1967). *The guaranteed income—Next step in socioeconomic evolution?* Doubleday.

USGS. (2022). *Volcanoes can affect climate.* [Online] Available at: https://www.usgs.gov/programs/VHP/volcanoes-can-affect-climate. Accessed 2.9.2022.

Wilson, D. J., De Flierd, T. D. F., McKay, R., & Naish, T. R. (2022). Chapter 10—Pleistocene Antarctic climate variability: Ice sheet, ocean and climate interactions. In F. Florindo, M. Siegert, L. De Santis, & T. Naish (Eds.), *Antarctic climate evolution* (pp. 523–621). Elsevier.

Chapter 12
How to Implement Gesell's Solution in Today's World

12.1 Governmentally Imposed Hoarding Tax

12.1.1 Pure E-money System

The simplest way to introduce a hoarding fee without stamping money is to simply abolish cash, i.e., physical coins and bills (Buiter, 2009; Fuders, 2009, 2010a; Rogoff, 2014; Assenmacher & Krogstrup, 2018, 2021). Instead of sticking tax stamps on physical bills, the hoarding fee would be charged on bank balances. Hoarding cash would no longer be possible at all, and all money would then be subject to devaluation through bank deposit fees.

If the hoarding tax is set high enough savers would be willing to offer their deposits for loans free of interest in order to avoid the hoarding fee. Banks will continue to serve an intermediary function for the interest-free loans as envisioned by Gesell (1949: 269 f.). The hoarding tax could be set by governmental decree (maybe on recommendation by the central banks). In order to protect small savers from devaluation, accounts might be exempted from deposit fees up to a certain amount (Fuders, 2009, 2010a). Note that the money charged, as with any tax, would be handed over to the government. It would not represent revenue for the banks.

12.1.2 Cash Cards for Anonymous Payments

In the above described pure e-money system banks register when, where and at what price we buy goods and services. Many people might prefer to be able to go shopping anonymously. To address this preference, we could consider a pure electronic but anonymous system with prepaid cash cards such as those used by German savings banks (Sparkassen) in the mid 1990s, and which today are offered

© The Author(s), under exclusive license to Springer Nature Switzerland AG 2023
F. Fuders, *How to Fulfil the UN Sustainability Goals*,
https://doi.org/10.1007/978-3-031-37768-6_12

internationally by a couple of firms as "prepaid debit cards", "prepaid credit cards" or "gift cards" (see, e.g., Nerdwallet, 2022). Such "cash-on-a-chip cards" could be used for anonymous electronic payments (Buiter, 2009: 24; Mikalauskas, 2020). In contrast to conventional cash-on-a-chip cards for our purpose a hoarding fee would be applied to the balances.

A similar system could be introduced as a mobile phone app. Even though any app on a mobile phone can be tracked to a specific phone number and thus is not *per se* anonymous, still, engineers I have spoken with have ensured me that anonymous payments could in theory be made possible via cellphone apps.

12.1.3 What About the Security of Such Cards?

Some might question the security of prepaid cash cards, since nowadays it is relatively easy to clone cards and/or credit them with monetary units using hacking software. The same applies to mobile phone payment apps. Today, the acceptance of electronic payment systems is still high because card providers usually reimburse losses due to fraud with relatively low administrative burdens. However, it would also be possible to implement a sophisticated system that makes cloning or illegal charging of balances impossible, using the fact that each card has a unique number. The POS (Point of Sale) card reading equipment could be connected to a centralized computer that registers and controls all transactions.[1]

12.2 Effective Central Bank Negative Interest Rate Policy

12.2.1 Gesell's Hoarding Fee Is a Negative Interest Rate

One way to practically apply Gesell's proposal would be through an **effective negative central bank interest rate** policy (Buiter & Panigirtzoglou, 1999, 2003; Buiter, 2005, 2009; Fuders, 2010a, 2015, 2016b; Fuders & Max-Neef, 2014). Indeed, the Gesellian hoarding fee can be viewed as a negative interest rate. For the first time ever in the history of central banks, the Swedish central bank introduced negative rates for deposits of commercial banks at the central bank in June 2009. The ECB followed in June 2014. A member of the ECB's executive board had compared this policy to Gesell's proposal (Cœuré, 2014, 2016).

However, what the ECB has done so far (2014—March 2022) was not the application of Gesell's proposal (Fuders & Löhr, 2014, 2016; Fuders, 2015, 2016a, b; Fuders & Louge, 2020). The reason is that the negative rate is not being applied to

[1] For example, if the sum on the card in not the sum registered in the system, then the card will be blocked.

cash (physical paper money), which is an obstacle to the implementation of an effective negative interest policy. Cash "by design" (Assenmacher & Krogstrup, 2018: 4) yields a nominal rate of zero. Cash therefore can be seen as the "zero bound" (Goodfriend, 2000) or "zero lower bound" (Buiter, 2009; Agarwal & Kimball, 2015) of monetary policy. Any policy that causes commercial banks to charge negative rates on deposits brings along the incentive to withdraw deposits and hoard money as cash (Assenmacher & Krogstrup, 2018, 2021).

12.2.2 The Problem of the "Zero Lower Bound"

Most commercial banks have been "shielding" (Agarwal & Kimball, 2015: 14) their customers from the negative rates, i.e., they do not pass them on to their customers. As we discussed above, we need **either positive or negative incentives** as a reward to abstain from liquidity. Since at this point negative interest does not apply to cash, it is always possible to 'invest' in cash and thus circumvent negative rates on deposits. This hampers the efficacy of negative rates. So, until March 2022 in the western economies we neither had clear positive nor negative rates. Therefore, money did not flow adequately, and the central banks, especially ECB and US Fed tried to compensate by printing more and more money nowadays called *Quantitative Easing*.

The policy of so-called *Quantitate Easing* can be seen as the contrary to what Gesell proposed (see Fuders & Löhr, 2016; Fuders, 2016b, 2017, 2020a, b). Gesell proposed money that circulates and not the printing of ever more of the 'sticky' money that does not flow well and will later be stored in ever larger amounts under the mattress. That "even the mildest form of quantitative easing" is inconsistent with a negative interest rate policy was shown by Buiter (2009: 6 ff.) in a sophisticated mathematical model. No amount of quantitative easing can make up for the inability of the monetary authorities to set negative nominal interest rates (Buiter, 2009), not to mention the risk that all of the newly created money could one day appear on the consumer goods markets causing incontrollable hyperinflation (Fuders, 2008, 2009, 2010a).

To recapitulate, we need a monetary policy where negative interest rates are not only charged on deposits of commercial banks at the central bank, but also apply to cash (meanwhile cash exists). Different ways have been discussed to achieve this (Buiter & Panigirtzoglou, 1999, 2003; Goodfriend, 2000; Buiter, 2005, 2009; Seltmann, 2010; Agarwal & Kimball, 2015; Assenmacher & Krogstrup, 2018, 2021), which will be discussed in the following paragraphs.

12.2.3 Pure E-money System

What is true for a government-imposed hoarding tax is also true for negative interest rates set by the central bank. Negative rates would theoretically be as easily implemented as positive ones (Buiter, 2009; Assenmacher & Krogstrup, 2018, 2021), although it may not be legal in all jurisdictions (Buiter, 2009). In a pure electronic system bank customers cannot circumvent negative rates (or deposit fees) by shifting into cash. A potential problem is that banks might try to 'shield' their customers by not passing along the negative rates to them if they fear that they would otherwise lose customers, as was seen in the eurozone in the years 2014–2022.

12.2.4 *Cash Cards for Anonymous Payments*

In the very same way as above described for a government-imposed hoarding tax, prepaid cash cards could be used to make anonymous payments possible. The only difference would be that a hoarding tax is now not the domain of the government but rather that of the central bank. Maybe new laws would be necessary to require all prepaid card providers to apply the negative interest rate on the cash card set by the central bank.

In this context an interesting question arises. Could banks not try to attract new customers by offering a deposit fee on demand deposits that is less than the negative rate charged on cards? The answer is yes, they could. In order to avoid this, the government must ensure, probably via a new law, that banks charge the same negative interest rate on all demand deposits and cash cards. This would have the advantage that market distortions would be avoided, since all banks would be required by law to charge the same rate, as set by the central bank.

12.2.5 *Cash Cards and Stamp Scrip Money Working in Tandem*

Even though electronic and bank-account-based means of payment could be implemented easily and without any "significant transitional problems, permanent efficiency loss or adverse distributional implications" (Buiter, 2009: 24), we might still consider a system of prepaid cash cards and stamp scrip money circulating in tandem. Why is this desirable? One advantage is anonymity. Indeed, 100% anonymity can only be assured with paper money. In addition, older people who are not familiar with cards or other electronic applications would want to pay with cash. Another important advantage of stamp scrip paper money is that in the event of an electricity outage we would still be able to pay.

To maintain physical money is "not beyond the realm of the possible" points out Buiter (2005: 192). Instead of stamps on bills, we could consider bills that expire (Buiter, 2009; Seltmann, 2010), such as were used in a regional money system in Köthen, Germany (see above Fig. 10.1). After expiration each bill could be exchanged for a new one by paying a hoarding fee (the fee can be set higher the longer the bill is expired).

12.2.6 The Eisler Proposal

Eisler (1932) proposed an interesting idea to make hoarding cash unattractive. Basically, the idea is to have two currencies: one for ordinary transactions, to perform as a unit of account to express prices of goods and services, and one (virtual) currency to pay taxes, fines or other payments required by the government. Also, government bonds could be denominated in this virtual currency. The government would establish an exchange rate between the two currencies. Unbundling the function of money as medium of exchange (payment) and the function of money as unit of account (Buiter, 2005) is quite difficult. At the end, money serves as medium of exchange precisely because of its function as unit of account, which makes possible to assign a numeric value to goods and services. Gesell's solution is better and more practically than Eisler's proposal (Buiter, 2005).

12.2.7 Decoupling Cash from Bank Deposits

Quite similar to Eisler's proposal is the idea of decoupling cash from electronic money. Ruchir Argawal and Miles Kimball (2015) worked this idea out in detail, and Katrin Assenmacher and Signe Krogstrup (2018) improved it. Decoupling cash (physical bills and coins) from electronic money would entail a split of the domestic money supply into two different local currencies: cash and central-bank reserves (Assenmacher & Krogstrup, 2018). A "time-varying paper currency deposit fee" for commercial bank deposits at the central bank can be used to eliminate the option to circumvent the negative rates by withdrawing and storing paper currency (Agarwal & Kimball, 2015: 7 ff.). Basically, the idea is that a negative interest rate can be accompanied by a deposit fee for cash that ensures the value of cash and the value of funds in electronic accounts will move in tandem (Agarwal & Kimball, 2015). The fee is time-varying, which means that it is higher the longer the cash is held before depositing it at the central bank.

Commercial banks, in turn, would then pass on these fees to their customers when physical cash is deposited. In practice this would effectively establish a sort of "exchange rate" (Agarwal & Kimball, 2015: 15; Assenmacher & Krogstrup, 2021: 70) between electronic money and cash. But not all banks would necessarily charge these fees, or not to all customers, or not the same fee to all customers, as was the case with the negative interest policy of the ECB or the central banks of Sweden, Denmark and Switzerland until recently. Agarwal & Kimball (2015: 15 f.) see this problem of the "pass-through" of the "exchange rate" between electronic and paper money. Therefore, ways would have to be found to ensure that the electronic currency becomes the relevant unit of account (Assenmacher & Krogstrup, 2018), the "real thing" as Agarwal and Kimball (2015) put it.

In this context another question arises. Would this exchange rate truly prevent people from hoarding physical cash at home? Or would it, on the contrary, create an incentive to never deposit cash in order to avoid the "exchange rate" between physical money and electronic money? Supermarkets could set two prices for all products, an e-money price and a cash price, the cash price being higher. This could in turn accelerate the phasing out of cash. The good thing is that such a policy falls within the existing statutes of central banks and would require no new regulations (Agarwal & Kimball, 2015; Assenmacher & Krogstrup, 2018, 2021).

12.3 Some Further Reflections

12.3.1 Stamp Scrip Has Advantages Over a Pure Electronic System

Even though it seems somewhat cumbersome, physical stamp scrip money as proposed by Gesell (1949) has several advantages. Firstly, we would not need to supervise commercial banks in order to apply the hoarding fee. Any bank that receives stamp scrip from its customers would automatically charge the cost of the stamps to its customers. On the contrary, any pure electronic system (using cash cards or otherwise) would necessitate new regulations requiring banks to charge the hoarding fee uniformly to all customers (and all cash cards). Market distortions are happening right now (as of spring 2022) in the eurozone due to the ECB's negative interest rate policy being passed on to customers by some banks but not by others. Secondly the velocity of money circulation could be easily regulated by increasing or lowering the stamp fee. In all likelihood, this tool would be sufficient by itself, thereby making other tools of monetary policy unnecessary. And thirdly, it makes truly anonymous payments possible.

12.3.2 Hoarding Fee Only for Demand Deposits

At this point it is of the utmost importance to highlight the fact that the hoarding fee/negative rate would only be charged on cash and demand deposits (checkbook accounts, current accounts), not on time deposits. Remember that the idea is to create an incentive for holders of excess money to lend that money. In practice this means that bank customers could convert their demand deposits into time deposits in order to avoid being charged the hoarding fee. If the hoarding fee was applied to time deposits, people would have no incentive to do so, hence money would not circulate via loans. In a Gesellian natural money system people would only be able to avoid the "rusting" of money (Gesell, 1949: 238) by spending or lending. The latter means that we save the money on time deposits where, in contrast to today's system, the saver will not receive any interest, but the other way around will be freed from paying the hoarding fee.

12.3.3 How to Improve Confidence

Any monetary system requires trust. People will trust the system if money is broadly accepted. Tying the currency to gold can certainly help to establish such trust. However, a gold-backed currency does not solve the problems that arise from the unnatural storability of money. Sect. 13.4.6 provides more detailed reflections on this aspect.

Nevertheless, pegging **natural money** to gold is possible. Since bank deposits would diminish with time so would the amount of gold for which those deposits could be exchanged. In the case of physical stamp scrip money, the quantity of gold for which each unit of money could be exchanged would diminish with the amount of time the money is held. In a natural money system, backing money with gold would help build confidence without being counterproductive.

Granted, there is a risk that people will exchange Gesell money for gold coins and pay directly with gold. Therefore, rather than gold, Gesell money could be backed by land, i.e., land owned by the public sector could be used to back the Gesell currency. For example, money could be exchanged for land at set prices. Since land can't be used to buy groceries in the supermarket, money would be indispensable. It would be impossible to avoid the hoarding fee since money is always needed for daily expenses. Only larger amounts could be exchanged for land. However, larger savings will probably not exist in a Gesell money system since the hoarding fee creates a disincentive to save the money. A similar proposal was made by Bernard Lietaer (2002).

12.3.4 Gesell's Proposal Is Still Not Properly Understood

I am, as the reader may already have noticed, convinced that without the 'zero lower bound', i.e., by implementing Gesell's natural money we would not need (or at least there would be much less need for) monetary policy since there would be no inflation and no deflation, no financial crises, and no (or much less pronounced) economic cycles. Furthermore, we could remedy ever-increasing income inequality as well as the growth imperative and thus the constant and ever-increasing overexploitation of natural resources. True sustainable development can only be achieved by implementing Gesell's money.

Unfortunately, these advantages of Gesell's proposal are still not properly understood. On the contrary, the current discussion (until March 2022) about how to break through the zero lower bound in monetary policy is focused on promoting economic growth. Business as usual. As Assenmacher & Krogstrup (2021: 99) clearly state, it is all about "conventional monetary policy tools without the constraints of the lower bound to stabilize the economy". Agarwal & Kimball (2015: 3) put it this way: "The zero lower bound has proved to be a serious obstacle for monetary policy, as shown by the recent efforts of central banks to stimulate economic growth in the wake of the Global Financial Crisis". It is all about "economic stimulus" (Agarwal & Kimball, 2015: 1).

12.3.5 *Outlook*

Negative central bank interest rates are easy to implement. They do not require new laws or the alteration of existing laws. If negative rates (the rate commercial banks pay for deposits at the central bank) are deep enough, they could already be effective because banks then will have to trespass these negative rates to their clients (Fuders, 2016b; Fuders & Löhr, 2016). Holding cash to avoid negative rates is a feasible solution only for relatively small sums. Larger sums will still be deposited at banks for security reason (Fuders, 2009, 2010a). Cash "is bulky and costly to store, it can be stolen, and it can be destroyed by fires and other man-made or natural disasters", as Buiter (2009: 11) correctly put it. Even though cash in the current system is not subject to an official hoarding fee, carrying costs still apply (Buiter, 2009). That is to say, it might not even be necessary to apply the hoarding fee to cash (or cash cards) or to switch to a completely electronic system.

However, we can only speculate at this point since no central bank has pursued a deep negative rate policy thus far. In any case, a better solution would be to enact laws that require all commercial banks to apply the same negative rate (hoarding fee) on all demand deposits and cash cards and to apply the same hoarding fee on cash. This would enable us to avoid market distortions caused by some banks passing on negative rates to their customers while other banks 'shield' their customers from negative rates.

12.4 Discussion of Some Myths on Negative Rates

12.4.1 *Do Negative Interest Rates Expropriate Holders of Money?*

Money is supposed to be a means to vastly expand the scope of possible exchanges beyond what would be possible via barter. Money should therefore be thought of as a public good and should be treated as such. In many countries it is therefore unlawful to destroy money. For example, in the US burning money is punishable by up to 10 years in prison (Murse, 2019). Similar laws exist in many countries. Money as a means of exchange is probably the most essential tool that we have as a modern, highly collaborative society. Cash must therefore not serve as a long-term store of value. Instead, it must circulate in the economy in order to fulfill its primary purpose of facilitating the exchange of real goods and services. A positive interest rate creates an incentive to not hoard money under the mattress, despite our *liquidity-preference* (Keynes, 1936: 168), but to lend it out instead. Remember that interest is a *reward for parting with liquidity* (Keynes, 1936: 167). We save our money in banks, which in turn lend it to others.

However, positive interest rates cause serious problems, many of which are not recognized in economics. In particular, as discussed throughout this book, they are the main reason for the ever-increasing and above all unfair unequal distribution of wealth, and for the destruction of the environment, which ultimately results from the compulsion of the economy to constant growth. Silvio Gesell's idea was to design money in such a way that it will circulate by itself without the need for (positive) interest as a reward for foregoing liquidity. To this end, he proposed banknotes that slowly lose their value if they are hoarded under the mattress for too long. And as outlined above, this hoarding fee can be viewed as negative interest on cash.

Negative interest does not represent expropriation as some believe (see. e.g., Seibel, 2021; Tagesspiegel, 2021; Ferber, 2021), but should rather be viewed as a measure to prevent the hoarding of money from harming the economy. It creates an incentive to keep money flowing. As a result, Gesell's hoarding fee could also be referred to as a "holding fee", or "carrying costs" as Keynes referred to them when discussing Gesell's proposal, since a holding fee is *de facto* a counterbalance to the Keynesian liquidity premium (the reward for parting with liquidity).

Expropriation is "the act of a government claiming privately owned property" (Kenton, 2020). However, as discussed money should be viewed as public, not private property and thus negative interest does not fit the definition of expropriation. Another definition of expropriation is the confiscation of tangible or intangible assets (see, e.g., Gabler Wirtschaftslexikon, 2022). However, money is not an asset, but a medium of exchange which allows assets to be exchanged more efficiently. Again, I highlight the conflation of money with capital, i.e., with goods that have intrinsic value. Money is neither an apple nor a pair of shoes nor any other real good. In classical economics, capital is the third production factor alongside human labor and land. It refers to machines or other means of production. Unlike machines or apples, money has no intrinsic value, it does no work and has no offspring as stated by Aristotle, who rightly considered interest from lending money to be "unnatural" (Aristotle, [384 BC] 1995: 1258b).

As discussed above, it is not unnatural to make profits via real economic activity but earning money from lending money is unnatural. It may be reasonable for owners of real, tangible goods to earn rent for lending them out since they temporarily relinquish the use of those goods and incur depreciation. However, it seems downright "insane", as Rudof Steiner (1979: 50) once put it, to rent out vouchers that represent goods instead of the goods themselves. Here it is opportune to once again recall the metaphor Keynes used to explain that money has no intrinsic value: saying that money has intrinsic value is the same as saying that a movie ticket is equal to the film the ticket entitles you to watch (Keynes, 1983).

12.4.2 Do Negative Interest Rates Cheat Savers?

Another erroneous and repeatedly expressed view on negative rates is that savers are being cheated out of their interest (see, e.g., Schieritz & Uchatius, 2014; Hüfner, 2015). Again, this view results from the misunderstanding that money is an asset or capital. This view fails to account for the fact that it is the savers themselves who ultimately pay the interest. This is true even if they are not in debt, since interest is included in the prices of all products (Creutz, 1993; Kennedy, 1990). In addition (as discussed in Chap. 9), interest is an unfair form of income, because it is not based on the effort of the money lender, but rather on the effort of the borrower who employs the money productively and assumes the entrepreneurial risk (Fuders, 2017). Only what has been previously produced can be bought. It is the goods and services available in an economy that give money its value. Therefore, as we noted in Chap. 9, the money holder should thank borrowers rather than charge interest, since it is borrowers who give value (purchasing power) to their money through productive investment (Fuders, 2017).

Those who confuse the medium of exchange with useful real goods will hardly recognize the problems inherent in our financial system, neither the economic nor the moral-ethical ones (Fuders, 2010b, 2017). It is not that negative rates are cheating savers but the other way around: a **positive lending rate cheats the borrower**, because the borrower must pay interest for something that itself has no value but only represents value (Fuders, 2010b, 2016b). Imagine, for example, that at a car rental company instead of the rented vehicle, you were only given the vehicle registration document that represents the car. Most people would probably feel cheated (see already Sect. 9.4).

12.4.3 Could Natural Money Cause Efficiency Losses?

Critics claim that with negative interest rates the central bank would manipulate the money market, and this could cause efficiency losses. However, as explained above (Sect. 5.5), the central bank is really only a trend follower, not a trend setter. In fact, interest should be negative for cash holdings and zero for loans. The reason why it is not negative for cash holdings is precisely because the unnatural design of our money makes hoarding free of charge. Savers have the ability to avoid negative rates charged by banks by saving money under the mattress. Without this unnatural zero lower bound, markets would automatically impose negative rates on cash holdings. Leading economists such as Larry Summers and Ben Bernanke have long emphasized that there is a global surplus of savings over investment.

Economic efficiency losses have therefore occurred for centuries due to the unnatural design of our money. We can intuitively understand it this way: any economic activities have been **artificially exacerbated** due to interest. Throughout this book we have used the metaphor of money as the blood of our economy, blood that has a design flaw, it is too sticky and needs constant injections of drugs to make it more liquid. This drug is interest. Gesell (1949) used a similar metaphor: Markets are like a road on which goods are exchanged, and money is a gate where we have to pay a toll, which is interest. Entrepreneurial undertakings which would otherwise be economically viable, but which cannot also pay the "toll" in addition, will not be undertaken or will fail.

In this context it is worthwhile to mention an interesting conference on entrepreneurship I participated in at Villa Gesell in Argentina,[2] a village founded by Silvio Gesell's son Carlos Gesell. The conference was dedicated to him. People as well as organizers had no idea who Carlos' father was. Carlos Gesell had been a very successful entrepreneur and forestry engineer. He invented a revolutionary method to fructify deserts and planted a forest on sand dunes using this special technique.[3] Villa Gesell is today one of the most famous seaside resorts in Argentina. Listening to the talks at the conference where successful local entrepreneurs shared their experiences, I identified a common thread running through all the talks: The most difficult part of being a successful entrepreneur in Argentina (and probably everywhere) is first to access credit and to then pay the interest on debt. The conference participants were surprised when they learned from my speech that Carlos' father Silvio was as much a genius as his son and had invented a natural form of money that would make entrepreneurship easier. Such money would make getting a loan easier, and the lack of interest would make paying a loan back easier as well.

With natural (and neutral) money we can finally achieve the free-market equilibrium discussed in textbooks. The fact that in the current system equilibrium is not stable is recognized in textbooks and mentioned as one reason for market failure. However, it is not recognized in mainstream economic theory that interest is a main reason for why this equilibrium is not achievable.

The unnatural (Aristotle, [384 BC] 1995) storability of money, is probably the most significant, but at the same time least recognized reason for market failure. An effective negative interest rate, which is also levied on cash, would remove the unnatural special position of cash vis-à-vis real goods. Only then would money be truly neutral, as is described in textbooks. With such a truly neutral money, the wealth-enhancing effects of the model of perfect competition could finally be realized, something that is not possible today.

---

[2] "1er congreso de Emprendedores: Carlos I. Gesell: emprendedor de emprendedores", organized by Rotary Club Villa Gesell (Argentina) 04/05/2019.

[3] This technique is similarly revolutionary to the monetary theory of Carlos' father Silvio. It consists of enveloping the roots of seedlings with bags in order to prevent rainwater from draining away too quickly. Once the trees have grown and developed stronger roots, they can reach ground water themselves while the netting of the roots serves as a reservoir. Carlos planted thousands of trees during his lifetime. This method is not very well-known even among forestry specialists, similar to Silvio Gesell's ideas to reform our monetary system are so far not well-known among economists. Regarding SDG 15 (life on land) a method to fructify deserts and/or to bring desertification to a halt is of the utmost relevance and we could learn a lot by studying Carlos Gesell's technique.

12.4.4 *Could Natural Money Affect Productivity, and Thus Economic Growth and Unemployment?*

In the current system, money circulation is achieved via a positive interest rate. In addition, money circulation is stimulated by inflation, which in turn results from interest, which causes the money supply to grow and gradually decouple from the real economy. That a positive interest rate causes the money supply to grow is unfortunately not understood in monetary and price theory. But it is precisely for this reason that it is so important for the European Central Bank to achieve its inflation target of at least 2%. Otherwise (with no inflation) the circulation of money will slow down even more. If money does not circulate sufficiently via interest, as was the case for the last 14 years in the western economies due to low interest rates combined with cash that can be easily stored under the mattress, then the central bank finds itself in a dilemma (Fuders, 2011; Fuders et al., 2013) and is compelled to print money (nowadays called *Quantitative Easing*) to keep money circulating adequately. However, achieving the circulation of money via positive interest rates and inflation is destructive. Its consequences are, among other things, the constant compulsion for growth, economic cycles—which can be interpreted as major or minor financial crises (Fuders et al., 2013)—as well as the redistribution of wealth from the majority of hard-working people to the interest rentiers, which in the long run causes an ever-growing concentration of wealth, which in turn threatens democracy.

Empirically, it has been observed that, contrary to conventional theory (see e.g., Stiglitz, 1998), interest and economic growth move synchronously in the medium term, i.e., a higher level of interest coincides with higher GDP growth rates, and *vice versa* (Fig. 5.1; see also Fuders et al., 2013; Fuders, 2020a). And whenever the GDP growth rate fell, unemployment increased. But since perpetual economic growth destroys the natural environment in the long run it cannot be considered a desirable goal. The only way to sustainably remove this growth imperative from the economy and transition to a post-growth (or degrowth) economy without creating unemployment and poverty is to let the interest rate fall to zero (see also Sect. 13.14.1). A *steady-state economy* (Daly, 1991, 1993) will only exist in an economic system where there is no positive interest on credit.

The criticism sometimes put forth that low and negative interest rates unnecessarily keep unproductive "zombie banks" and "zombie companies" alive can hardly be surpassed in its sweeping cynicism (Fuders & Bockting, 2019b). Even in wealthy economies such as Germany, one in seven companies cannot cover its borrowing costs with its profits (Hillmer, 2021). If interest rates were to rise, many companies would immediately become insolvent. In many of these companies, meaningful work is being done; small and medium-sized companies that produce goods and services that help people satisfy their fundamental human needs. In southern Europe, a rise in interest rates would have even more dramatic consequences (Fuders & Bockting, 2019a).

12.4.5 Could Natural Money Threaten Pensions and Retirement Plans?

When it is argued that low interest rates threaten people's retirement savings (OECD, 2015; Barrington, 2022), what is overlooked is the fact that for every penny of interest we receive, we have usually paid three times that amount in "hidden" interest. Interest is part of the price of any product or service. Even if we don't borrow money, we still pay interest whenever we buy goods or services (Kennedy, 1990; Creutz, 1993). Only if we have a large amount of savings (like a million USD[4]) do we come out ahead in terms of interest. Otherwise, we are a net payer of interest.

The negative interest rates (hoarding fee) in the Gesellian natural money system apply only to demand deposits (current accounts) and cash. Time deposits would not be subject to the negative rates, but they would not earn interest either. Since inflation would most probably not exist, saving for retirement or major purchases in the future would still be possible on time deposits. Those who hope to earn positive returns and who are willing to accept the corresponding risk can, in the same way as today, invest in the real economy, buy stocks, participate in companies, cooperatives, buy shares of investment funds, etc., as outlined in Sect. 10.2.4.

Today we have positive interest rates on deposits but also inflation. The higher the interest rates, the higher the inflation. This is because inflation is a secondary phenomenon of positive interest. Interest makes deposits and thus the money supply (composed of cash plus bank deposits) grow. Since interest causes deposits to grow not in a linear fashion but exponentially (first very slowly and then faster and faster), the money supply gradually starts to decouple from the real economy and thus causes inflation (more money in relation to products and services). As previously discussed, it can be empirically shown that inflation and interest move synchronously in the long run (Fuders et al., 2013). In a Gesellian natural money system, i.e., negative rates which abolish the zero lower bound, there will no longer be inflation.

12.4.6 Could Natural Money Put the Stability of the Monetary System at Risk?

Concerns about bank profitability, and thus the stability of the financial system, play a major role among those who criticize negative interest rates. However, effective negative interest rates, which also include cash as proposed by Gesell, reduce the risk of bank failures rather than increasing it. With the ECB's current policy (as of spring 2022—negative rates for holdings of commercial bank funds at the central

[4]We can only approximate this figure. The exact amount where we cross over to receiving more interest than we pay depends on what we consume and with how much debt these products and services have been produced.

bank but no negative interest on cash), it is mainly small banks that are passing on the cost to customers and charging negative interest on demand and even savings deposits, while larger banks or those that have a solid private equity cushion as 'buffer' can hold out longer without having to charge customers.

Many customers of smaller banks therefore have an incentive to withdraw their funds and move them to larger banks that do not charge negative interest on deposits. The resulting market distortion increases the probability of bank failures. Charging interest on bank deposits only but not on cash therefore does not only not have the desired effect of keeping money circulating well but may on the contrary increase the risk of bank failures which could in turn lead to a dangerous bank run. If, on the other hand, a negative interest rate (hoarding fee) was also applied to cash as proposed by Gesell, banks would automatically charge this interest rate on deposits as well. This would have the advantage that the negative interest rate would be the same for all banks, thus avoiding market distortions.

In this context it is crucial to again point out that in addition to cash, only demand deposits (current accounts) should be subject to this charge if we want to ensure that money is loaned out and continues to circulate. After all, people are supposed to spend their money, invest it, or lend it out interest-free. The incentive for savers to ask banks to lend their money interest-free is precisely to avoid the negative rate that is charged on demand deposits.[5]

Regarding the stability of financial markets, there is another argument why with natural money the system would be more stable and not less: Normally, banks pay higher interest rates the longer customers promise not to withdraw their money. For banks, such time deposits have the advantage that they cannot be withdrawn unexpectedly, which helps them plan lending. In recent years, however, less and less money has been deposited on a long-term basis. On the contrary, long-term financial deposits have sharply fallen while the volume of overnight deposits has literally exploded. This, in turn, has led to a further massive increase in systemic risk (Fuders & Bockting, 2019a). The introduction of effective negative interest rates that include cash holdings would encourage bank customers to use time deposits in order to avoid the negative interest rate charged on cash and demand deposits. This would significantly reduce the banks' maturity transformation risk, which would greatly improve the stability of our banking system (Fuders & Bockting, 2019a). Also, negative rates on cash would remove an incentive for bank customers to withdraw deposits on a large scale, which *per se* strengthens the stability of the system (Assenmacher & Krogstrup, 2018, 2021).

Regarding the argument that negative rates would affect the profitability of banks (see, e.g., Agarwal & Kimball, 2015), which could in turn negatively impact the stability of the system, remember that banks can charge commissions for the brokerage of loans. Therefore, they would not necessarily earn less. There is no

[5] In Gesell's natural money system, banks will lend customers' money only if the customer expressly designates it for this purpose, in order to avoid the hoarding fee (the negative interest rate) that would otherwise apply. Such customers would agree to abstain from withdrawing their money for a fixed period (months or years) and in turn would not be charged negative interest.

evidence that bank profitability has been negatively affected by negative interest rates *per se* (Ball et al., 2016; Assenmacher & Krogstrup, 2018, 2021).

12.4.7 Could Natural Money Cause 'Capital' Flight?

An interesting question is whether negative rates could cause capital flight. The answer is: it would not if the negative rates are implemented as Gesell proposed. In a Gesellian natural money system, in which the negative interest rate is applied to cash too, the feared outflow of funds to other economies, where positive interest rates may still be available, would in fact be very limited. On the foreign exchange markets, a currency to which a hoarding fee is applied would be priced with a corresponding discount compared to foreign currencies that are not subject to a negative rate (hoarding fee). In other words, the hoarding fee would also apply when the currency is exchanged for a foreign currency, which would reduce the incentive to move money abroad. In the euro zone, at present (spring 2022), many banks charge a negative rate on demand deposits but not on cash. Money can therefore be easily converted into other currencies without paying the hoarding fee. As long as the negative rate does not apply to cash, it can be avoided by 'investing' in cash or converting cash in foreign currency and investing funds abroad. For this reason, too, cash should be subject to a money holding fee (Fuders & Bockting, 2019a).

Since a currency with a negative rate applied to it is less prone to become a target of speculative investment, nor does it serve as reserve currency for foreign central banks, it will not be over evaluated. The overvaluation of a currency can reduces exports and thereby harm the economic development of a country. Conversely, the devaluation of a currency can have positive effects on exports. In this context, Rogoff (2016) assumes that the foreign trade of a country that introduces deep negative interest rates would benefit greatly from a devaluation on the foreign exchange markets.

Domestically, no loss of purchasing power is to be expected. As mentioned, there would be no inflation in a natural money system. The money supply would not be forced to grow via interest-driven money creation (see Sect. 5.2) and would therefore not decouple from the real economy. We would finally achieve true monetary neutrality. It can be expected that this would lead to a stable currency and an economy without (or with greatly reduced) economic cycles (Fuders et al., 2013; Fuders, 2016a, b; Fuders & Bockting, 2019a).

12.4.8 Could Natural Money Cause a Loss of Confidence in Central Banks?

Another interesting question is whether negative interest rates could cause a loss of confidence in central banks and the monetary system, creating the potential for severe social and political crises. The financial system has already been

Fig. 12.1 M1 (blue) and M3 (yellow). (Source: ECB, 2023)

experiencing a loss of confidence for many years due to the bursting of the 'dotcom bubble' in the year 2000, and the 2008 global financial 'subprime' crisis. The ECB tried to counteract the threat of deflation by printing money. That is, it tried to offset the decreased money multiplier (caused by low lending) by expanding the central bank money supply M0. As a result, the money supply M0 and M1 grew at a faster pace than the money supply M3 (see Fig. 12.1). In other words, if the ECB had not printed money, the money supply M3, which includes the money supply M0 and M1, would have hardly grown at all.

However, only a small part of the huge amount of fresh money made it to the real economy. Instead, it was (and is) being invested in the stock and real estate markets, or even in gold or Bitcoin, where it is causing ever larger speculative bubbles. If these bubbles burst and all of that money is suddenly available for transactions in the real economy at some point, hyperinflation cannot be ruled out (Fuders, 2016b; Fuders & Bockting, 2019a) and the latest developments (as of June 2022) in industrialized economies seems to confirm this risk.

In the euro zone, the ECB's quantitative easing measures, despite relieving the financial system in the short and medium term, are ultimately ineffective because people can avoid negative interest on demand deposits by hoarding cash at home. In fact, this is probably precisely why more and more cash is being hoarded. The volume of euro bills "in circulation" (inverted comas because these bills are not really in circulation but in ever-higher sums stored under the mattress) has increased more than fivefold since 2002, to more than 25 trillion euros (ECB, 2022). Accordingly, each inhabitant of the euro zone should have an average of 3540 euros in his or her wallet. Since cash is secure central bank money that is basically not subject to any risk of default, the German Bundesbank estimates that around 20% of it is hoarded in Germany alone. 70% of the banknotes issued by the Bundesbank are either in other euro countries (20%) or outside the euro zone (50%). Only around 10% is used for payment purposes (Thiele, 2017).

So, instead of fixing the flaw in the monetary system and giving money, especially cash, a built-in motivation to circulate, the markets, especially in Europe, the US and Japan, are flooded with ever more of the money that does not serve its purpose well. This is the opposite of what Gesell proposed. Gesell proposed that money should circulate, not that hoarding should be counteracted by printing money, i.e., artificially expanding the money supply. This policy will hardly stop the loss of confidence.

12.4.9 *Could Natural Money Cause a Shift into Gold and Cryptocurrencies?*

Keynes (1936) discussed the possibility that a Gesellian natural money system might people lead to invest in alternative coins such as gold and silver and in this way avoid the Gesellian hoarding tax. Nowadays many would also include Bitcoin or other cryptocurrencies, which are supposedly safe (on this see Fuders & Schreiber-Martens, 2018). However, it would not be necessary to invest in gold in order to avoid the hoarding fee on cash and demand deposits. In a Gesellian natural money system time deposits would not be subject to the hoarding fee, and since inflation would not exist, it would be possible to save money without loss.

And anyone who wants to invest their money in gold or cryptocurrencies is welcome to do so. This would not affect the stability of the monetary system. As in the current system, in a natural money system too there is nothing bad about buying gold coins, bitcoins, jewels, valuable paintings or whatsoever one would believe to be a good investment. This would not affect the amount of natural money in circulation. If Fritz sells an ounce of gold to Hans for, say, 1000 natural money units, these 1000 natural money units move from Hans to Fritz. The natural money that Hans no longer possesses is now possessed by Fritz. The total money supply has not changed. People can buy as much gold (or any other asset) as they want. The functioning of the natural money system would not be affected.

It would become problematic, however, if gold were to become established as currency. Then people who didn't have gold would have to borrow it. And since gold is not perishable and is easily storable (there is no negative interest rate attached to it), those who owned gold would only lend it out in exchange for interest. Since gold would be necessary to buy essential goods like food and shelter, owners of gold would have the same ability to extort those who do not possess gold as holders of money have under the current system (Fuders, 2014; Fuders & Schreiber-Martens, 2018).

But why gold should change its status from an asset into a currency? What we use as currency depends strongly on what the government defines as legal tender for the payment of taxes. If the legal tender were natural money, then people would pay taxes with natural money. There would be nothing to prevent them from building up reserves in gold or other assets. And there would be nothing bad about it either.

12.4.10 Could Natural Money Lead to the Surveillance State?

One might ask if a Gesellian natural money system would lead to total surveillance of the consumer. Others might ask if a natural money system would not require a great deal of effort in terms of communication, organization and technology.

This question regarding the control of citizens is not so far-fetched. Indeed, if one wanted to implement effective negative interest rates by abolishing cash, then all payments would be recorded on bank statements. Abolishing cash would be an easy solution for the implementation of negative rates as explained above, but one casualty would be our freedom to pay anonymously. The consequence would indeed be total surveillance. For this very reason, it would be better to maintain cash, but to find a way to apply negative interest to it, which would be technically feasible as outlined above.

The effort required to apply negative interest to cash would be significant even if a high-tech alternative to physical stamps was adopted (Buiter & Panigirtzoglou, 2003), but the effort would be worthwhile, considering it would allow us to solve a variety of problems including inflation, deflation, financial crises, increasing wealth inequality and the growth imperative. And we do not necessarily need bills and coins in order to preserve anonymity. Prepaid cash cards or electronic bills would preserve anonymous payments while negative interest could be easily applied. In fact, administering such a system might even be cheaper than our current system.

References

Agarwal, R., & Kimball, M. (2015). *Breaking through the zero lower bound* (IMF Working Paper WP/15/224). IMF.

Aristotle. (1995). Politik 1st Book. In *Aristoteles Philosophische Schriften in sechs Bänden. Band 4*. Felix Meiner Verlag.

Assenmacher, K., & Krogstrup, S. (2018). *Monetary policy with negative interest rates: Decoupling cash from electronic money* (IMF Working Paper WP/18/191). IMF.

Assenmacher, K., & Krogstrup, S. (2021). Monetary policy with negative interest rates: De-linking cash from digital money. *International Journal of Central Banking, 17*(1), 67–106.

Ball, L., Gagnon, J., Honohan, P., & Krogstrup, S. (2016). *What else can central banks do?* International Center for Monetary and Banking Studies.

Barrington, R. (2022). *Lower interest rates could ruin your retirement.* [Online] Available at: https://www.moneyrates.com/investment/how-lower-interest-rates-ruin-retirement.htm. Accessed 22 June 2022.

Buiter, W. (2005). Overcoming the zero bound on nominal interest rates: Gesell's currency carry tax vs. Eisler's parallel virtual currency. *IEEP, 2*, 189–200.

Buiter, W. (2009). *Negative nominal interest rates: Three ways to overcome the zero lower bound* (NBER working paper series). NBER.

Buiter, W., & Panigirtzoglou, N. (1999). *Liquidity traps – Gesell's solution*. Bank of England.

Buiter, W., & Panigirtzoglou, N. (2003). Overcoming the zero bound on nominal interest rates with negative interest on currency: Gesell's solution. *The Economic Journal, 113*, 723–746.

Cœuré, B. (2014). *Life below zero – Learning about negative interest rates.* [Online] Available at: https://www.bis.org/review/r140911a.htm. Accessed 25 May 2022.

Cœuré, B. (2016). *Assessing the implications of negative interest rates.* [Online] Available at: https://www.ecb.europa.eu/press/key/date/2016/html/sp160728.en.html. Accessed 6 July 2022.

Creutz, H. (1993). *Das Geldsyndrom. Wege zu einer krisenfreien Marktwirtschaft*. Wirtschaftsverlag Langen Müller/Herbig.

Daly, H. (1991). *Steady-state economics* (2nd ed.). Island Press.

Daly, H. (1993). Steady-state economics: A new paradigm. *New Literary History, 24*(4), 811–816.

ECB. (2022). *Banknotes and coins circulation.* [Online] Available at: https://www.ecb.europa.eu/stats/policy_and_exchange_rates/banknotes+coins/circulation/html/index.en.html. Accessed 17 Oct 2022.

ECB. (2023). *Monetary developments in the euro area: February 2023.* [Online] Available at: https://www.ecb.europa.eu/press/pr/stats/md/html/ecb.md2302~952f671d72.en.html. Accessed 4 Apr 2023.

Eisler, R. (1932). *Stable money: the remedy for the economic world crisis : a programme of financial reconstruction for the International conference 1933*. London: The Search.

Ferber, M. (2021). *Geld aus dem Nichts – doch gratis ist es nicht. Die Sparer bezahlen die Zeche.* [Online] Available at: https://www.nzz.ch/meinung/enteignung-der-sparer-geld-aus-dem-nichts-ist-nicht-gratis-ld.1648941?reduced=true. Accessed 21 June 2022.

Fuders, F. (2008). Eine Wirtschaftsordnung ohne Zinsen. *Zeit-Fragen, 17*(23), 3–4.

Fuders, F. (2009). Die natürliche Wirtschaftsordnung als Option nach dem Zusammenbruch. *Aufklärung & Kritik, 16*(2), 128–145.

Fuders, F. (2010a). Alternative concepts for a global financial system – An answer to the present world financial crisis. *Estudios Internacionales, 166*, 45–56.

Fuders, F. (2010b). Warum der Zins auch moralisch nicht zu rechtfertigen ist. *Humane Wirtschaft, 2*, 26–29.

Fuders, F. (2011). Wie Zerstörung von Sachkapital dem Finanzsystem in die Hände spielt. *Humane Wirtschaft, 2*, 34–38.

Fuders, F. (2014). Indexierte Währungen und Recheneinheiten als Mittel gegen Inflation und Finanzkrisen? *Zeitschrift für Sozialökonomie, 51*(180/181), 15–24.

Fuders, F. (2015). Auch das Bargeld braucht einen Negativzins. *Fairconomy, 11*(Special Edition: Wirtschaft Natürlich), 10.

Fuders, F. (2016a). Smarter money for smarter cities: How regional currencies can help to promote a decentralised and sustainable regional development. In *Decentralisation and regional development – Experiences and lessons from four continents over three decades* (pp. 155–185). Springer.

Fuders, F. (2016b). Negativzinsen ja – aber richtig. *Fairconomy, 4*, 6–11.

Fuders, F. (2017). Neues Geld für eine neue Ökonomie – Die Reform des Geldwesens als Voraussetzung für eine Marktwirtschaft, die den Menschen dient. In *Finanzwirtschaft in ethischer Verantwortung – Erfolgskonzepte für Social Banking und Social Finance* (pp. 121–183). Springer-Gabler.

Fuders, F. (2020a). *The 'tragedy of the commons' and the role of the money interest rate*. Springer.

Fuders, F. (2020b). Freigeld und Freiland: für eine wirklich freie, nachhaltig funktionierende und gerechte Wirtschaftsordnung. *Humane Wirtschaft, 2*, 16–24.

Fuders, F., & Bockting, B. (2019a). Negativzinsen werden noch nicht richtig verstanden. *Fairconomy, 15*(1), 12–16.

Fuders, F., & Bockting, B. (2019b). Von der Zombie-Ökonomie zur Fairconomy. *Fairconomy, 15*(4), 12–13.

Fuders, F., & Löhr, D. (2014). Hat die Europäische Zentralbank ein Tabu gebrochen? *Fairconomy, 10*(4), 4–5.

Fuders, F., & Löhr, D. (2016). Die EZB und Silvio Gesell – Eine neue Partnerschaft? Eine Einschätzung zwei Jahre nach der bemerkenswerten Rede Benoît Cœurés. *Humane Wirtschaft, 6*, 4–9.

Fuders, F., & Louge, C. (2020). Effektive Negativzinsen statt Corona-Schuldenwachstum. *Fairconomy, 16*(2), 4–9.

Fuders, F., & Max-Neef, M. (2014). Dinero, deuda y crisis financieras. Propuestas teórico-prácticas en pos de la sostenibilidad del sistema financiero internacional. In *Economía Internacional. Claves teórica-prácticas sobre la inserción de Latinoamérica en el mundo*. LATIn.

Fuders, F., & Schreiber-Martens, A. (2018). Bitcoin – eine Alternative für das Finanzsystem? *Fairconomy, 14*(3), 16–17.

Fuders, F., Mondaca, C., & Azungah Haruna, M. (2013). The central bank's dilemma, the inflation-deflation paradox and a new interpretation of the Kondratieff waves. *Economía (U. de los Andes), 36*, 33–66.

Gabler Wirtschaftslexikon. (2022). *Enteignung.* [Online] Available at: https://wirtschaftslexikon.gabler.de/definition/enteignung-35922. Accessed 21 June 2022.

Gesell, S. (1949). *Die natürliche Wirtschaftsordnung durch Freiland und Freigeld* (9th ed.). Rudolf Zitzmann.

Goodfriend, M. (2000). *Overcoming the zero bound on interest rate policy*. The Federal Reserve Bank of Richmond.

Hillmer, H.-J. (2021). *Niedrige Insolvenzzahlen als Trugschluss?* [Online] Available at: https://www.creditreform.com/. Accessed 22 June 2022.

Hüfner, M. (2015). *Niedrigzinsen: Betrug am Anleger?* [Online] Available at: https://www.boerse-online.de/nachrichten/aktien/niedrigzinsen-betrug-am-anleger-1000444657. Accessed 21 June 2022.

Kennedy, M. (1990). *Geld ohne Zinsen und Inflation – Ein Tauschmittel, das jedem dient*. Permakultur-Verlag.

Kenton, W. (2020). *Investopedia – Expropriation.* [Online] Available at: https://www.investopedia.com/terms/e/expropriation.asp. Accessed 21 June 2022.

Keynes, J. M. (1936). *General theory of employment, interest and money*. Hartcourt.

Keynes, J. M. (1983). Economic articles and academic correspondence. In D. Moggrigde (Ed.), *The collected writings of John Maynard Keynes* (p. 402). Macmillan.

Lietaer, B. (2002). *The future of money – Creating new wealth, work and a wiser world.* Random House.

Mikalauskas, E. (2020). *The ultimate guide to safe and anonymous online payment methods in 2022.* [Online] Available at: https://cybernews.com/resources/the-ultimate-guide-to-safe-and-anonymous-online-payment-methods/#:~:text=Prepaid%20debit%20cards%20are%20an,with%20some%20sort%20of%20website. Accessed 6 July 2022.

Murse, T. (2019). *Why burning money is illegal in the United States.* [Online] Available at: https://www.thoughtco.com/is-burning-money-illegal-3367953. Accessed 16 Jue 2022.

Nerdwallet. (2022). *5 best prepaid debit cards.* [Online] Available at: https://www.nerdwallet.com/article/banking/best-prepaid-debit-cards. Accessed 6 July 2022.

OECD. (2015). *Business and finance outlook.* OECD.

Rogoff, K. (2014). *Costs and benefits to phasing out paper currency* (Working Paper 20126). National Bureau of Economic Research.

Rogoff, K. (2016). *Der Fluch des Geldes – Warum unser Bargeld verschwinden wird.* Finanzbuchverlag.

Schieritz, M., & Uchatius, W. (2014, November 13). Mark Schieritz/Wolfgang Uchatius, Der stärkste Trieb der deutschen – Ausgespart! *Die Zeit*, pp. 13–15.

Seibel, K. (2021). *Die ungesunde Zinssituation treibt einen Keil in die Gesellschaft.* [Online] Available at: https://www.welt.de/wirtschaft/article232290653/Strafzinsen-Die-ungesunde-Zinssituation-treibt-einen-Keil-in-die-Gesellschaft.html. Accessed 16 June 2022.

Seltmann, T. (2010). Umlaufsicherung von Banknoten – Wie sich die Einführung der Liquiditätsgebühr bei Geldscheinen systemkonform realisieren lässt. *Humane Wirtschaft, 2*, 10–16.

Steiner, R. (1979). *Die soziale Grundforderung unserer Zeit – In geänderter Zeitlage, Zwölf Vorträge, gehalten in Dornach und Bern v. 29.11. bis 21.12.1918.* Rudolf-Steiner-Verlag.

Stiglitz, J. E. (1998). *Macroeconomía* (2nd ed.). Ariel.

Tagesspiegel. (2021). *Rechtsgutachten hält EZB-Negativzins für verfassungswidrig.* [Online] Available at: https://www.tagesspiegel.de/wirtschaft/enteignung-der-sparer-rechtsgutachten-haelt-ezb-negativzins-fuer-verfassungswidrig/27389236.html. Accessed 16 June 2022.

Thiele, C.-L. (2017). *Die Bedeutung des Bargelds als Wertaufbewahrungsmittel.* [Online] Available at: https://www.bundesbank.de/de/presse/reden/die-bedeutung-des-bargelds-als-wertaufbewahrungsmittel-613736. Accessed 22 June 2022.

Chapter 13
Alternatives to Gesell's Solution?

13.1 "100% Money" and "Sovereign Money"

13.1.1 An Old Idea in Vogue Again

Nowadays it has again become very much *in vogue* to blame the fractional reserve banking system for many of our problems (Huber, 1998, 2017; Huber & Robertson, 2008; Jakab & Kumhof, 2015; Gross & Siebenbrunner, 2019; Positive Money, 2022; Sovereign Money, 2022; Vollgeld-Initiative, 2022; Monetative, 2022). There is a growing awareness that many problems are connected to our monetary system. Unfortunately, the aforementioned authors do not see these problems as inherent in the unnatural design of money but, instead, attribute them to the fractional reserve banking system. According to their interpretation, these problems could be solved by changing the fractional reserve banking system to a system with 100% reserve requirements for demand deposits. Banks would thereby be prevented from lending money held in demand deposits. The idea of such a "100%-Money" system, sometimes also referred to as "full reserve banking", is not new but was envisioned, as explained in Chap. 2, by Irving Fisher (1936a) and by the so-called Chicago Plan in the 1930s. In German this idea is referred to as "Vollgeld", which the inventor of this term Joseph Huber (1998) translates as "sovereign money" (Huber, 2017).[1] Even though the "sovereign money" proposal is not the same as full reserve banking (the 100%-Money proposal), the two proposals only differ in terms of minor details, which for the purpose of the following analysis are not relevant.

Most people who advocate for this concept believe in the mistaken idea that banks create money out of nothing (*ex-nihilo*), and that money creation is limited only by reserve requirements. Therefore, they believe that out of control money supply growth and the problems associated with it, especially financial crises and

[1] In an earlier publication he had translated it as „plain money" (Huber & Robertson, 2008).

© The Author(s), under exclusive license to Springer Nature Switzerland AG 2023

F. Fuders, *How to Fulfil the UN Sustainability Goals*,
https://doi.org/10.1007/978-3-031-37768-6_13

inequality, could be solved or strongly mitigated by simply not allowing banks to create money (Huber, 1998, 2017; Huber & Robertson, 2008; Jackson & Dyson, 2013; Positive Money, 2022; Sovereign Money, 2022; Vollgeld-Initiative, 2022). However, the idea that banks create money is wrong and is based on an erroneous interpretation, as was explained in Chap. 2. Banks do not create money endogenously, but the money supply grows in the banking sector as a whole according to the so-called *money multiplier* effect: deposits keep circulating via loans.

13.1.2 100% Money Can Avert Bank Runs

It is noteworthy that not all supporters of the Chicago Plan (100%-Money proposal) were followers of the idea that banks create money. Even Irving Fisher himself who initially believed in the erroneous idea of money creation *ex-nihilo* (Fisher, 1936a) later published a text with the same title "100% Money", but in this text he did not repeat the idea of money creation *ex nihilo* anymore (but still advocated for a 100% reserve requirement on sight deposits, see Fisher, 1936b). As previously discussed in Chap. 2, a 100% reserve requirement on demand deposits could be viewed positively since it would prevent bank runs. Banks would have to deposit 100% of their customers' demand deposits physically in their safes or as deposits at the central bank. Loans could then no longer be funded by demand deposits, but only from time deposits or inflows from selling obligations or other debt securities. This would prevent situations in which bank customers demand more money than the bank holds in its vaults or at the central bank. A 100% reserve requirement could thereby significantly improve the stability of the financial system. Here it is worth noting that a Natural Money system is automatically (without the need for further regulations) a full reserve banking system since loans would be issued when customers specifically ask banks to loan out their money in order to avoid the hoarding fee on demand deposits. Funds not expressly released by customers for lending would therefore not be lent out. Bank customers would therefore understand that their money has been lent out and would not expect to have access to it until the agreed upon loan period has elapsed.

13.1.3 The Problem Is Not Solved

Even though a 100% reserve requirement for demand deposits might avoid bank runs, it will not solve the main problem that results from the unnatural design of money: the existence of interest as a "*reward for parting with liquidity*" (Keynes, 1936: 167). Banks will continue to grant loans from time deposits and charge interest on them. Hence, if customers save money in time deposits (or buy bank bonds, obligations or other debt securities) banks can lend that money and charge interest.

And as long as loans are subject to interest, deposits will grow, and banks will be forced to keep expanding the volume of loans. Even if loans can only be funded by time deposits, the money supply will still grow; after all, time deposits are included in the definition of the money supply M3. Hence, the money supply would still grow in a system with 100% reserve requirements.

However, banks would oppose a 100% reserve requirement on demand deposits. To grant loans from demand deposits enables banks to offer loans at particularly low interest rates or to earn higher margins, since demand deposits earn little or no interest. This is why in the eurozone today about half of the total debt volume stems from demand deposits (Bundesbank, 2017). And this way, the collapse that occurs when banks can no longer find borrowers (or when borrowers default on a large scale), can be delayed. The fractional reserve system is thus a secondary phenomenon of the compulsion to steadily expand the money supply. Here we see a main reason why reserve requirements are usually only lowered but never increased by central banks.

13.1.4 "100% Money" vs. "Sovereign Money"

The term *sovereign money* is the English translation of the German expression "Vollgeld" coined by Joseph Huber (1998) and has become quite popular in recent years, including outside the German speaking countries. In Switzerland there was a plebiscite held in 2018 on whether Switzerland should switch to a full sovereign money system. According to supporters, full sovereign money differs from a 100% reserve requirement (full reserve banking). However, I cannot see a big difference. Huber (2022) himself states that both concepts differ only in terms of technical issues regarding accounting policies.

The term "sovereign money" is supposed to express the idea that only the central bank can create money, while commercial banks "only" function as financial intermediaries between depositors and borrowers (Huber, 1998; Sovereign Money, 2022). The sovereign-money proposal is focused on the false idea that commercial banks create money endogenously. In Chap. 2 it was explained why the idea of such money creation *ex-nihilo* is wrong. Banks do not create money endogenously. The central bank creates the monetary base from which the money supply, defined as cash plus deposits, starts growing within the **banking sector as a whole** (*money multiplier* effect). This dynamic will **always** exist as long as commercial banks exist. Yet, the existence of commercial banks is not challenged by adherents of the full-sovereign-money proposal (Monetative, 2022). The only difference from the current system is that banks—just as in the 100% Money proposal—cannot fund loans from demand deposits, but only from time deposits or from selling obligations or other debt securities. Hence, full sovereign money is in its essence just a different expression of the proposal of a 100% reserve requirement.

13.1.5 'True' Sovereign Money

'True' sovereign money would be a so-called Central Bank Digital Currency (CBDC) as is being studied by several central bank. A Central Bank Digital Currency completely replaced the traditional monetary system, i.e., private commercial banks would not exist anymore (or they would be converted into branches of the central bank), and the central bank is the only institution that grants loans to private persons. Then we could truly speak of a full sovereign money system, a nationalization of money creation for in this case there would be no *money multiplier* effect anymore.[2] Such nationalization of money creation is nothing new. It was proposed by Silvio Gesell (Gesell, 1892) even before he wrote his masterpiece "The Natural economic Order".

Silvio Gesell did not emphasize the nationalization of money in "The Natural Economic Order" for good reason. The problems in the current system are **not caused by the money multiplier itself but the non-neutrality of money**, which results from interest. If money was neutral the money multiplier effect would, as explained in Sect. 5.4, make Fulltarton's *law of reflux* work and the money supply would adapt to the performance of the economy. The nationalization of money creation would not solve any of the problems associated with the unnatural storability of money. People would hoard ever larger amounts of money physically at home, and the central bank would have to issue ever more loans to keep money circulating. But this new money injected into circulation would also end up in large part under the mattress or in the stock, real estate or other investment markets causing more price bubbles and ever-increasing risk.

And if in such a sovereign money system, the central bank refused to issue more loans, deflation would result. The precarious situation caused by deflation will then lead people to offer their savings (held physically at home). Then again, money will keep flowing by credit, the money multiplier effect will start working. And since these loans would surely not be free of interest, we would soon have the same problems as we have today.

13.1.6 Neither Full Reserve Banking, Nor True Sovereign Money Are Forms of "Positive" Money

Those who advocate for a full reserve banking system (100% Money) usually believe that in such a system we would have money that is not based on debt anymore: "positive money" (Jackson & Dyson, 2013; Positive Money, 2022). However, we already concluded in Sect. 2.6.2 that the fact that money is the counterpart of debt is neither a problem, nor is it possible to avoid. Even if we went back to paying

[2] Note that this was not the original idea of Joseph Huber (1998) who coined the term "sovereign money".

with gold coins, this would still not be positive money. As soon as a gold coin is deposited in a bank, it becomes a liability for the bank while the deposit is an asset for the account holder. In fact, strictly speaking cash—even before depositing it in a bank account—can be seen as the counterpart of debt, since on the central bank balance sheet it appears as liability. Money is **always** the counterpart of debt. This fact cannot be avoided. Therefore, even if we changed to a system of "true" sovereign money where all banks are branches of the central bank money would still be the counterpart of debt.

13.1.7 Summary

Neither a 100% reserve requirement nor a "true" full sovereign money system would solve the problems inherent in our monetary system. To solve these problems money needs some sort of hoarding fee. This is true for a fractional reserve system as well as for a system with 100% reserve requirements or a "true" sovereign money system. Otherwise, interest will exist. Interest makes deposits and credit volume grow according to an exponential function, slowly at first and then faster and faster. This is a mathematical law. The problem is not that in our current monetary system money is the counterpart of debt. The problem is that banks are forced to constantly expand the credit volume because the interest payable on deposit, savings accounts or on debt instruments requires it. But an expanding credit supply just means an expanding money supply. The idea of 100% reserve requirements as well as a "true" full sovereign money is not a solution to the problem.

13.2 Modern Monetary Theory (MMT)

13.2.1 A Noble Objective

Today another concept which is very much *in vogue* as a cure for financial crises and unemployment, and in general as a justification for debt-based expansionary fiscal policy is so-called "Modern Monetary Theory" (MMT), sometimes also called "Modern Money Theory" (Mosler, 1997; Forstater & Mosler, 2005; Wray, 2012; Ehnts, 2017; Kelton, 2021). It is considered a branch of post Keynesianism since it strongly promotes State interventionism and countercyclical fiscal policy. MMT mixes classic Keynesian concepts with the theory that banks create money out of nothing, and in this sense MMT can be seen as related to the 100%-Money and sovereign-money approaches. The first text that propounded this theory was published by Mosler (1997); the most comprehensive treatment is, as of 2022, probably that of Dirk Ehnts (2017). Quite innovative is the use of the double-entry accounting system to analyze (credit) money creation, central bank instruments and fiscal operations, among other things. The primary and noble objective is to fight unvoluntary unemployment.

However, since a cornerstone of MMT is the erroneous theory of money creation *ex-nihilo* (banks can endogenously create money through lending) this theory is built on an incorrect foundation. Regarding the classic Keynesian concepts contained in MMT, strong arguments exist (as analyzed in Sect. 7.3) that fiscal policy, if not debt-based, does not have any impact on economic growth at all and that the so-called *Keynesian multiplier* is, in fact, being confused with the *money multiplier*. Hence, MMT is based on erroneous premises.

13.2.2 Keynesianism Serves the Recipients of Interest

As was explained in Sect. 7.3.3, (only) debt-based expansionary fiscal policy can counteract deflationary tendencies and foster economic growth. Even though such policies cannot work in the long run, Keynesianism has provided and still provides policymakers with a scientific rationale for expanding debt. In the same way as private debt does, sovereign debt-financed social security programs contribute to a steadily widening gap between rich and poor, even though these programs aim to achieve the opposite. More debt always means growing deposits (savings), namely for those who collect interest. Any interest leads to growing inequality between those who are engaged in producing real wealth (goods and services with real intrinsic and subjectively perceived value) and those who earn by lending money. There is no such thing as a fair interest rate and consequently no socially just debt (Fuders, 2017, 2018).

In this context, it is noteworthy that Keynes advocated for countercyclical fiscal policy, but it is often forgotten that he correctly saw in interest (he called it the "marginal efficiency of capital") the "cumulative oppressive power of the capitalist to exploit the scarcity value of capital" (Keynes, 1936: 376), and in lowering the interest rate to zero "the most sensible way of gradually getting rid of many of the objectionable features of capitalism" (Keynes, 1936: 221). Unfortunately, MMT does not, with some exceptions (Forstater & Mosler, 2005) deal with the role of interest, and especially not the problems inherent in it. Sustainable development and long-run economic growth are two contradictory goals as explained in Chap. 6. The only way to develop sustainably in terms of improving the satisfaction of fundamental human needs (Max-Neef et al., 1991) without the need to constantly grow is a zero-interest environment.

13.2.3 A Dangerous Proximity to Marxism

The idea proposed by advocates of MMT that the State can practically (through its central bank) print as much money as it wants and in this way provide a "job guarantee", acting as "employer of last resort" (Mosler, 1997; Wray, 1998a, b; Mitchell,

2000) means that the central bank would no longer be an autonomous institution and suggests a kinship with Marxism. Marxism does not offer any solution to the problems inherent in our monetary system but adds new problems. This will be outlined in the next paragraph.

13.3 Marxism, Communism and Socialism

13.3.1 Another Noble Objective

"Marxism" is a catch-all term for various theoretical approaches and policy proposals dating back to the teachings of Karl Marx (1872) and Friedrich Engels, [1882] 2019. The main objective is to overcome "class differences", which could be translated into today's language as income inequality. However, Marx did not see that income inequality fundamentally stems from interest. Marxism today is frequently referred to as "communism" even though this term goes back further than Marx and includes works of Marx's followers such as Lenin. The term "socialism" is often used interchangeably with Marxism, but they are not the same. Many people think of socialism as a sort of "communism light" or as a mixture of communism and social market economy. However, studying literature on this topic in English, German and Spanish I could not find convincing, clear and generally accepted definitions of the three terms. In the following I will use the term communism.

13.3.2 Little Freedom and Heavy Welfare Losses for People

A planned economy leaves little freedom to market participants (that is to us, the people: consumers and producers) and workers. Sometimes communist governments decide whether someone has the right to study at the university, as was the case in former GDR "German Democratic Republic" (Adenauercampus, 2022). The inverted commas here refer to the fact that today we know that the GDR was not democratic at all. For example, elections were falsified (Bundesregierung, 2022). Good reasons exist to assume that the same is happening today in communist countries such as North- Korea, Cuba and Venezuela. The history of communist nations (ex-Soviet Union, North-Korea, Cuba, Venezuela), tells us that communism delivers worse results in terms of social welfare than a competition-based market economy. We discussed previously (Sect. 10.4) that competition, when functioning properly[3] (getting as close as possible to the ideal of perfect competition which includes

[3] A prerequisite for competition to produce benefits for society is that it is organized properly, especially morally. On this aspect with many more references see Schachtschneider (2002, 2005, 2007).

monetary neutrality), offers an ideal amount of economic freedom (Fuders, 2009b, 2011) as well as, in all likelihood, a maximum of well-being; well-being defined in terms of the greatest possible satisfaction of fundamental human needs.

13.3.3 The Origin of the Conflation of Money with Capital

In addition to the fact that planned economies leave less freedom to individuals compared to the free-market system in terms of how individuals can choose to lead their lives, Karl Marx and his followers apparently did not recognize the problem emanating of interest. Marx was subject to the same confusion that still characterizes economics today. In fact, it was Marx who played a major role in giving rise to the conflation of money with capital which now characterizes all schools of economic thought, including those who are fierce opponents of Marxism. The first chapters of Karl Marx's (1872) main work "Das Kapital", contain a sound analysis of many economic problems we still see today. But from that point he confuses concepts and defines money as capital, as if money was a production factor. Part II of volume I of "Das Kapital" is even titled (in the English translation) "The Transformation of Money into Capital".

We find the same concept in all communist texts. As an example, I would like to mention a famous Marxist textbook: Bogdanoff (1925). Similar to Marx, Bogdanoff delivers in the first 100 pages a more or less adequate description of the production and exchange processes similar to the analysis we find in neoclassic textbooks of that time. But then money and the production factor capital are confused and used as synonyms. The author speaks of earning interest from "capital", but should have said "money" (Bogdanoff, 1925: 140) and this, even though he correctly observed some lines earlier that "the usurer has nothing to do with the process of production" (Bogdanoff, 1925: 140).[4] The complete conflation and confusion then comes on page 141 f.: "Means of production and money, which represent the general form of value of the former, become capital...". Bogdanoff emphasizes here that production factors as well as money are "capital". Indeed, the term "capitalism" coined by Marxism as the contrary to communism also contains this confusion, since it refers principally to the private accumulation of money, not machines or tools.

13.3.4 A Most Efficient "Smoke Grenade"

This conflation of money and capital, which can be found today in all branches of economic theory, blinds us to the problems inherent in interest. If we believe that money is just another production factor along with human labor, machines, tools and natural resources, then of course we won't see why it should make any

[4] "Usurer" refers to a person who receives interest for lending money. We see here, as mentioned in Sect. 9.12.3, that in former times interest for lending money and usury used to be synonyms.

difference whether we earn money from lending money or by producing real goods and services. Money is not a production factor. The real reason for the division of society into classes, both then and now, is our monetary system. Since Marx didn't recognize the problems inherent in our monetary system, it is not surprising that his proposed solution of nationalizing the means of production was aimed in a completely different direction.

Class differences do not arise from the competition-based market economy. On the contrary, competition reduces class differences (see e.g., Friedman, 1966). We can intuitively understand this: the more competition there is, the lower prices tend to be, which is beneficial for households, but not for producers. As we discussed previously, in the ideal situation of perfect competition, company profits would not exceed the opportunity costs of the production factors employed. Monopoly rents would not be possible (see e.g., Frank, 2015). Even though "perfect competition" describes a utopian situation which will never exist in its pure form, **it does serve as a compass** that shows us in which direction to move. Competition leads to a more equal distribution of income between households and firms and thus to **greater social justice** (the so-called distribution function of competition).[5] The distribution function refers not only to the relationship between households and firms, but also to the relationship between competitors. Since in the model of perfect competition each market participant earns just as much as he could earn elsewhere with the same production factors, free market competition ensures that competitors are remunerated according to their performance (Herdzina, 1999; Tolksdorf, 1994; Frank, 2015). Hence, competition prevents the "establishment of entrenched positions of power" (Kantzenbach, 1967: 16 f.), which threaten the freedom of all (Emmerich, 2001: 2). Marxism eliminates these positive aspects of competition.

Again, the enormous differences in wealth that we observe today in almost all countries, where less than 20% of the population owns more than 80% of the wealth, are not caused by competition but by our monetary system. It is interest that causes the gap between rich and poor to constantly grow according to mathematical logic. And it is not only that the wealth gap grows, but the resulting income inequality is also unjust. We discussed above the fact that interest represents the "right to the foreign labor yield" (translated from German *Das Recht auf den fremden Arbeitsertrag*) (Gesell, 1920: 210), because the recipient of interest does not work productively, but instead it is the borrower who works. We also observed that the lending of money does not require one to renounce or abstain from consumption. The recipient of interest does not perform any useful function or renounce any real good. Marx correctly viewed the an unfair distribution of wealth. However, since he did not recognize the true cause of the problem, his solutions do not solve the problem at its roots but, on the contrary, cause many new problems.

[5] On the power-distribution function between the economy and society, see Böhm (1961). He calls competition the "greatest and most ingenious instrument of disempowerment in history" (p. 22, original in German – author's translation). See also Herdzina (1999).

13.3.5 *Communism Strengthens the Current Monetary System*

Social legislation and extensive state transfer payments can hardly compensate for the injustice caused by the current monetary system. On the contrary, exaggerated social legislation regularly requires the State to incur in further debt. This is true even for countries that do not pursue Marxist policies. It is probably no coincidence that social welfare expenditures in Germany for many years have been roughly equal to the amount of increase of sovereign debt. If generous social welfare legislation in, e.g., Germany requires the state to take on additional debt (or prevents that debt from being reduced), this favors interest recipients and thus ultimately increases income inequality. As we discussed in the Chap. 9 on the moral-ethical aspects of interest, **every loan** (even those destined for productive investments or social welfare programs) unjustly increases the wealth gap and can therefore never be socially beneficial. **There is no such thing as socially justified sovereign debt**.

Communism strengthens the current monetary system for a second reason. Communist countries are particularly prone to debt because they produce inefficiently (see next two paragraphs) but, of course, people still need to consume. The lack of domestic productivity is then compensated by credit-financed imports.[6] In the end it does not make a substantial difference if it is private debt or State debt (in the name of its citizens, who then pay interest indirectly via taxes). Viewed from this perspective, the age-old debate about whether the free-market or communist system is better for the people may only serve as a distraction.

13.3.6 *Old Class Differences Are Being Replaced by New Ones*

Communism does nothing more than address symptoms, while Silvio Gesell gave us a solution that cures the disease (inequality and the necessity to grow which leads to the destruction of the environment). And regarding the "class problem" communism may not even cure the symptoms. Experience shows that in communist countries those who are well positioned vis-a-vis the government, i.e., work in or in connection with the government live extraordinary well compared to others. The old class differences are thus replaced by new ones: **you either belong to the "government class" or you work for them (and will be poor)**. This is what we saw in the former GDR and what we currently see in North Korea, Cuba and Venezuela. In addition, the problems associated with our unnatural monetary system and the resulting phenomenon of interest also apply in communist countries. Communism does, however, deprive us of freedom through a planned economy. In fact, a planned economy can be viewed as a form of forced labor, **governmentally organized slavery**. Since everyone earns the same regardless of their efforts (except for those who

[6] If more is imported than exported, the deficit in the trade balance must eventually be balanced by credit in the balance of payments.

belong to the "government class"), there is little incentive for hard work. Supervision is necessary so that people work properly. In a correctly working free market economy people work because they want to, not because they are forced to. And what they earn is theirs, not the government's.

To prevent people from emigrating to countries with free market economies, communist countries often impose travel restrictions. In Venezuela, Cuba and North Korea it is very difficult to obtain a passport. To prevent people from leaving the country, the former GDR fenced off the whole country, effectively converting the whole country in a prison, the biggest prison in the history of mankind with 17 million innocent innates. What we need is a truly free market economy, free most importantly from interest which requires growth and produces (unjust)[7] inequality via mathematical logic. We need a market economy that is not perverted by its monetary system.

13.3.7 We Are the Market

Some believe that markets are to be blamed for all economic problems and should therefore be banned. In reality, markets are not bad and cannot be banned. Markets consist of consumers and producers. In other words, markets consist of us, the people. **We, the people, are the market**. The marketplace where we meet to negotiate and trade can be a physical place for example the classic marketplace at the center of a city or virtual space on the internet. We can regulate markets. But we cannot ban markets.

What should not and in the long term cannot be regulated are prices since prices will always be determined by market equilibrium (Frank, 2015). Let me explain. In communism, in the case of the (relatively few) privately supplied products, prices are frequently dictated by the government. If the price is lower than the equilibrium price between demand and supply, suppliers will adapt (lower) the quality of the product. This way a new equilibrium price is found. If it is not possible to adjust the characteristics of the product to the price (e.g., regulations require y certain standard or quality), producers will simply not produce. This results in shortages and a flourishing of illegal markets, a fact that can be easily corroborated in countries such as Venezuela, Cuba or the former GDR. Supermarkets have little variety and low-quality products. Yet, those who are willing to pay higher prices (usually in US-dollars or in case of the former GDR in Western-German Deutsche Marks) will be able to acquire products with higher quality also in Venezuela or Cuba on the black market.

For example, in Venezuela you can either buy gasoline almost for free at the subsidized gas stations or you can buy at a private gas station paying in US-dollars.

---

[7] It is crucial de bear in mind that unequal income distribution is not necessarily unjust, however the income inequality produced by interest is always unjust (see Chap. 9 on the moral-ethical aspects of interest).

Why should someone opt for the latter? The answer is that the line of people waiting at the subsidized gas station is so long that it takes about 24 h to get gasoline. People often have to sleep in their cars while waiting in such lines. So, it is inaccurate to say that the government provides gasoline for free. Instead of paying higher prices in money, people pay with their time.

13.3.8 Communism Is Perfect Monopolization

Since in communism the State manages everything and since of course it makes little sense for it to compete with itself, all production is performed by monopolies. Monopoly is the opposite of perfect competition. Monopolies work inefficiently. While in perfect competition price equals marginal costs and marginal revenues (the last unit produced has the same value to the consumer as it costs to produce), this is not the case in monopolized markets. Inefficiency means we need more input and energy to produce a given product or service, which works counter to the ideal of sustainability. This can be confirmed by looking at communist countries where we usually observe devastating environmental degradation, such as in the former GDR and North Korea.

The inefficiency of publicly owned monopolies is usually even worse than in the case of private monopolies due to bureaucracy (Leibenstein, 1966). As the profits belongs to everyone (or no one), civil servants have no incentive to economize. In Eastern Germany where from 1945 to 1990 the communists governed under Soviet protection, the State-owned distribution system and retail shops ruthlessly abused their legal monopolies and charged a 300% margin on their (mostly poor quality) products (Walker, 1995). In fact, communism can be thought of as the **perfection of monopolization of markets**: all markets are pure monopolies, and even worse, State-owned monopolies.

Many people in Chile and elsewhere are against neoliberalism. This is understandable since in Chile as well as in other countries, especially in Latin America, the concept of free markets has been misinterpreted as markets freed from any regulation. Most markets in developing countries are strongly concentrated (monopolized), and this is one of the main reasons why these countries are less developed. At University Austral de Chile we have conducted a study using data from 100 cities and regions. We found a **negative correlation** between market concentration (monopolization of markets) and the degree of development of the respective country, defining development by the UN Human Development Index (Namatama et al., 2018). That means, the roadmap towards more development requires more competition, not less: "Small is beautiful" as Schumacher (1973) correctly put it. The best way to fight "gigantism" Max-Neef (1986: 58, 63, 136, 149, 151 f., 184) is to get as close as possible to perfect competition.

13.3.9 Communism Is Governmentally Organized Robbery

The term "allocative efficiency" as one of the outcomes of perfect competition is indeed a very academic concept, and readers not trained in economics might have trouble understanding it. But intuitively we can understand what it means. Since monopolists have no competitors, they have no need to improve products. Therefore, in monopolistic markets we usually find few products (since there is only one producer) low quality and **high prices**. This enables the monopolist to earn an **unjust income**, so-called "monopoly rent". It is to be considered an unjust income since it is not based on the monopolist's effort (investment of production factors) but in taking advantage of the dominant position. In other words, the high price that the monopolist charges is not justified by the quality or characteristics of the product, yet the consumer has no choice but to pay the price, because there are no alternatives. Hence, monopoly rent is, for all intents and purposes, comparable the blackmailer's profit, or robbery. Since in communism all production is organized by State-owned monopolies, communism is governmentally organized robbery and the most efficient way to exploit people, especially the working class. This brings us to the next point.

13.3.10 The Most Serious Flaw of Communism

The most serious flaw of communism is that all production factors are commonly owned, which means they belong to everyone (or no one). One reason why markets fail to efficiently allocate natural resources—and perhaps next to the growth imperative the most significant failure regarding the overuse of natural resources—is that many natural resources are non-excludable goods (for a detailed overview of this problem see, e.g., Daly & Farley, 2011).[8] In communism this market failure (also known "free rider" problem[9]) **occurs intentionally**, and not only regarding natural resources but all production factors.

What has been said can also be understood if we remember from Sect. 10.4.6 that the positive welfare effects attributed to competition eventually result from the fact that **competition limits our selfishness**. Competition obliges producers to offer the best products at the best prices (what in economics we refer to as allocative efficiency) in order to avoid losing market share. In other words, competition compels us to behave **as if we loved one another**.[10] Of course, it would be even better if we

[8]This does not mean that all goods provided by nature, and which are naturally open access regimes (air, sea, eco-systemic services), should be private goods, as discussed above.

[9]The name is derived from fare evasion ("free ride") on the running boards of streetcars or other public transportation systems.

[10]This reasoning is unfortunately not well understood in conventional economics.

truly did love one another and offered the best products at the best prices not because competition forces us to do so but out of a desire to contribute to the well-being of all. But as long as the majority have not yet reached this deeper understanding of the purpose of life, we need competition (Fuders, 2017; Fuders & Nowak, 2019). Since in communism all markets are monopolies and competition is intentionally abolished, this curb on selfishness is also eliminated. Hence, communism could only work efficiently and deliver the desired results, if we were all angels and behaved altruistically (if we perfectly loved one another).

Unfortunately, supporters of communism usually argue in its favor on the basis of justice. They believe that communism produces a more just society when the contrary is true. In fact, recipients of interest, monopolists and free riders all share a common trait: they wish to earn without working, i.e., they want to earn at the expense of others. And they find all kinds of arguments to justify their behavior. Of course, there is nothing wrong with helping the poor and a market economy is not inconsistent with a social security system (Frank, 2015). We do not need communism in order to help the poor. And we do not need the government to implement social welfare programs. Anyone can do so on their own. If I see poverty nothing prevents me from sharing with the less fortunate. Have you ever seen a politician use his or her own money to help the poor? It is easy to advocate for socialistic reforms at the expense of others, to promulgate socialistic policies funded by the taxpayer's money.

13.3.11 Communism Is Even More Unnatural Than the Interest Economy

We have discussed above that our monetary system is based on the unnatural storability of money and that this gives rise to the unnatural phenomenon of interest. We also observed that the major problems the world faces today are consequences of interest. Communism does not solve these problems. Instead, it just creates more problems through further deviation from nature (Fuders, 2018). We argued above that it is natural for private property to exist for all man-made products. The more our lifestyle, laws and behaviors deviate from nature (from the original equilibrium that existed many thousands of years) the unhealthier it is for society. As I discussed earlier (Fuders, 2010a), neither interest-capitalism nor communism are natural economic forms, that is, they are not consistent with human nature, meaning that in the long-run both are destined to fail. Communism leads to laziness and interest-based neoliberalism to greed. In communism, people are enslaved through the forced labor inherent in a planned economy. In an interest-based economy, people are enslaved through the burden of interest.

13.4 Gold Standard or Other "Backed Money"

13.4.1 Gold Standard

Since the 2008 Great Financial Crisis, many have been arguing for a return to the gold standard, especially since former World Bank president Zoellick advocated for this in 2010 (Beattie, 2010). "Gold standard" means that we either pay directly with gold coins or with paper money that is backed by gold, i.e., the central bank is required to hand over a certain amount of gold in exchange for each bill or coin. Originally, all paper-money currencies were backed by gold or Sterling silver and each bill stated the amount of gold or silver it represented. In the US a gold standard existed until 1971. However, a gold standard does not address the unnatural storability of money and the resulting phenomenon of interest as a "*reward for parting with liquidity*" (Keynes, 1936: 167). On the contrary, gold is perfectly storable.

Therefore, a gold standard does not cure the main problem of our monetary system. Money would still have an advantage over real goods due to its imperishability and, therefore, there would still be an incentive to hoard gold money. Those who needed to borrow money would still have to pay interest as a reward for parting with liquidity, i.e., to incentivize the holder of money to lend. Interest invariably increases the money supply (defined as cash plus deposits) and debt since there is no interest without debt. As the amount of gold in the world cannot be increased as needed, a gold standard cannot function effectively in the long run. The sum of monetary units in bank accounts and the existing amount of gold will start to decouple and the central bank will not be able to meet its obligation to redeem bills for gold. This is the reason why any gold-standard currency **must** eventually break down and why today there are no longer any industrialized counties using the gold standard.

13.4.2 Tangible Wealth-Backed Money

Some authors propose overcoming the problems of our monetary system with money that is tied to tangible wealth (see e.g., Reichmuth & Reichmuth, 2001). Every unit of such currency would represent a share of a basket of real goods, such as energy resources, land, etc. Holders of such money would have the right to claim their share of this basket of goods. Such a basket of goods would be comparable to an investment fund whose shares could be purchased at banks. The investment fund would invest in real assets such as real estate, commodities, and precious metals. Shares in these funds could theoretically be used as a medium of exchange, as money. Proponents of this idea believe that money backed by real goods would not devalue or be prone to financial crises since it would be backed by real tangible wealth (Reichmuth & Reichmuth, 2001). However, this is not the case. Such currency would be comparable to gold-backed money. A gold standard system can be thought of as an indexed currency. Like the gold standard, currencies that are backed by other forms of tangible wealth do not represent a solution to the problems of our financial system (Fuders, 2014a, 2016).

It may be beneficial and profitable to invest in funds that hold real assets. Nevertheless, a share of such a fund would only maintain (or increase) its value over time if shares of the fund **are not themselves used as currency**, and if their value is specified in a third currency (that actually will experience inflation). For example, if a fund unit that had a cost of 100 € some years ago today costs 150 €, then this value may correspond approximately to the former purchasing power. But the possibility of maintaining the original purchasing power by investing in such funds only works if the value is measured in a **third**, "virtual" (Soddy, 1933, 1934: 24 ff.) currency, e.g., Euros. If the shares of such an investment fund were used as currency **themselves**—i.e., if supermarkets designated prices in this currency, bank accounts were denominated in this currency and banks granted loans in this currency—it would give rise to the same problems that afflict our current monetary system. Without a hoarding fee, a (positive) interest rate would result. Even a low interest rate would, in the long run, force banks to constantly expand their lending in order to be able to pay interest on deposits. The quantity of currency units (i.e., of fund shares) in bank accounts would grow throughout the banking system due to interest, just as the supply of Euros or US-dollars grows today.

In the long term, the guarantee to redeem each monetary unit for the promised amount of real goods would not be able to be maintained, just as in the case of the gold standard (Galbraith, 1983). Backing a currency with real goods such as real estate, commodities or any other non-perishable goods cannot be maintained in the long run because as long as interest exists the money supply will multiply exponentially. This is, as explained above, a mathematical law, and it is true for every currency that can be lent for interest, regardless of whether it is backed by tangible assets or not. Money, in the long run, cannot simultaneously function as a means of exchange and a means of storing value. This contradiction results in the existence of interest and thus in the unavoidable increase of the money supply over time, which in turn causes inflation, financial crises, price bubbles (partial inflation) and ever-growing debt as explained above (Chap. 4).

In the long run, if a fund share is supposed to serve as a store of value, it cannot function at the same time as a medium of exchange. This may be possible in the short run, but not in the long run. As soon as investors and borrowers denominate their contracts in fund shares (or fractions of fund shares) instead of national currencies, these fund shares thus become currency.[11] However, fund shares, if used as a means of payment, would be subject to the same interest-related self-multiplying problem. Like gold-based currency, a time will come when the number of fund shares in deposit accounts can no longer be covered by real assets in the original proportion.

[11] Regarding this proposal see Reichmuth and Reichmuth (2001).

The solution to the problems inherent in our financial system can only be found in a financial system where loans are free of interest and deposits and cash naturally decay (pay a hoarding fee or negative interest rate), regardless of whether we back money with real assets or not. As has been described in Sect. 10.2.4, private equity investments remain possible in such a natural money banking system. Investment funds, which invest in stocks or other real assets represent a way to avoid the hoarding fee that applies to cash. However, if the fund shares *themselves* become currency, they will be subject to the same self-multiplying effect as any other easy-to-store currency.

13.4.3 Is a Tangible Wealth-Backed Currency Not Natural Currency?

Reichmuth and Reichmuth (2001) are convinced that if money and real goods are pegged together can humanity find its way back to unity with nature. Indeed, in the state of nature, a couple thousands of years ago, people exchanged goods for goods. Only later was money invented as a universal medium of exchange to overcome the limitations of barter. However, if money is not subject to natural decay (hoarding fee), it can be easily hoarded, unlike goods and services. People then begin to hoard their money instead of using it as a medium of exchange. This results in the problem described throughout this book. If, however, money was linked to perishable goods, then this would indeed represent a natural money system. Of course, such banknotes, which would then be classic vouchers (each representing a specific quantity of a certain good), would naturally also devalue at the same rate as the goods they represent. Such money, linked to perishable goods, would therefore not be suitable as a long-term store of value.[12] For this very reason, of course, hardly anyone would propose linking money to perishable goods. The purpose of backing money with real assets is precisely to ensure the preservation of value. However, linking money to non-perishable goods such as precious metals, non-perishable raw materials or real estate leads people to hoard money rather than using it as a medium of exchange. The problems inherent in our monetary system would therefore remain.

As mentioned throughout this book, Aristotle and later St. Thomas Aquinas considered interest to be unnatural.[13] The constantly growing total money supply and debt induced by interest is the cause for the growth imperative of the economy and the environmental problems associated herewith. We can conclude, the unnatural phenomenon of interest promotes the destruction of nature. Finding a way to

[12] Incidentally, it would probably also be unsuitable as a medium of exchange, since it would not be possible to ascertain from the banknotes the condition of the underlying assets. It would therefore be more convenient to directly exchange the goods themselves.

[13] See also Frederick Soddy (1934) who points out that interest violates natural laws, especially those of physics. Rudolph Steiner found that "There is something highly unnatural in the social order today. That is that money increases when you merely have it", author's translation, original wording in German: "Es gibt heute etwas höchst Unnatürliches in der sozialen Ordnung. Das besteht darin, daß das Geld sich vermehrt, wenn man es bloß hat", quoted from Seminar für freiheitliche Ordnung (1984: 35).

prevent people from excessive hoarding of money without the need for interest as a "*reward for parting with liquidity*" (Keynes, 1936: 167) will not only solve many of the problems the world faces today at their roots. It will also allow us to live in greater harmony with nature.

13.4.4 The Impossible Circular Reasoning

The fact that it is not possible to have a form of money that functions sustainably and at the same maintains its purchasing power can also be illustrated by the following reasoning. Value-preserving assets are, e.g., real estate. Imagine a real estate property with buildings that are rented out. The investment will only hold value if the rental income is used for regular maintenance of the buildings. Otherwise, the buildings will deteriorate over time and lose value. Let us now imagine that our money is backed by a fund that invests in such real estate, i.e., assets that only retain their value as long as regular maintenance investments are made. In what currency, then, do the tenants of the properties contained in the fund pay their rent? If shares of this fund were money, then the tenants would pay the rent with monetary units representing shares of the same real estate they occupy. In other words, they would pay the rent from the substance of the same property in which they live (Fuders, 2014a).

The reason that in our current monetary system a real estate fund can be a value preserving investment works because, as was described above, the real estate and the rental payments for tenants are valued in a third currency, such as the Euro. Here, this problem of circular reasoning does not occur because the real estate or the fund shares are valued in a third (virtual) currency. Inflation is not prevented with this currency, though. The value preserving function of an investment in such a fund will only exist as long as the fund shares are not used as currency themselves. There is no such thing as a perfectly purchasing power preserving currency. Either it is subject to some sort of hoarding fee (Gesell's solution), or there will be inflation and financial crises.

Even money backed by tangible assets has **no intrinsic value**, even though some might think otherwise (Reichmuth & Reichmuth, 2001).[14] "Money in itself is worth nothing, since it is incapable of generating any value", as the car manufacturer Henry Ford correctly put it (Seminar für freiheitliche Ordnung, 1984: 35). Of course, this applies all the more to our current fiat money system, which is not directly backed by goods. But it also applies to banknotes that are backed by a precisely defined quantity of goods. Money is then a classic voucher representing a specific quantity of specific goods. These goods have an intrinsic value, conferring a subjectively perceived benefit or value, but not the voucher (money bill)

[14] "The source of the stability of the currency, however, is its intrisnic value", author's translation from German „Quelle der Wertstabilität hingegen ist der innere Gehalt" (Reichmuth & Reichmuth, 2001: 79).

representing them. We remember, to confuse a banknote with the goods that can be bought by it would be the same as confusing the cinema ticket with the movie itself, as John Maynard Keynes (1983: 402) correctly put it.

13.4.5 The "Rentenmark" Was Not a Long-Term Solution Either

The "Rentenmark", which was issued in 1923 in Germany to tackle the highest hyperinflation the world had ever seen, can be understood as an indexed currency. It was indexed to or backed by 4% of Germany's real estate used for agriculture, forestry, industrial or commercial purposes (ODR[15] § 6 and 9). The real estate was, without asking the owners for permission, burdened with mortgages, which had priority over all other mortgages and did not have to be entered separately in the mortgage register (ODR § 6 in conjunction with § 10). This measure was intended to limit the number of Rentenmark bills and coins in circulation to 3.2 billion Rentenmarks (ODR § 2). The Rentenmark was quickly accepted by the people and inflation was eliminated. However, the Rentenmark existed as the sole currency for less than a year. As early as 1924, the "Reichsmark" was introduced in parallel.[16] Until the end of World War II, the Rentenmark circulated alongside the Reichsmark.

Even though the introduction of the Rentenmark initially solved the inflation problem immediately, a currency backed by land, like all other currencies backed by tangible wealth, is not a long-term solution to the real problem. The Rentenmark only remained stable in terms of value because shortly after its introduction the Reichsmark, which was subject to inflation, was introduced in parallel. Even the mortgages, which secured the Rentenmark, were not denominated in Rentenmark, but in "Goldmark". One Goldmark was defined as 1/2790 kg of fine gold (ODR § 6 par. 1 and 2). If, however, the Rentenmark had remained the only valid currency, and consequently all bank accounts and prices had been denominated in Rentenmarks and loans had been granted in Rentenmarks, the Rentenmark money supply would also have grown as a result of interest-driven money creation and the *money multiplier*, even though the quantity of Rentenmark money units printed would still not have exceeded the 3.2 billion stipulated by law. But when loans are granted on the one hand, debt securities (which increase the money supply) are booked on the other. The interest to be paid on the deposits would also have forced the banks to steadily expand credit volume and thus money creation via the *money multiplier*. The share of German real estate represented by each Rentenmark would thus have steadily declined.

[15] ODR, Ordinance on the Establishment of Deutsche Rentenbank (Verordnung über die Errichtung der Deutschen Rentenbank) of October 15, 1923 (RGBl. I. pp. 963–966). The value of the land was calculated on the basis of the so-called military contribution value in gold marks (§6 par. 1–3).

[16] Coinage Act (Münzgesetz), RGBl. II 1924, p. 254 ff.

13.4.6 Backed Currency with Hoarding Fee?

Theoretically, currencies backed by tangible wealth could incorporate a hoarding fee (demurrage fee, carrying cost, Gesell tax). Such a proposal was made by Bernard Lietaer, who termed such a currency "Terra" (Lietaer, 2002). Unlike other backed currencies, the Terra is subject to a hoarding fee of 3.5–4% annually, because storage costs for the goods used to back the currency are passed on to the money holders. Long-term hoarding is thus prevented. A currency backed by tangible wealth, which cannot be hoarded eternally, is a variation on Silvio Gesell's natural money that deserves serious consideration. The backing of the currency with tangible wealth would increase confidence in such a currency. The negative effects of our existing form of money would be absent since this kind of money would be non-hoardable. The interest on loans issued in such a currency would tend toward zero.

Especially suitable for such a real values-backed currency could be seeds. These are somewhat storable but not indefinitely and the hoarding requires space and dedication and is not free of costs. However, the backing by **tangible wealth only has a psychological effect**. What is decisive for the proper functioning of such money is its non-hoardability, which ensures that it would circulate consistently and reliably. Even without the backing of tangible wealth, each monetary unit indirectly represents a share of economic productivity. The problem with our current monetary system is not that money has no backing, but that the number of monetary units grows over time, first slowly and then more and more rapidly. This can only be prevented by lowering the interest rate to zero. The only way to do this, is to deprive money of its unnatural property as a means of hoarding.

13.4.7 Conclusion

In summary, we can state that fund shares could function as a store of value in a Gesellian natural money system. If people want to save for the future, they could do so through private equity investments (as explained above in Sect. 10.2.4) and also, of course, through investments in funds. This way, as is the case with any equity investment, the investor "naturally" shares in the entrepreneurial risk. Any positive return on the investment ("interest") would correspond to the real economic profits of the companies. The money supply would therefore not decouple from the real (productive) economy. Naturally, equity investments have the disadvantage that they do not generate guaranteed interest income.

If one tried to use fund shares as a means of payment, such money would be subject to the same flaws as the current monetary system as would any form of money not subject to some sort of hoarding fee. The quantity of fund shares held in bank accounts would multiply via interest. However, because in reality there would not be an unlimited number of fund shares, lending and thus the circulation of fund share as money would soon come to a halt. The result would be devastating deflation. The only possible solution to this problem would be a reduction of the quantity of tangible wealth backing each fund (money unit) share, as has happened frequently under the gold standard. The quantity of gold that each unit of money represented in the various gold standard systems was regularly reduced until eventually the gold standard was completely abandoned. Again, money cannot, in the long run, function simultaneously as a medium of exchange and a store of value.

13.5 Inflation Adjusted Accounting Units Like the Chilean UF

In some South American countries, it is common to express prices in an inflation adjusted accounting unit, particularly in the financial sector. Chile was a pioneer in this practice, using the inflation adjusted Unidad de Fomento (UF) since 1967. The UF is an accounting unit (*unidad de cuenta*), whose value is calculated based on a hypothetical basket of products—*i.e.*, using the same method through which the consumer price index is calculated. The UF was originally used to ensure that the interest rates on mortgage loans were always higher than the inflation rate. Today, the UF, whose value in Chilean pesos is calculated daily, is principally used in the financial sector. All Chilean banks use UF to express interest rates for loan contracts, and often bank fees as well. Since most people finance the purchase of real estate through mortgage loans, real estate prices in Chile are usually also expressed in UF. Prices expressed in UF have remained relatively stable, recording almost no inflation.[17] Expressed in Chilean pesos, however, property prices have indeed risen with inflation, because the UF is becoming increasingly expensive. In this way, the UF has served its purpose of giving banks a reliable basis for calculating interest rates and ensuring that they are always set above the inflation rate. Employees whose employment contracts are expressed in UF are protected against inflation. So, one might therefore ask, why not use the UF instead of the Chilean peso as a general method of payment?[18]

The fact that the UF is relatively stable in value, and therefore not subject to inflation, is probably due to the fact that it is not a currency, but rather an accounting unit whose value is expressed in a third currency—the Chilean Peso—that is subject to inflation (Fuders, 2014a). It is not the case that prices expressed in UFs have not increased. In fact, real estate prices, for example, have risen significantly. Consider, for instance, a property that cost 5000 UF 10 years ago and has the same UF value today. 10 years ago, this value corresponded to 50 million Chilean pesos, whereas

[17] With exception of the last 10 years, but this is because the real estate market became subject to speculation, which in turn is a secondary phenomenon of an ever-growing money supply, as explained throughout this book.

[18] The owner of a Swiss private bank, Karl Reichmuth, asked me this question. He is co-author of the book: "Der RealUnit" (Reichmuth & Reichmuth, 2001).

today it corresponds to 160 million pesos, since the UF 10 years ago was equivalent to about 10,000 pesos and today it is equivalent to 32,000 pesos.[19] If we were to use the UF not only as an accounting unit for a third currency but as money itself, the UF would probably also have recorded inflation. That is to say, if all businesses established prices in UFs, and if bank accounts were denominated in UFs and loans were issued in UFs, then the amount of UFs in UF bank accounts would grow via interest and banks would be forced to expand lending in UF. The problem of an expanding money supply would then also affect the UF, which is why there would also be UF-inflation (Fuders, 2014a).

13.6 Time Banks and Time Vouchers

13.6.1 Time as Currency?

Time-based currencies work as follows: people work or provide services for, let us say, 1 h and thereby get 1 h of time credited to their accounts at the Time Bank. This credit can be redeemed for 1 h of work done by another person. For example, if a craftsman repairs someone's bathroom, and this service takes 3 h, the craftsman will get 3 h credited to his account. The person who received the service will have these hours deducted from his or her account. The Valdivian Time Bank is an example of such a system (BDTI, 2020).

While time banks today usually work purely electronically, older systems used physical time money bills. One of the first and largest time-based currency systems using physical bills was Ithaca HOURs, where each bill represents 1 h of time (Kennedy et al., 2012). The system, which no longer operates, functioned between 1991 and 2015. "Time vouchers" are a similar system as, e.g., the "Minuto" (Fig. 13.1) in Germany. With these vouchers, which are issued in denominations of 1 min, 15 min, and an hour, recipients of services pay in terms of the time it takes to provide the services they receive. On the backs of these vouchers the recipient of the service states the service they can offer in return, and anyone who holds these vouchers can demand this service.

The idea is that these vouchers start circulating as money backed by concrete services. For example, a craftsman fixes the air conditioning at a podiatrist's home which takes, let us say, 1 h. The podiatrist then issues a "Minuto" voucher of 60 min. On the back of the "Minuto" voucher the doctor states his service as podiatrist (Minuto-regional, 2022). On the front of the bill two other people guarantee and sign for the service. The craftsman can then demand 1 h of the doctor's services or, alternatively he can pass the voucher on to someone else who can use it to pay for other services.

[19] The daily values of the UF since 1990 can be downloaded from the website of Chile's tax authority. See: http://www.sii.cl/pagina/valores/uf/uf2012.htm.

Fig. 13.1 Minuto Zeitgutscheine (time vouchers). (Photo: Felix Fuders)

13.6.2 Does It Make a Difference What We Call the Bills?

The time-based money systems facilitate the exchange of services. On the website of the Valdivian Time Bank, which belongs to a nationwide network of time banks in Chile, it is explained that this system is supposed to be an alternative to money (BDTI, 2020). On the website of the "Minuto" we read similar thing. According to its founder, the minuto system even makes interest-free minuto loans possible (Minuto-regional, 2022). To answer the question whether such time-based barter systems represent a true alternative to the current monetary system we need to examine whether or not the time-based currency can be hoarded indefinitely. After all, what difference does it make if we call our monetary units dollars, euros, pesos or hours? The answer is: it doesn't matter. As long as the currency (be it electronic or physical currency) can be hoarded indefinitely, the system will not differ fundamentally from our existing monetary system since the same basic flaw exists in both systems.

To expand upon this point, let's suppose that we pay with time units instead of euros or pesos; the Time Bank is the legal payment system. All goods and services are priced in units of time. Bank accounts are denominated in time-money instead of conventional euro or peso accounts. Anyone who ran out of time units would have to borrow them. The same applies to companies planning investments, which would have to be paid for in time units. As long as such a time-based money system does not include a hoarding fee or other disincentive to hoarding, some people will hoard time money, while others will need it urgently. As with the conventional monetary system, in a time-based money system those who have accumulated currency have no incentive (other than charity) to lend them out, unless interest is offered as an incentive. Therefore, a time-based money system with hoardable currency would in the long run be plagued by the same problems as our conventional monetary system.

13.6.3 Only Non-storable Time-Based Money Is a True Alternative

The Ithaca Hours, time money bills did not incorporate any restrictions on hoarding which is why they are **not a true alternative** to the current system. Time units at the Valdivian Time Bank are, during a person's lifetime, indefinitely storable. So, they **are not a true alternative either**. Again, any Time Bank system can only be a true alternative to the current monetary system if the currency is not indefinitely storable.

The aforementioned "Minuto" time vouchers are not indefinitely storable. They expire after 5 years. This creates a strong incentive not to hoard for too long. Therefore, they can indeed be considered a true alternative to the current system.[20] However, it must be highlighted that the distinguishing feature compared to the current monetary system is non-storability, not the fact that they are backed by concrete services.

13.6.4 Time-Based Money Is Not "Positive" Money

At different conferences I have heard proponents of time-based money pointing out that such money is "positive" money, not debt-based money, and that this is why it would function in a sustainable manner. It would be "positive money", so the argument goes, since the money is paid only after services are provided. Each voucher or time unit in a Time Bank account represents real work: behind every time-money unit, there is a service that has already been performed. However, the same is true in the current system where we usually first work and then get paid. The holder of the time units in a Time Bank account cannot force anyone to exchange these time units for goods or services, just as how in the current system we cannot force anyone to take our euros or pesos in exchange for goods and services. And what is even more important: for the person who provided the service the time voucher or time unit at a Time Bank account is an asset, while for the Time Bank (or time voucher community) it means a debt. The time bank community is in debt with the person who provided a service. In fact, each time voucher can be thought of as a bearer bond (Kennedy, 2011). Or, to put it the other way around, the people who provide a

[20] Maybe 5 years is a too long period. In case people started to charge interest for loans the "best-before" date of each minute bill could be shortened to a year or even months.

service give credit to the system because they trust ("credit" comes from Latin "credere" = to believe, to trust) to receive services worth the same amount of time. That is, time money too is the counterpart of debt. As explained in Chap. 2, the fact that money is the counterpart of debt cannot be avoided, nor is it a problem.

13.6.5 Some Problems Associated with Time-Based Money

Is an hour of a dentist's labor worth the same as that of a gardener? Some time-based money systems such as Ithaca Hours, therefore, permit negotiations and let physicians and dentists charge more "hours" per hour. But most time-based currencies value participants' contributions equally: 1 h equals one service credit, as is the case, e.g., at the Valdivian Time Bank. In these systems, one person performs an hour of work for someone and is thus credited with 1 h, which they can then redeem for an hour of work from someone else. The idea is, as stated on their website (BDTI, 2020), that every person's time is worth the same.

However, one could argue that such a system generates inequality since study time is not accounted for. If, for example, 1 h of services is provided by a person who needed 15 years of university studies (or 15 years of practical experience) to be able to provide the service and only results in being credited with 1 h in the Time Bank, the 15 years of study are not taken into consideration. These are 15 years during which students typically do not earn any income. If university studies are not taken into consideration, this is fundamentally unfair and disincentivizes the participation of such people in the system.

And here is another problem. A time-money system may encourage people to perform less efficiently than they theoretically could. In fact, they might even try to artificially overstate the amount of time necessary to complete a job, since this would result in receiving more time units. This incentive structure would disadvantage those who perform most efficiently.

13.7 Social or Local Money

13.7.1 Promoting Local Commerce

Social money, sometimes also called local money, regional money or complementary money[21] (in former times in the US also "scrip", see Fisher & Cohrssen, 1933; Champ, 2008), are currencies created and used in communities to facilitate the

[21] These terms are often used synonymously. However, some authors try to distinguish between regional and complementary currencies (Kennedy et al., 2012) or to distinguish between complementary currencies based on their nature and purpose (Blanc, 2011). On the emergence of complementary currencies see also (Gelleri, 2020, 2022).

exchange of goods and services alongside the legal tender. These currencies are usually only accepted in a certain region and intend to avoid the leakage of money to other regions and in so doing to boost local commerce and optimize the use of local resources (Lietaer & Belgin, 2004). In most cases it circulates alongside the legal tender currency. There are countless examples of such initiatives worldwide. As far back as 2006 in a paper published by the German Bundesbank 65 local money initiatives were identified (Rösl, 2005, 2006). In Spain there are currently more than 100, and some estimate that worldwide there are about 5000 such initiatives (BDTI, 2020). The idea is simple: Most poor or underdeveloped regions are not inherently poor, i.e., it is not that people have less abilities or that there are less resources that can be exploited, but rather money, which is needed to exchange goods and services, does not stay in the region. Development attracts development, and money flows from less developed regions to more developed ones, causing unequal and uneven development. Wealth and development concentrates in specific regions while others get progressively poorer. Local money intends to avoid this leakage of currency and to promote a decentralized local development.

13.7.2 The Extraordinary Example of Banco Palmas

The "Palma" is an example of how a local currency can promote local development. This local currency helped a formerly very poor suburb of the Brazilian town Fortaleza, the Conjunto Palmeira, to stimulate development. The Palma, first issued in 2001 by a neighborhood community bank called "Banco Palmas" (founded 1998), is probably one of the most successful local currencies in terms of promoting local development. It has been estimated that as of 2012 as many as 1800 jobs were created as a result of this initiative (Kennedy et al., 2012). Since the Palma currency cannot be spent elsewhere and converting it back to the Reais incurs a 1% fee, it has a greater tendency to continue circulating within the community. One of the co-founders once marked a Palmas bill with a pencil and reported that the same bill returned to his till, a local grocery store, five times. By contrast, a marked bill of the official Brazilian currency, the Reais, was never seen again (Kennedy et al., 2012). The initiative has achieved international recognition, gained the **UN Millennium Development Goals Award**. For more on the Palmas and how local currencies can contribute to local development see Kennedy et al. (2012), Fuders (2016), and Banco Palmas (2022). According to a representative of Banco Palmas at a conference held at Universidad Austral de Chile, as of 2019 there have already been 113 other such community banks inspired by the extraordinary example of Banco Palmas.

13.7.3 Only Local Money with a Hoarding Fee Is Natural Money

Even though local currencies have the power to promote local development they are **not a true long-term working alternative** to our current monetary system if they are not subject to some sort of hoarding fee or other incentive to circulate. Imagine if this system were implemented not alongside the legal tender currency but was itself the legal tender currency. All prices and bank accounts would be denominated in this currency. If there is no disincentive to hoarding, some people will hoard, while others will need it urgently. As is the case of the conventional monetary system, someone who has money saved will have no reason (apart from charity) to lend it, unless interest is paid as an incentive to do so. This means, the phenomenon of interest will arise, leading to the exponential growth of bank accounts and debt.

Fortunately, some local currencies do, in fact, incorporate hoarding fees, for example the German **Chiemgauer**, a local currency that has been in circulation since 2003 in the Bavarian region Chiemgau, mainly in the cities Rosenheim and Traunstein. Chiemgauers can be acquired 1:1 against the legal currency, the euro, but changing them back into euros incurs a 5% fee, providing a powerful incentive to spend them rather than exchanging them back into euros. Currently about 500 businesses and 300 charitable foundations, associations and clubs participate (Gelleri, 2009, 2022). The hoarding fee is 3% and must be paid if the bills are held longer than 6 months. The fee therefore amounts to 6% per year. According to Christian Gelleri, the founder of Chiemgauer, on average, each participant pays only one euro of hoarding fees per year (Gelleri, 2018). But the psychological effect is huge, because every Chiemgauer user makes sure to spend the Chiemgauer quickly.

The very small financial disadvantage is therefore offset by a major advantage for the community: The circulation of the Chiemgauer is consistently high and leads to stable sales (Gelleri & Stodder, 2021). On Chiemgauer bank accounts, which today can be held at most German Banks in the region,[22] this fee is charged automatically according to a specific algorithm. With its physical bills Chiemgauer applies the classic Gesellian stamp scrip method (see Fig. 13.2).

The hoarding fee is not pocketed by the Chiemgauer bank but donated to one of the participating charitable organizations. This is the "trick" how the Chiemgauer rapidly got accepted in the region. The organizations have an incitive to accept the local currency since only then they are prone to receive the donations. In addition, when someone changes its euros in Chiemgauer to participate in the system 3% goes right away as donation to one of these charitable organizations. If, for example,

[22] Chiemgauer accounts can be opened at the cooperative banks (Genossenschaftsbanken) and at mutual savings banks (Sparkassen). Those two types of banks together hold about 70% market share. That is, most people have a bank account at one of these two banks.

Fig. 13.2 Chiemgauer bill (front and back side). (Source: Chiemgauer, 2022)

someone changes 100 € into 100 Chiemgauers the Chiemgauer issuing cooperative Regios eG transfers 3% to one of the connected charitable organizations. The participants when acquiring the local currency can decide which of the connected charitable organization should receive the donation (Chiemgauer, 2022). Since only organizations with charitable purpose—in Germany there are many of them—can receive these donations, people feel less uncomfortable should they forget to spend the local money at time and must afford the hoarding fee. Loans in Chiemgauer are generally interest free meanwhile the loan is duly amortized.[23] For businesses and charitable organizations, the Chiemgauer represents an effective marketing tool as it promotes customer's loyalty, customer acquisition as well as a form of club sponsorship since all members of the network are potential customers.

13.7.4 Argentina's Patacones & Co.

One of the most successful recent experiences with local money took place in Argentina after the 2001 financial crisis, when a bank run occurred, and banks were not able to repay customers' deposits. Some provinces started to issue their own money. This money was exchangeable 1:1 with the Argentinian peso but it was not possible to deposit it in banks. Even though this local money was not explicitly based on Gesell's theory, it had similar effects. It was not possible to hoard it for a longer period at home because inflation in Argentine pesos was high and the local money was 1:1 bound to the peso. Therefore, it circulated rapidly and helped pull the country out of the deep economic depression (Louge, 2015). The most successful provincial money was the "Patacones", which was issued by the province of Buenos Aires and was also accepted in other parts of the country. Other provinces

[23] For those who do not amortize on time, a penalty rate will be charged.

Table 13.1 Local money in Argentine 2002

Province	Bono (local money)	Number issued	Acceptance in %
Nación	Lecop	3300	100
Corrientes	Cecacor	200	<50
Entre Ríos	Federales	318	60–70
Chaco	Quebracho	50	No information
Buenos Aires	Patacones	2.695	100
Tucumán	Bocade	168	85–90
Formosa	Bocanfor	50	70%
Catamarca	Bono Público	48	100
La Rioja	Bocade	10	100
Córdoba	Lecor	800	100
Mendoza	Petrom	80	100

Source: Own work, based on Louge (2015)

also issued their own money although they were not as successful as the Patacones (see Table 13.1).

Despite this positive example of how to bring an economy out of deflation caused by a banking crisis, the Patacones only circulated well because they were pegged 1:1 to the Argentine peso and thus linked to inflation. Inflation has the same effect as a hoarding fee. However, the idea of natural money is precisely to avoid inflation by using a hoarding fee and not to avoid the need for a hoarding fee via inflation. In order to become true natural money and represent a true non-inflationary alternative to the Argentine peso, local money needs to incorporate a hoarding fee.

13.7.5 Some Legal Issues

One might ask, why in the case of Banco Palmas or Chiemgauer, the central bank does not intervene as it did in Wörgl (see Sect. 14.2.2), the classic example of a very positive natural money experiment? The answer is that there is a difference between the stamped money in Wörgl and all the modern regional money initiatives. Neither the Chiemgauer nor the Palma were implemented by municipalities, as was the case in Wörgl, but instead are run by community organizations, which pursue narrowly defined social or environmental missions (Gelleri, 2020). This is the reason why Palmas and Chiemgauer as well as many other social currencies around the world have not been prohibited so far. It is noteworthy that in the case of Banco Palmas, the Brazilian central bank did initially fight against the initiative. The Central bank sued Joaquim Melo, one of the local organizers of Banco Palmas. In the eyes of the Brazilian central bank, he had committed fraud, issuing this private "false" money. But Joaquim Melo won the lawsuit. The court did not see any reason why private people, enjoying the constitutionally guaranteed freedom of contract, should not be allowed to use such vouchers to facilitate the exchange of goods and services. The judges also affirmed that access to finance is a constitutionally guaranteed right (Kennedy et al., 2012). Since then, the Brazilian central bank changed its attitude towards Banco Palmas and even started to promote the establishment of community banks (for more on Banco Palmas see Banco Palmas, 2022; Fuders, 2016).

13.7.6 Outlook: Confederation of Local Currencies

We could also imagine a sort of "confederation" of local complementary currencies (Fuders, 2016). This could be a way how money based on Silvio Gesell's proposals could be implemented throughout a national territory via private initiatives, i.e., without government involvement. The exchange of goods with other regions would still be possible, either by exchanging local currency into legal tender, e.g., the euro, and then exchanging the euro into other local currencies. In the event that there was no longer any legal tender (perhaps due to a severe financial crisis causing people to lose confidence in the euro and stop using it), a clearing system could be set up to exchange local currencies directly, comparable to how the foreign exchange markets work today.

In the context of this book, it is crucial to note that the key feature of social money is not the fact that it circulates locally and fosters the local economy but that it is subject to a hoarding fee. The positive impact in terms of promoting local development is just another positive feature of local currencies. Of course, these effects would disappear the moment such a currency is introduced nationwide. But even then, a parallel currency operating nationwide would have the power to promote the national economy instead of that of a certain region. This is because producers of imported products, unlike domestic producers, would have to convert the national complementary currency into legal tender in order to be able to exchange it for the respective currencies of their home countries on the foreign exchange market. The hoarding fee, which also must be paid exchanging the complementary currency back to the legal tender, they would have to add to the price. This would make imported products relatively more expensive. In contrast, domestic producers would not necessarily be disadvantaged by customers paying with non-storable complementary currency, because they would be able to pay wages and buy intermediate products with that currency. In the long run, this would encourage supermarkets and shopping centers to offer more domestic products.

13.8 Credit Cooperatives and Trustee Savings Banks

13.8.1 The Idea Behind Credit Cooperatives

The first credit cooperatives, also called credit unions, were founded in Germany about 160 years ago. Generally, in cooperatives people unite in order to achieve a common goal. The first cooperatives that came into existence were not credit but farmer's cooperatives. Farmers united to buy seeds, since seeds can be acquired cheaper using economies of scale. In Germany, today most farmers belong to at least one such cooperative. Nowadays, cooperatives are becoming ever more popular. For example, in Spain in the city of Barcelona there have been 61 cooperatives formed in just 1 year (Catà, 2018).

To combine efforts is likewise the purpose of credit cooperatives. The first credit cooperatives were founded in Germany by Herman Schulze-Delitzsch and Friedrich Raiffeisen. Originally the idea was to permit persons access to credits more easily. Interest rates were in many cases lower than at commercial banks. In Germany up to the present day these banks are called Volks- und Raiffeisen banks.[24] They hold about 27% market share in the German banking sector (Statista, 2020). The idea spread around the world. As of 2020 there were 86,451 credit unions in 118 countries with a total of 375,160,065 members which issued a total loan volume of USD 2 trillion (World Council of Credit Unions, 2021). Unlike the first credit cooperatives, today interest rates at cooperative banks are not generally lower than at commercial banks.

The same idea that led to the creation of credit unions prompted municipalities in Germany, Austria and Switzerland, and later in many other countries to form publicly owned trustee savings banks, called "Sparkassen". In Spain we find the "Caja/Caixa de ahorros" and in Chile the "Banco Estado". Germany is the country where such institutions have the highest market share; the Volks- und Raiffeisen banks and Sparkassen together hold about 65% market share (Statista, 2020).

13.8.2 Evaluation

Unfortunately, the idea of offering interest free loans, or at least loans at significantly reduced rates, has never really worked for prolonged periods. Some credit unions granted (and still grant) loans at lower than market interest rates. However, this is not the case for most credit unions or cooperative banks nor for trustee

[24] For example, the biggest cooperative bank in the region of Aachen, Germany, is the Volks- und Raiffeisen Bank. I am very happy to mention that it was my great-grandfather, Anton Josef Fuders who played a leading role in the founding of this bank (VR Bank eG, 2013).

saving banks in Germany, Austria and Switzerland.[25] The same is true, e.g., in Chile with the Banco Estado. Even the Swiss WIR bank, founded 1934 as a cooperative **and expressly based on Silvio Gesell's proposals** and the idea that money should flow without interest, today charges market-based interest on loans (WIR Bank, 2022).

Why has this happened? The answer can be found in the fact that money which is not subject to some sort of hoarding fee simply does not circulate properly. Such money will inevitably be hoarded physically at home "under the mattress". For the holder of the money who has no interest in an interest-free loan (perhaps because he or she already owns a fortune), there is no incentive to participate in a credit union. For such people, there is no incentive to lend the money except for interest. Now imagine that some members of a credit union are unable to properly repay their loans, then the credit union has a problem. What should it do then? The answer is it must borrow from the money market to make up for the lack of inflow of the defaulting loan. In the current system with money that has a design flaw and gets stuck "under the mattress", the money market requires the payment of a positive interest rate. The credit union must pay interest in order to borrow money and refinance the defaulted loan. Consequently, it must start to charge interest on its loans.

13.8.3 The JAK Cooperative: An Extraordinary Example

A very positive example is the Swedish JAK[26] credit cooperative founded in 1965 that was able maintain very low interest rates of less than 2% on loans over the course of decades (Kennedy, 2011). Currently the rate is 1.95% for mortgage loans (JAK, 2022). From the beginning, the guiding principle was the idea of offering loans interest-free as much as possible, as interest was rightly seen as the cause of many social problems and financial crises (Wikipedia, 2022). The "trick" that made and still is making it possible to offer loans with very low rates is to

- allow only members of the cooperative to be granted loans and to have savings accounts,
- JAK bank does not refinance in the capital markets, i.e., all loans are financed solely by member savings,
- savings accounts do not pay any interest and
- default rate is extremely low

[25] The credit cooperatives as well as the trustee savings banks in Germany and Austria still offer building-loan agreements, where people agree to deposit money in savings accounts paying lower than market interest rates in exchange for being offered mortgage loans at lower than market rates.

[26] JAK is an acronym for *Jord Arbete Kapital* (Land Labour Capital), which are the classic production factors.

The latter is due to the fact that all borrowers are members of the cooperative and have a vested interest in the success of the cooperative. This promotes solidarity among members, who feel like one big family; the bank offers help to members who cannot pay, while those who have received loans do everything possible not to default. We might refer to this as *careholder value* instead of shareholder value (Kennedy, 2011).

Since JAK does not refinance via capital markets and savings accounts do not bear interest, the interest rate charged on loans is **solely directed toward the operating costs** of the bank (JAK, 2022). In fact, since the bank finances loans solely from members' interest free savings accounts, no liquidity premium applies. And if defaults are kept to a minimum, then there is no risk premium either (on the components of interest rates see Sect. 4.2). If the only reason to charge "interest" is to finance operational costs, then this "interest" is, in fact, **not interest at all but rather an administrative fee** for the service of brokering credit. In Sect. 10.2.2 we discussed the fact that it is not always easy to differentiate between commissions and interest. Indeed, in some situations a commission could be thought of as a camouflaged interest rate. In the case of JAK, however, it is the other way around. What the credit cooperative itself terms "interest" is in fact a commission. A fee that has neither a liquidity, nor a risk premium component is not really interest.

13.8.4 Outlook

Even though JAK is an extraordinary and positive example, it is still **not a true alternative** to the introduction of Gesell's natural money. This is because even if we could incentivize all credit cooperatives to go back to their roots and operate in the same way as JAK, these credit unions would still not be able to guarantee the circulation of money—and serve as the "heart" of the economy, to return to the metaphor of money as the blood of the economy—as they only hold a small share in the credit market. To be a real solution to the problems inherent in our monetary system, all banks in an economy would have to operate according to this principle. However, **the latter is impossible as long as money is easily storable**. Remember, not all people need loans. Those who don't, would have no reason to become members in such credit cooperatives engaging in savings contracts where they abstain from liquidity for many years without receiving interest.

What would happen if the government simply banned conventional banks and only allowed cooperatives like JAK? Then people with no need for loans would likely prefer to hold their money at home. Therefore, money would not circulate well, and the economy would eventually plunge into recession or even depression. Soon, conventional banks with interest-bearing savings accounts (and loans) would reappear. The only way to prevent this is to charge a hoarding fee on money. Then the liquidity premium will disappear by itself, even in banks that do not operate as cooperatives.

13.9 Cryptocurrencies

13.9.1 What Is Bitcoin?

There has been a lot of coverage in the international media about the digital currency "Bitcoin", introduced in 2009, and other cryptocurrencies inspired by it, such as "Ethereum", "Litecoin", "Ripple" (Krempl, 2017; Fuders & Schreiber-Martens, 2018; Andriotis et al., 2019; Irrera & Wilson, 2020; Marcus, 2020). Bitcoin has not only attracted interest in the media, but also in academia. A search for "Bitcoin" in the web-based database of scientific articles "Science direct" yields (as of May 2022) 4575 search results. Governments in some countries are considering officially introducing digital currencies (Simon, 2018), and taxation of profits from trading in digital coins is also being considered and has already been introduced in some countries, e.g., Chile. Central banks around the world are trying to assess the growth potential, implications on banking and price stability of cryptocurrencies (ECB—European Central Bank, 2015; Bank of England, 2021; Liao & Caramichael, 2022), and blockchain, the technology behind Bitcoin, has not only been proposed as the basis for a central bank cryptocurrency (Danezis & Meiklejohn, 2016), but some central banks have already begun to officially consider launching central bank digital currencies (Bank of England, 2022). For some, Bitcoin has become a kind of "ideology" and some "believers" think that cryptocurrencies are the "answer" to many of the world's problems (Wash, 2022).

One of Bitcoin's primary characteristics is that it does not require a central settlement authority to function, because it is based on so-called "blockchain" technology, a decentralized peer-to-peer application (Kroll et al., 2013). This technology supposedly makes it impossible to counterfeit because the necessary information is stored in a distributed manner on millions of computers worldwide. Another selling point of this technology is its anonymity. While every computer connected to the internet has an IP address that is not anonymous, blockchain technology supposedly allows anonymous buying and selling and thus anonymous payments. Many buyers are also attracted by the prospect of speculative profits. In US dollar terms, the bitcoin price has risen from USD 0.09 in 2009 to almost USD 69,000 at the end of 2021.

13.9.2 Blockchain Currency as Legal Tender?

The reason why Bitcoin is so expensive when measured in US dollars or euros is not because Bitcoin is good money but rather because there are more and more US dollars and euros in circulation. Remember that the money supply of any storable and thus interest-bearing currency grows exponentially. Graph 5.3 and 5.5 show this exponential growth of the US dollar and euro money supply. All these US dollars and euros need to be invested somewhere, and this is where price bubbles come from. We previously discussed the fact that speculation and price bubbles are a symptom of the current monetary system. Bitcoin is perceived by many to be a good investment. The fact that its value has risen so much when measured in US dollars or euros creates the impression that Bitcoin is a "safe haven" (Umara et al., 2021), similar to investing in gold.

This makes it conceivable that entire countries might choose to make Bitcoin legal tender, especially in response to financial crises. In 2010 the Greek finance minister openly discussed such an idea. It is conceivable that the US dollar or the euro might eventually be converted to blockchain technology in order to create the impression of a safe, crisis-free currency (Simon, 2018; Mühlbauer, 2018; Bank of England, 2022). After all, blockchain is supposedly a safe technology (for a more critical view, see e.g., Kroll et al., 2013). Especially in the event of a collapse of the financial system—and this is to be expected soon—attempts could be made to convince the public that such digital currencies represent safer alternatives. It is argued that financial crises are less likely with this new technology because it prevents expansion of the Bitcoin money supply, which is seen as an advantage over a classic fiat currency (Bank of England, 2021).

13.9.3 Evaluation

The idea of replacing our current monetary system with one based on blockchain technology is similar to the call for the reintroduction of the gold standard, which is a common refrain but would not solve the problem inherent in our flawed monetary system, as discussed above. We would then have a purely electronic but still hoardable currency. This is actually not very different than what we have today. The existing national currencies are already booked, transferred and traded electronically. As long as a currency is easy to hoard, the same problems will arise. Let us imagine an economy where all prices are denominated in a new legal tender blockchain currency. Just like conventional currencies, people would need it to buy food and other essential goods. Some people would hoard the currency, while others would need it urgently.

Interest would thus arise as a non-hoarding premium (liquidity premium), which would cause both deposits and debts to self-multiply just like conventional currencies do. One might object that blockchain technology promises that only a certain amount of digital currency can be mined. The answer is that they will eventually have to change the technology so that more units can be mined. Alternatively, the quantity held in bank accounts and the number of blockchain currency units that actually exist will be decoupled, in the very same way as was regularly done with gold-based currencies, where the amount of gold that backed each currency unit was repeatedly reduced until the gold standard was eventually eliminated altogether. Eventually we would see the same price bubbles, financial crises and, above all, the compulsion for ever more economic growth, ever-growing deposits and ever-growing debt as is the case in the current system, in other words **business as usual**.

13.9.4 The Solution: Crypto Currency with Hoarding Fee

Despite everything discussed above, a digital currency could indeed be a solution to our current monetary problems if it incorporated a hoarding fee or some other mechanism to prevent long-term hoarding. Then we would have digital Gesell money. Such non-hoardable cryptocurrencies already exist: "Freicoin" (Freicoin, 2021) and "Bitgesell" (BitGesell, 2022) are two examples. The names of the two cryptocurrencies refer to Silvio Gesell's natural money, which Gesell himself had called "*Freigeld*" (free money). Both are based on the same technology as Bitcoin, but they are subject to hoarding fees. In this respect, Freicoin and Bitgesell represent a real advance over Bitcoin. Unfortunately, as with any means of payment subject to hoarding fees, it will not be easy to find supporters. Today, people and institutional investors buy Bitcoin precisely because it is storable.

Particularly interesting in this context is the result of a study conducted in Austria in 2021, which found that such a Gesell cryptocurrency has a "significantly higher potential for general acceptance compared to regular cryptocurrencies" if the local economy can be supported in this way (Janisch & Stapleton, 2021). For the study, a non-hoardable cryptocurrency was developed, the "Circle Coin". This study provides hope that Gesell money could be widely accepted, even though it is not infinitely hoardable. Crucial to acceptance is the communication strategy. People need to be aware of the positive impact on the local or national economy. Furthermore, if we can communicate to people that through such a currency we can fix the flaw in the design of our money and thereby address the most important challenges facing the world today, then I am convinced that people will prefer Freicoin or Bitgesell to Bitcoin.

13.9.5 What About 'Stablecoins'?

So-called "stablecoins" are cryptocurrencies based on blockchain technology that are pegged to a third fiat currency, usually the US dollar. One unit of such currencies are usually backed by one US dollar. Examples are the "Tether", "USD coin", "Biance USD" or "Dai". The idea behind this form of indexed blockchain currency is to reduce Bitcoin's price volatility. Bitcoin valued in US dollars or euros or any other third currency is very volatile. Many people are reluctant to use Bitcoin because they fear that it could lose a significant portion of its value. In the last 2 years stablecoins have been gaining ever more followers. USD-pegged stablecoins circulating on public blockchains have seen 500% growth in 2021 compared to 2020, with a combined circulating supply of nearly USD 130 billion as of September 2021 (Liao & Caramichael, 2022). The Bank of England as well as the US Fed have published discussion papers on this issue, trying to assess the impact of this development on credit intermediation and price stability (Bank of England, 2021; Liao & Caramichael, 2022).

Why would someone want to use this money? What is the advantage over traditional money? The answer is that people might prefer this money over classic fiat money since blockchain technology only allows a certain amount of currency units to be produced (mined) thereby allegedly preventing uncontrolled expansion of the money supply, in contrast to what happens in the current monetary system. At the same time, it eliminates price volatility when measured in third currency since the stablecoins are tied 1:1 to the USD or euro. However, to believe that blockchain-based cryptocurrencies are not fiat money is wrong. In the very same way as our current money, blockchain-based currencies are not backed by any real assets.[27] A so-called stablecoin is a **fiat currency backed by another fiat currency**. In any case, as we have discussed throughout this book the fact that money is not backed by real assets is not a problem. Instead, the problem—and indirectly the major challenges facing the world today—arises from the legal tender currency's hoardability. This flaw in our current monetary system is not solved by any cryptocurrencies, with exception of the aforementioned "Freicoin" and "Bitgesell".

13.9.6 High Energy Consumption

Regarding the sustainability aspect of blockchain-based cryptocurrencies, there is another significant problem. The electricity consumption of Bitcoin mining increases with the number of Bitcoin units generated, because the computing processes for each transaction become more complex as more units enter circulation (Mühlbauer, 2018). A single Bitcoin transaction equals several hundred thousand credit card transactions (Best, 2022). As of October 2021, Bitcoin consumed an estimated 180 terawatt hours, which is equivalent to the annual electricity consumption of Finland (de Best, 2021). From this point of view traditional money, even if not subject to a hoarding fee, is still preferable to blockchain technology.

13.10 Prohibition of Interest/Islamic Banking

13.10.1 What About a Simple Ban on Interest?

It is worth considering whether Silvio Gesell's relatively complicated stamp scrip money is actually necessary in order to ensure the adequate circulation of the currency without the necessity for interest to be offered as a reward for refraining from liquidity. Would it not be easier to simply ban interest by law? Indeed, the result

[27] The former president of the Dutch central bank compared the hype over cryptocurrencies with the first ever recorded speculative bubble, the tulip mania in Holland in the years 1634–37. But he concluded, and I can only agree with this analysis, that the Bitcoin bubble is even worse than the tulip mania, since "at least then you got a tulip, now you get nothing" (Hern, 2013).

would be similar. If there were no interest, saving money at banks would not be free. This is because banks would not receive an interest margin. Banks thus would have to charge fees for their services. Of course, people would still have the option to hoard money at home. But since doing so creates the risk of being robbed, larger sums would still be saved in banks, as is currently the case in countries that follow Sharia rules.

So, while Silvio Gesell's proposal seeks to reduce the interest rate towards zero by means of a hoarding fee, a ban on interest would create the need for a hoarding fee in the form of bank fees (Fuders, 2009a). The fees would then create an incentive for holders of money to pay banks for the service of intermediating lending so they could avoid paying fees on idle bank balances. In other words, just as in a Gesellian monetary system, money held in demand deposits would incur hoarding fees, while time deposits would not earn interest but would also not incur fees.

13.10.2 Avoiding Hoarding Fees by Saving Abroad?

One could argue that a ban on interest would lead people to save abroad in countries where interest is not banned. However, to invest money abroad, one must first buy the foreign currency on the foreign exchange market. Money that does not earn interest and is even subject to bank fees would certainly be less in demand on the foreign exchange markets, i.e., it likely trades at a discount relative to other currencies. This would make investing abroad less attractive.

Some banks offer accounts denominated in foreign currencies. For example, most private banks in Chile offer US-dollar denominated accounts. Now let us think about what would happen if the US government prohibited interest. Could people transfer their money to Chilean US-dollar denominated bank accounts and thereby avoid fees or even earn interest? The answer is, probably not. This is because the foreign bank will offer a fee-free foreign currency account (or even pay interest) only if it can do business with the foreign currency, i.e., if it can lend the money. However, who would pay interest for loans dominated in US dollars in Chile if one can get interest-free loans in the US? So, if Chilean banks cannot earn interest margin, they will be forced to charge fees on foreign currency accounts. Thus, the hoarding fees cannot be avoided.

13.10.3 Prohibitions Are Not Necessarily a Constraint on Freedom

The prohibition of interest on loans, as was actually enforced in Europe until the height of the Middle Ages (Weitkamp, 1993; Fuhrmann, 2003), is therefore worth considering. Prohibitions are not necessarily a constraint on freedom

(Schachtschneider, 1994, 2006, 2007), as some might believe. Rather, people achieve freedom, which is "born with the humanity of man", through respect for the laws that "protect each from the coercive arbitrariness of the other", as Immanuel Kant (1868a: 34) put it. Laws may restrict individual autonomy, and in fact many laws do, but such restrictions are sometimes necessary in order to ensure broader freedoms. For example, if we had the freedom to kill each other at will, and everyone had to worry about being killed by their neighbors at any moment, would this represent real freedom? It thus makes sense to prohibit homicide by law. The law, as Wolfgang von Goethe ([1800] 1987) put it in his famous Sonnet, makes us free. Jean Jacques Rousseau ([1769] 2012) in a similar vein spoke about the "social contract" (*contrat social*), a sort of compromise between individual freedoms and the collective general will aimed at protecting the common good. Laws thus have the power to create freedom, at least insofar as the laws are just and promote righteous behavior among citizens and do not run counter to morality, i.e., if they are in line with Kant's categorical imperative (1868b).[28]

So, prohibitions are not necessarily incompatible with a free society nor with the principles of a free market economy. For example, Milton Friedman, one of the most celebrated advocates of liberalism, believed the State is necessary to set the rules of the game within each participant in the economy can move freely. He compared the market economy to a game. There are rules and a referee. Within the confines of the rules, the players can behave freely (Friedman, 1966). A ban on the greatest evil in the economic system, interest, seems to me perfectly justifiable. To clarify, I am referring to interest on loans, not to the generation of profits in the real economy, which is often also referred to as "interest" in academia.

In summary, prohibitions do not contradict the free market model. Conversely, they may be necessary for economic freedom (Fuders, 2011). In a market economy, in which the market players are forced to perpetual growth by a constantly growing interest burden, many are not necessarily much freer than they would be in a government-imposed planned economy. What we need is a truly free market economy, that is free from interest. It must not be forgotten that Gesellian natural money would also be enforced by law, as is our current legal tender.[29] For example, some people would perhaps prefer to pay with precious metals instead of dollars, euros or pesos. But these are not currently accepted by any State as legal tender or for the payment of taxes.

---

[28] Which is basically a paraphrase of the biblical commandment to love one's neighbor (Fuders & Nowak, 2019). On the requirement of laws to be moral in order to promote freedom see, e.g., Schachtschneider (1994, 2007).

[29] For Germany: § 14 para. 1, sentence 2 BBankG; for Chile: Decreto Ley 1123 from 30.07.1975.

13.10.4 Islamic Banking

In many Islamic countries, there are at least some banks or financial institutions that operate without interest since they adhere to the principles of Sharia, a Koran-based body of religious Islamic law (on this see, e.g., Saeed, 1999). Some countries such as Iran, Pakistan and Sudan even expressly prohibit the charging or payment of interest. The often repeated claim that the prohibition of interest is circumvented by tricky business arrangements in these institutions (e.g., Hucko, 2009; Ginsburg, 2011) may be a misinterpretation resulting from the fact that, as explained in Sect. 5.6.5, our (Western) academic texts label all investment returns as "interest", regardless of whether they are profits generated from productive investment in the real economy or from lending money. To give just one example of such a misinterpretation, an article in the Financial Times (Hucko, 2009) stated that Mercedes-Benz in Dubai, in order to increase financing of its vehicles and to circumvent the ban on interest, formed a separate company that buys the cars and resells them to their customers on installments at a higher price. The basic thrust of the article is that this is ultimately the same thing as a normal loan, but concealed by complexities so that it can be claimed that interest is not being charged.

However, this financing model is not a money loan and the overall higher price that is paid does not represent interest. Even though the customers pay more than they would with a cash purchase, the surcharge is not interest for borrowing money. Rather, it is compensation for arranging different payment terms. The surcharge applied for different payment terms—contrary to a common misinterpretation in textbooks on banking theory (Borchert, 2010; Drukarczyk, 2003; Stiefl, 2005)—is not interest, but rather the rental fee for lending real goods (Fuders, 2008, 2010b). Here, **not money but products are lent**.

The surcharge paid by customers in Dubai through the Mercedes-Benz affiliate is not interest for a money loan but rather the leasing of cars until final payment is made, at which point ownership is transferred to the buyer. The product (the car) and the money are **not separated**, unlike in the case of a typical money loan. The leasing company bears the risk, e.g., that the customer finds a problem with the vehicle and does not continue paying. A conventional bank would not care whether the product the customer buys with the borrowed money has defects. The bank would collect interest regardless. If all "loans" functioned as in this example, the money supply would not decouple from the productive economy. A money supply that does not grow faster than the (productive) output of the economy produces a stable currency without inflation, price bubbles or financial crises. The real productive economy would not be forced to grow and there would be no redistribution of wealth via interest.

In the German literature on Islamic banking, this form of credit, which is common in Islamic countries, is correctly referred to as "material means loan" (*Sachmittelkredit*). This designation is apt, because material resources, i.e., goods, and not money, are lent. It is therefore not a tricky circumvention of the prohibition of interest.

13.10.5 A Ban on Interest Compared with Gesell's Natural Money

At first glance, an interest ban has two advantages over Gesell's natural money. First, an interest ban would probably be viewed as more palatable than Gesell's natural money. While lobbyists for interest recipients try to convince people that natural money is a bad idea (in German they pejoratively refer to it as *Schwundgeld*, or "shrinkage money"), the elimination of interest would not negatively impact most people.

However, charging deposit fees while money is easily storable encourages hoarding of money at home. We have been able to study the impact of near-zero interest rates on deposits while maintaining cash (physical notes and coins) which is easily storable at home in Japan, Europe and the US in recent years. Cash hoarding in these economies increased dramatically and the respective central banks had to counter the risk of falling into deflation (and deep economic depression) by printing money (*Quantitative Easing*). In other words, the central banks counteracted money hoarding by injecting new money into the economy, a situation that creates the risk that all the hoarded money could at some point be spent suddenly and unpredictably, which would then lead to hyperinflation and ever larger price bubbles in asset markets (Fuders, 2010a), a phenomenon that we are already seeing play out.

For this reason, holding large amounts of cash in Islamic societies is considered immoral, while large amounts deposited in banks which operate according to Sharia law must be invested in the real economy (Schoon, 2016: 43 ff.). Since Gesell's proposal does not rely on people's moral insight, it is preferable to the prohibition of interest.

Another argument in favor of natural money, is that it gets to the **root of the evil**. We have discussed throughout this book the fact that interest is the root of the major problems the world faces today. The root of interest, however, is the fact that money can be hoarded easily. Since it is always better to treat the cause of a disease rather than just the symptoms, one should start with the cause of interest: the unnatural design of money.

13.11 Ethical Banking and Microcredits

13.11.1 "Ethical" Banking

Banks pursuing the concept known as "ethical banking"[30] commit themselves to transparency. Their customers can see, or in some cases even help decide, what investments the banks use their deposits for. This means that customers can be sure

[30] Sometimes the concept is also referred to as "social", "alternative", "civic" or "sustainable banking".

that their money is not being invested in such things as environmentally harmful technologies, wars, child labor, drug trafficking or other ethically questionable activities. This of course is an improvement on conventional banking.

However, the concept is neither a solution to the problem of wealth inequality, nor to the problem of environmental destruction, and it also distracts us from the real problem, which is interest. As far as I know, none of the many banks pursuing this goal question interest. However, as we have already discussed, the real reason for environmental destruction is the growth imperative resulting from interest. Since the ethical banking concept does not challenge interest, investments in these banks do not truly foster sustainability. As pointed out above (Sect. 6.4), there is no such thing as sustainable growth. Since the redistribution of wealth due to interest is not addressed, the term "ethical banking" may be misleading.

13.11.2 Microcredits

In 2006, Muhammad Yunus, founder of the Grameen Bank received a Nobel Peace Prize for promoting microcredits. Micro entrepreneurs are creditworthy, and they should not be denied access to credit and thus the opportunity to start businesses (Yunus & Weber, 2010). Micro entrepreneurs refer to members of the Indian lower class. However, it is questionable if microcredits really serve the lower class. Particularly in countries with large and growing, but predominantly poor populations like India, this has allowed lending to expand greatly. In the case of the Indian bank "SKS Microfinance Limited" alone, the number of microcredit clients had risen abruptly from 80,000 to 7.3 million (Blume, 2010). However, it is not only Indian banks that have flocked to this market since 2006. Many international banks are also serving this market, and Yunus has been openly supported by the International Monetary Fund and the U.S. government (Blume, 2010; Hein, 2011).

One could interpret the awarding of the Nobel Prize for the idea of microcredits in a different way: Banks are so compelled to increase loan volume due to the ever-increasing obligation to pay interest to depositors that they seem to be willing to use any means possible to expand credit. Even the poorest people in the world, who have no resources of their own, are now offered such loans, and we are told that such activity is beneficial by the awarding of the Nobel Prize, while previously contributing to the indebtedness of the poor was seen as unethical. The fact that these microloans have not brought welfare gains to many borrowers but, instead, have in many cases lured the poor into the debt trap, is highlighted in the literature (Wolff, 2010; Blume, 2010). Everywhere where the microcredit model has reached critical mass, it has in fact undermined and prevented the achievement of sustainable and equitable 'bottom-up' development and meaningful poverty reduction (Bateman, 2017). In October 2010, the Indian state government of Andhra Pradesh banned new lending after a series of suicides (Blume, 2010), and Yunus was removed from the chairmanship of the bank he founded by the Reserve Bank of India in February 2011. Prime Minister Sheikh Hasina accused Yunus of exploiting the poor (Hein, 2011).

13.12 Bailouts, "Bad Banks" & Co.

13.12.1 Bailouts and Bad Banks

During and after the 2008 financial crisis (the so-called subprime crisis), which was probably small in magnitude compared to the financial cataclysm that is to come, many countries 'rescued' banks with bailout programs. In so doing they were able to avoid (or, perhaps more accurately, postpone) a chain reaction of bank failures similar to what happened in the 1929 crisis.[31] A special form of bank bailouts are so-called "Bad banks" (Afhüppe, 2009). Bad Banks were used for the first time during the 2008 subprime crisis in the US and some European Counties. A Bad Bank is a publicly owned institution, established to acquire "bad" (defaulted) debts or specific assets seized in foreclosures. Beleaguered banks unload bad assets to a Bad Bank, and this way they can clear their balance sheets (Wagemann, 2022).

These bailouts had two purposes aimed at preserving the economic status quo **without curing the root of the problem**. Firstly, a bank bailout is an excuse to use taxpayer money to backstop bank liabilities, which is ultimately nothing more than a gift to the banks' owners. Secondly, the bailouts required significant expansion of public debt. But who is the creditor of this debt? The main creditor of public debt is the private banking system. In other words, the financial system was rescued with money that the government borrowed from the same financial system they rescued.

This second effect of the bailout programs, i.e., governments acting as customers of banks and allowing them to expand the loan volume to creditworthy borrowers was no less important than the first one. Remember, deposits grow via interest and therefore credit volume must grow too. But the more indebted households and businesses already are, the more difficult it becomes to expand credit volume by lending them more money. Banks are eventually forced to issue loans to people—and to countries, e.g., Greece—with questionable (subprime) credit ratings. Banks do this not only due to greed, as many would believe, but because they must grant loans at all costs if they want to survive since interest must be paid on ever-increasing bank deposits.[32] And when these increasingly risky loans fail, the banks are at risk of bankruptcy themselves. That governments step in as solvent clients demanding huge sums of loans is enormously beneficial to the ailing banking sector.

The subprime crisis was triggered by the bursting of the real estate bubble. In the years prior, banks had made their calculations and granted loans based on real estate prices four or five times higher than prices were after the bubble burst. Before the

[31] The official name of these bank bailouts in the US was "The Emergency Economic Stabilization Act of 2008", which became part of Public Law 110–343. It created the USD 700 billion Troubled Asset Relief Program (TARP) to purchase 'toxic' assets (defaulted loans) from banks.

[32] Some banks may indeed have committed fraudulent actions to hide the fact that they were already insolvent, such as bundling toxic assets (loans that they knew would most probably go default) in investment funds and then selling them to customers. But these behaviors are symptoms of a monetary system that is close to collapse rather than the cause of the crisis.

bursting of the bubble, real estate prices seemed to grow forever. These ever-growing prices encouraged banks to lend money to customers with low credit ratings (subprime). In the event that borrowers were unable to keep up with their payments, the banks would simply sell the mortgaged real estate. This 'business plan' worked as long as prices were rising. But suddenly the bubble burst. Then, when foreclosed properties were sold at auction, they did not yield the sums necessary to cover the loans, and when this began happening on a larger scale, banks themselves were in danger of default. Rescuing the banks then became a double burden on taxpayers:

- Banks were then saved from collapse by taxpayers' money or by further expanding the national debt. In the latter case, the interest of this new public debt again has to be funded by the taxpayer. This interest generated by public debt, by the way, flows back to the same financial system which is being rescued because the State is predominantly indebted to private banks (Hauch-Fleck, 2004).
- The establishment of a Bad Bank transforms 'toxic loans' into triple-A claims against the public sector, in other words against the taxpayer (Fuders, 2009a).

In addition to this burden on the citizens, the rescue of the banks had another negative effect. It delayed the collapse and thus the rebuilding process for years. The quicker the inevitable collapse comes, the less damage it will do to the real economy.

13.12.2 Economic Recovery Programs

The economic recovery programs that were launched in the US in conjunction with the bank bailouts had the same effect, since they were also financed by debt. Bloomberg calculated that the cost of the bailout and economic recovery programs in the US amounted to USD 12.8 trillion, which is almost equal to the US GDP for 2008 (Pittman & Ivry, 2009). In other words, in the period of 2003–2011, the US government increased the total debt by an amount greater than the total of all of the debt that had previously existed in the history of the US.[33] Here we again see the logic of an exponential function. The total credit volume grows exponentially. Similar economic recovery programs were implemented in European countries, especially Germany. Here again, the public sector was a good and solvent customer for the banks.

The very same effect had now, 12 years later, the Covid-19 economic recovery programs enacted in 193 from 197 countries (IMF, 2021). Since most countries did not have sufficient funds for this exceptional spending, these expansive fiscal

[33] The total US public debt (debt held by public + intragovernmental holdings) as of September 1st, 2012, reached 15,236,506,139,986.86 USD according to the website of the US Bureau of Public Debt. In 2003 the total public debt was 6.8 trillion (US GAO, 2004: 6). The US net fiscal stimulus was modest relative to European countries (Aizenman & Pasricha, 2013: 397 ff.).

politics were financed by larger deficits (e.g., IMF, 2020; Snyder, 2020; Ahval, 2020; KPMG, 2020). The total debt of most industrialized economies doubled because of these programs, and we observe the associated exponential growth of the (credit) money supply. The Covid-19 pandemic served as the perfect pretext to again rescue the financial system, which was close to collapse at the end of 2019[34] (Fuders & Louge, 2020). Economic recovery programs took the place of bailouts.

13.12.3 Quantitative Easing

Most people are convinced that hyperinflation like that of Germany in 1923—the most severe inflation in recorded history—is impossible today, because today's governments would not repeat the mistake of out-of-control money printing. However, in 1923 the German government was forced to print money, not only due to the reparations required by the Treaty of Versailles, but also due to the interest-burden of war debts. The US Fed and the ECB have found themselves in a similar dilemma since 2008 and have responded by printing money (Steltzner, 2009). However, as in the past (Fisher, 1937: 212 f.) governments today do not openly admit this but instead describe it as the "buying back of national bonds" (e.g., Frühauf, 2009), otherwise known as "*Quantitative Easing*". The ECB has announced that it would embark on *unlimited* purchases of sovereign bonds from crisis-stricken countries (Reiermann, 2012; Kaiser, 2012).

When the ECB began to buy European government bonds, at first there was great indignation (Steltzner, 2009). However, as we discussed above, long-term interest rates trend downward over the long run. Having nearly reached zero interest during the past decade money did not flow adequately due to the fact that interest is needed as "reward for parting with liquidity" (Keynes, 1936: 167). The central banks in industrialized countries found themselves in a dilemma where they had to either risk plunging into deflation or print more and more money (see also Chap. 16; Fuders et al., 2013). Printing money has nothing to do with Gesell's solution. Gesell proposed a form of money that would automatically flow by itself, not printing ever more of the 'sticky' money that does not circulate well. This newly printed money was again largely hoarded physically at home (see Fig. 5.10 in Sect. 5.5.2).

[34] In an almost panicky reaction, the US Fed began in September 2019 to respond to liquidity shortages of commercial banks with a so-called "standing repo facility". Basically, this is a new instrument with which the US Fed can provide virtually unlimited liquidity against highly secure US government bonds if required (Cox, 2019; Cœuré, 2019)

13.12.4 European Stability Mechanism and Other Rescue Funds

The European funds EFSF (European Financial Stability Facility) and the subsequent ESM (European Stability Mechanism)[35] did not solve any of the fundamental problems, but instead added fuel to the fire. They serve to keep the system artificially afloat. This fund makes it possible for private banks to keep providing credit to countries with questionable solvency such as Greece.

In this context, we should note two things: Firstly, the fund does not actually have much money, but instead the promise of member States to lend money in the event that a 'country bailout' is necessary. Given that all European Union countries are already highly indebted, where will the money come from if a country defaults? The answer is that member States would have to increase their sovereign debts in order to fulfill their obligations to the EMS rescue fund. However, if Greece or any other European country should plunge into insolvency, it is unlikely that the rescue fund would be able to bail out the debt of an entire country. Secondly, even if the EMS was able to rescue a country, the bailout funds would not go to benefit the people in the recipient country but rather would be used to pay off old debts. In other words, the taxpayer's money would circulate within the European financial sector but would hardly reach the people.

13.13 Tobin Tax and Austerity

13.13.1 Tobin Tax

Speculation and price bubbles are a major problem for the real economy and the goal of achieving allocative efficiency because prices no longer reflect the true value of goods and the relationship between supply and demand is distorted. A measure employed in some countries in the aftermath of the 2008 subprime crisis was a tax levied on financial transactions that is supposed to decrease speculation. Such a tax was discussed as far back as 1929 (New York Times, 1929) and proposed by Keynes (1936: 160). Today it is known as the "Tobin Tax" named after James Tobin (1978).

[35] EFSF and ESM were created by the euro Member States. The EMS's mandate is to safeguard financial stability in Europe by providing financial assistance to euro area Member States. The EMS is authorized to use the following instruments: (1) Provide loans to countries in financial difficulties; (2) Intervene in the primary and secondary debt markets. Intervention in the secondary market will be only on the basis of a European Central Bank analysis; (3) Act on the basis of a precautionary program; (4) Finance recapitalization of financial institutions through loans to governments. To fulfill its mission, the EMS issues bonds or other debt instruments on the capital markets. EMS is backed by guaranteed commitments from 19 Member States and currently has a lending capacity of 705 billion euros (EMS, 2022).

However, speculation was not the cause of the financial crisis (neither of the subprime crisis nor of the 'euro' crisis) but rather is a symptom of the current monetary system that generates a steadily growing money supply. It is doubtful whether this measure really has the power to decrease speculation. Speculation can be seen, as explained above, as *partial inflation* in investment markets, such as real estate and stocks. It is impossible to avoid speculation as long as the money supply keeps growing. Such a tax does not solve the problem of an ever-growing money supply. Hence, it was predictable (Fuders & Max-Neef, 2014) that implementing this tax would not have the desired effect. Indeed, stock prices and real estate prices grow to ever new records in Europe, the US and elsewhere, e.g., Chile.

History teaches us that even tulip bulbs can be subject to speculation (Garber, 1989; Galbraith, 1993; Dash, 2001; Louge, 2015). In Holland in 1637 a box of tulip bulbs cost 30,000 guilders, which at that time was equivalent to the value of three houses. The bursting of the tulip bubble brought an economic crisis to Holland. Tulips were physically traded on the market square exchanging gold coins against boxes of tulip bulbs, bringing along the risk of being assaulted. Such trading conditions undoubtedly constituted a major obstacle to trade as a 1–2% tax on the financial transaction that we today pay buying stocks via mouse click. If even under such conditions and in the absence of sophisticated markets price bubbles were to be found, then we must conclude that it is not possible to prevent bubbles from forming as long as there is an ever-increasing amount of money circulating. The only way to prevent speculation without changing the underlying problem of ever-growing money supply would be to ban trading altogether. But this would be like a remedy that 'heals' patients by killing them.

13.13.2 Austerity

In the course of the 2010-euro crisis, which was not so much a crisis of the euro but a sovereign debt crisis[36] the former German chancellor Angela Merkel visited Greece and other European countries to try to convince the governments to 'tighten their belts' via austerity measures. The idea was that if we were able to pay back our debt there will be no crisis.

Unfortunately, few realize that the enormous national debt in most economies is not just due to the irresponsibility of governments, but rather follow a mathematical

[36] Even though the common currency, too, generates problems, especially imbalances in the trade balances of European economies. If they still had their own national currencies, the southern European countries would devalue against the currencies of the northern European countries which would help to increase exports. Exchange rates must fluctuate in order to compensate for differences in productivity and inflation. Now these differences translate into extremely positive (for the northern European countries) and negative (for the southern European countries) trade balances. A negative trade balance translates into more debt for the respective country. This is a significant reason for why the southern European countries are relatively more indebted than the northern ones (Fuders, 2014b).

logic, as was shown above. Because the value of bank deposits grows steadily through interest and compound interest, total debt must also increase at the same pace since, as we previously discussed, there is no interest without debt. That is a fact, which cannot be changed by anything except by abolition of interest. With the exception of the monetary base originally created by the central bank, every unit of money held by one person represents debt owed by someone else. Consequently, **if no one were indebted, there would be (almost) no money**.[37]

This is the reason why the wealthiest nations are also the most indebted ones (in absolute figures): U.S., UK, France and Germany (CIA, 2022). If governments actually managed to pay off a large portion of their nation's debts, this would inevitably result in the people becoming more indebted. This latter phenomenon is found in those (few) countries where the national debt has been, until recently, low, e.g., in Chile. Thanks to its income from copper mining, Chile managed to almost completely eliminate its net national debt (BCCh, 2012). Instead of the State, now the majority of the Chilean population finds itself highly indebted. This is because in our financial system someone has to be indebted.

Notwithstanding the fact that, as we had also understood above (Sect. 5.6.4), that in the current system it is mathematically impossible that everyone cancels their debt, it is also not desirable. Without expanding the total debt and thus money supply, we would see deflation, because if the quantity of goods is growing as a result of increasing productivity, but the amount of money in circulation does not, then every unit of money will have greater purchasing power. When people realize that their money is gaining value over time because prices are falling, they have an incentive to hoard their money even more in the expectation that its purchasing power will continue to increase in the future. But if all, or many, withhold their money from circulation, it can no longer serve adequately as a medium of exchange. This results in a situation where no one buys or sells anything, and the economy grinds to a halt.

13.14 Degrowth, Decarbonize and Circular Economy

We previously mentioned (Sect. 7.2 and 6.5.5) that neither degrowth strategies, nor decarbonizing the economy, nor circular economy can be seen as alternatives to monetary reform. In this chapter, we will discuss these ideas with some more detail.

[37] In this context one might argue that those who hold that money should be grateful to those who are indebted, for money is created by debt and money only has an exchange value if products and services exist that can be exchanged for it; but it is predominantly the borrowers who produce theses goods and services.

13.14.1 Degrowth and Less Consumerism

It was Nicholas Georgescu-Roegen (1975) who started the discussion on economic degrowth almost five decades ago (see also Latouche, 2003; Martínez-Alier et al., 2010). As long as interest exists as an opportunity cost of any productive investment any degrowth will result in business owners trying to save costs in order to maintain their return on investment at least at the level of the interest rate. Saving costs usually means replacing humans with machines. In economics we describe this as "improving productivity". But which is worse, destroying the environment or increasing unemployment? Only by eliminating interest can we get rid of this dichotomy which forces us to choose between economic growth (and destruction of nature) or increased unemployment.

Many blame excessive **consumerism** as the cause for environmental degradation (Traverso, 2015; Pérez, 2016; Becker, 2020; Higgs, 2021). We should, so they say, become more conscious regarding the damage unnecessary consumption causes. However, what is unnecessary consumption? And what would happen if we really consumed less? The answer is businesses would go bankrupt and the economy would fall into a depression. In our current monetary system, we must grow, or else businesses will close, and unemployment rates will skyrocket. Nobody wants this to happen. As we discussed above (Chap. 7) as long as we do not change our monetary system we must choose between economic growth or unemployment. If we do not achieve monetary neutrality, it will not be possible to abolish consumerism without incurring the cost of high unemployment.

13.14.2 Decarbonize and Circular Economy

It is of course desirable, and I am convinced that it is possible to significantly decarbonize our economy. For example, Chile could generate all of its energy from natural and clean resources. It is one of the countries with the most active volcanos and geothermal solutions would work in most parts of the country. Most regions in Chile are peppered with rivers. One of the 14 regions is even called "Los Ríos" (The Rivers). In ancient times even small streams were enough to run water mills. Such a watermill requires approx. 5 kW of energy. This is enough to power up to five houses (using a system with batteries or other solutions as energy "cushion"). In the same way we could, in fact, even power large parts of Europe with electric energy using the energy provided by the Rhine River, one of the biggest rivers in the world with a very powerful current. Installing wheels with a sufficient diameter (of approx. 50 m) on both banks of the Rhine running from its source in Switzerland to its mouth in the Atlantic would provide clean energy equivalent to the output of about 50 nuclear power plants.

However, even if we managed to draw 100% of our energy from clean and carbon-neutral resources, and even if we managed to achieve a 100% circular economy, if the productive output of our economies keeps growing, we will use more and more energy and other resources. Thus, we will keep destroying nature.

13.15 Doughnut Economy, Gemeinwohl-Ökonomie and Buen vivir

13.15.1 Doughnut Economy

The term "Doughnut Economy" was coined by Kate Raworth (2012, 2017). In her bestselling book with the same title, she uses the metaphor of a doughnut for an economy that functions in harmony with nature and society. Such a doughnut economy would have three layers or circles. The outer circle defines the limits of our ecosystem. The inner one defines regulations of a social system, e.g., those that provide help for the poor. In between the outer and inner circles, we have the space where humans may thrive. Raworth's approach has been celebrated as an alternative to orthodox and neoliberal thinking (Wolf, 2017; Ideo, 2018). However, the idea that the economy is part of a social system which itself is part of the biosphere and as such has to respect the limits of the biosphere is not new (see, e.g., Azkarraga et al., 2011 and in this book Sect. 6.3). In fact, a similar perspective is presented in the Brundtland report (1987) and at least indirectly in the 1972 Meadows study "the Limits to Growth" (Meadows, 1972). And the fact that the market economy and its notion of efficient resource allocation is not necessarily at odds with the desire to help the poor via social programs is agreed upon by textbooks on classical market economics (e.g., Frank, 2015). Later it became known, especially in Germany, as the social market economy (see, e.g., Fikentscher, 1992).

Raworth says that in the 1870s economists made an analogy between Newton's laws and economics: just as gravity pulled things towards it, prices pull the economy toward equilibrium. The problem is that while science advanced, economics remained stuck in the nineteenth century (Kayser, 2020). This comparison, which in a very similar way can also be found in Smith and Max-Neef's (2012) book "Economics unmasked" does not get to the root of the problem: our monetary system. She says it all comes down to design of regulations (DEAL, 2020). But this is incorrect. Yes, we need rules (for design rules see Daly & Farley, 2011). And yes, she is right in saying that many people are greedy and focused on material wealth.

However, it is not only about using rules to keep greed in check. Our money makes it possible to easily hoard surpluses and thus creates an incentive to hoard more than we would in nature to survive the next winter. Raworth explains that "an economy that is distributive by design is one whose dynamics tend to disperse and circulate value as it is created, rather than concentrating it in ever-fewer hands" (Raworth, 2017: 128). She does not see that it is our monetary system that hinders this circular flow and causes money to concentrate in ever-fewer hands. In microeconomic theory, one of the assumptions regarding consumer preferences is: "more is better" (Frank, 2015: 64). The very same theory teaches us the law of "diminishing marginal utility" (Frank, 2015: 181), which says that the more we have already consumed of a particular product the less additional utility we gain by consuming one more unit. So how can we reconcile these two perspectives? The reason for this apparent contradiction is that we accumulate money not physical goods. If, for some reason, we have more of a certain good than we need at the moment, we can sell it and accumulate money.

Our money, as it is designed today, fosters greed. And it does not only encourage such behavior but, as explained throughout this book, it requires it. As long as interest exists, we will have to grow, or unemployment will skyrocket. As discussed throughout this book, interest will continue to exist as long as money is easily storable. Raworth rightly criticizes materialism but does not address the underlying force that strongly promotes materialism: the possibility of accumulating wealth unnaturally via money.

The desire to live in harmony with nature is of course nice. This is what we all want and always wanted, even those economists that some would label as neoliberal. John Starts Mill's thoughts on utility (Mill, 1863) had exactly this in mind. However, living in harmony with nature is impossible as long as we are obliged to grow.

13.15.2 Gemeinwohl-Ökonomie

The book "Gemeinwohl-Ökonomie" by Christian Felber (2010) has attracted quite a bit of attention in German and Spanish speaking countries and generated a movement with the same name consisting of regional groups and 2000 supporting and 500 officially associated businesses (GWÖ, 2022). Gemeinwohl-Ökonomie can be translated into English as "economy for the common good". This auspicious name alludes to Daly's "For the common good" (Daly & Cobb, 1989). In 2017 a book was authored by Nobel prize winner Jean Tirole (2017) with a very similar title "Economics for the common good" and even though it was a completely different book, the overall message has a similar tenor to Felber's book. As with Raworth's "Doughnut Economy" in Felber's work the well-being of people and the natural environment are the primary goals of economic activity (GWÖ, 2022). Conceived as an "ethical market economy" (GWÖ, 2022), it is based predominantly on private companies, but instead of striving for financial gain in competition with each other; they would cooperate toward the goal of achieving the greatest possible common good.

However, as outlined in Sect. 10.4.5, the idea of classical (market) economic theory and the ideal of "perfect competition" does not preclude cooperation. In fact, any corporation involves people coming together and cooperating in order to produce a certain product or service. In the classical model of market economy, businesses are allowed to and are expected to cooperate. Businesses cooperate in production processes. What is prohibited is collusion and price fixing.[38] And for a good reason, since this form of cooperation works against the common good. Businesses form cartels to abolish competition in order to be able to sell their products at higher prices and/or lower quality than they would in an environment of perfect competition.

Felber says that his idea is consistent with the Aristotelian notion of "*oikonomia*", that is the efficient use of resources in the household (Aristotle, [384 BC] 1995a). However, this is exactly what economics is and always has been about, to find a way to most efficiently allocate resources. That this does not always work properly and that many seek material gain as an end in itself (and at the expense of others) had even been recognized already by Aristotle. He termed such behavior contrary to "*oikonomia*" as "*chrematistics*" (Aristotle, [384 BC] 1995a). As explained throughout this book, the main reason why chrematistics exist is the design of our money and the resulting phenomenon of interest, not greed alone. And this monetary system then promotes greedy behavior since money can be accumulated and even bears interest. This, too, Aristotle already had recognized. He considered making money with money to be the most immoral form of income (Aristotle, [384 BC] 1995a).

Of course, one might argue that our monetary system is a consequence of our greed. We invented this monetary system, and we find every excuse to not recognize the problems inherent in it, precisely because we like to accumulate and receive interest. If we were angels, we would also not need competition to incentivize us to produce good products at good prices. We would do so of our own volition. We would understand that it is part of the deeper sense of life, and part of neighborly love, to employ our talents (our vocation[39]) to serve each other. We would be happy and experience the Aristotelian notion of *eudaimonia* (Aristotle, [384 BC] 1995b), when we see people use our products and services (Fuders & Nowak, 2019). We would not abuse a dominant market position (monopoly) if we held one but would sell our products at the lowest possible price since we want to serve others with our products. For the same reason, we would not form price cartels to sell products at artificially high prices. We would not engage in unfair competition. And we would not charge interest on loans or accumulate much money under the mattress. We would not need monetary reform or competition law. In fact, if we were angels, we would not need any laws whatsoever. We would achieve what can be called an "economy of neighborly love" (Fuders, 2017; Fuders & Nowak, 2019). But if we were angels, we would already be in Heaven. As long as we are not angels, we need laws, that regulate markets, and above all we need money that is not unnaturally storable.

---

[38] Besides collusion and price fixing, there are other forms of cartels. To discuss each form of prohibited cooperation goes beyond the scope of this work. For European regulations see, e.g., Fuders (2009a).

[39] The expression 'vocation' derives from 'vocatio', which is Latin and means invitation. 'Vocatio' itself contains the expression 'voca' most probably from 'bucca' (latin = mouth); in other words, we are called to serve. In the same way, the German word for vocation is 'Berufung'. The word 'Berufung' itself contains the word 'Ruf', which means 'call'. The 'Berufung' is God's call for a special service (Kluge, 1989).

It must be mentioned that even though through Felber's book and movement the term "Gemeinwohl-Ökonomie" has attracted wider public attention, the term was not coined by Felber. As far back as 2001 a book with the very same term in its title was published but unfortunately gained less attention. The complete title is "Vision einer Gemeinwohl-Ökonomie auf der Grundlage einer komplementären Zeit-Währung (Sikora & Hoffmann, 2001). Translated into English this would be "Vision of an economy for the common good based on a complementary time currency". As discussed above, time-based currencies can be seen as viable alternatives to the current monetary system if they are designed according to Gesell's proposal, i.e., if they incorporate some sort of hoarding fee.

13.15.3 Buen vivir/Bem viver

In Latin America a concept has become popular in recent years called "*buen vivir*" in Spanish or "*bem viver*" in Portuguese, which can be translated into English as "good living" or "the art of living", *ars vivendi* as it was called in former times in Latin. The expression we find in most Andean indigenous cultures and languages from the Amazonas region to Tierra de Fuego, e.g., "*sumak kawsay*" (Quechua), or "*suma qamaña*" (Aymara) or "*küme mogen*" (Mapuche) means *buen vivir* in the respective languages. *Buen vivir* is supposed to be a sort of a proposal as well as opportunity to think of another reality in which human beings are part of a more harmonious whole with nature and with other human beings (de la Cuadra, 2015; Schlemer Alcântara et al., 2017; Cioce Sampaio et al., 2018; Mora, 2020; Loncon Antileo, 2022). But the concept is not attributable to any particular author or group, nor does it imply a magical formula to be adhered to. It is a "proposal under permanent construction" (de la Cuadra, 2015). Cioce Sampaio et al. (2019) define it as:

> Thus, the buen vivir, more than a material, socio-educational, psychosocial and health condition, is a particular state of happiness, where different cultural patterns vibrate, measured by human multiculturalism, as suggested by Max-Neef et al. (1991) in the thesis of development on a human scale [author's translation].

Essentially, it is a critique of capitalism, but, as the Spanish/Portuguese name suggests, it has less Eurocentric and more Latin American connotations (de la Cuadra, 2015). However, those who discuss this concept are mostly not trained in economics, which is why they do not know that what they are asking for is what we describe in economics as perfect competition: a strongly diversified market structure with small-scale production of small and medium-sized producers, producing good products at good prices for the well-being of all. Perfect competition is a market structure in which no producer has a dominant position that could be exploited, and everyone earns an income commensurate with their effort. In this sense, *buen vivir* is not a critique of capitalism but a critique of poorly implemented capitalism.[40]

[40] In this context, it helps to remember that the term "capitalism" contains the conflation of money with the production factor capital.

To better understand the notion *buen vivir*, it is important to keep in mind that most Latin American economies are far from the ideal of perfect competition. Companies are poorly networked, cooperate little and markets are highly monopolized (Soza-Amigo et al., 2021). In Chile, for example, the dominant sector is copper mining followed by forestry and aquaculture. The main companies are predominantly foreign, usually European or US companies or joint ventures between foreign companies and domestic ones. People in Chile have never experienced markets that function correctly. People erroneously believe that this is what competition-based market economics is about, when in fact such a monopolized market is the contrary of the ideal of perfect competition.

Hence, it is understandable that when authors discuss the concept of *buen vivir* and criticize the "exponential growth of resource exploitation" (de la Cuadra, 2015) they fail to understand that this is not (only) rooted in greed or the "capitalistic" model, but in a poor implementation of that model and, above all, in the growth imperative inherent in our monetary system. But in Latin America (as well as in other developing regions) it is not only a problem of the monetary system. Markets are highly monopolized because governments have not effectively combated collusion and have not encouraged the creation of small and medium-sized enterprises to diversify the economy and promote competition. Remember, income inequality mainly stems from the monetary system. Nevertheless monopolization of markets also contributes to the problem. Regarding income inequality, in Latin American economies we have this 'doble pack': in addition to the monetary system heavily monopolized markets are the norm (for Chile see Fuders & Michaelis, 2016). Apart from the reform of the monetary system, in Latin America competition should be promoted, e.g., through encouraging start-ups, small and medium-sized companies, cooperatives, etc.—some would call this "*economía solidaria*" (solidarity-based economy), see Cioce Sampaio et al. (2018)—in order to transition to a market structure that is less concentrated and less dominated by a few large companies.

It is understandable that people, especially in Latin America, long for less inequality. It is also understandable that Latin Americans wish to be free of European and US American paternalism and to find their own identity. *Buen vivir* reflects this desire as well as the universal desire to live in harmony and peace with each other and with nature. However, despite the Spanish/Portuguese branding and the regularly cited translations of this term in different indigenous languages (Mora, 2020; Loncon Antileo, 2022), it is doubtful that *buen vivir* is really a uniquely Latin American, or even indigenous concept, as the existence of *ars vivendi*, the ancient Latin term, reveals. The desire to live in harmony and peace with each other and with nature is a desire that probably all, or at least most people share worldwide and have always shared. Countless philosophers from all times and places have thought about how to achieve a good life. Our legal system, including laws that regulate markets, is based on the principle of "good living of all in freedom" (Schachtschneider, 1994: 569 f., 638, 978, 980, 996, 2006: 20). Indeed, economics is about finding ways to improve well-being, not about gaining money as many, especially in Latin America, believe today. *Buen vivir* is a vague notion that gives us no clear rules or road map for how to achieve a good life, as authors themselves admit (de la Cuadra, 2015).[41]

In this context some interesting questions arise. In Latin America we still find groups of indigenous people that have been able to maintain their traditional lifestyles and live in a more or less self-sufficient manner. If all people lived in such a way, natural resources would surely not be over exploited. So, can we learn from them? How were they able to preserve this way of life? Is it because they are more spiritual people, better connected to nature? Maybe. But it also might be that the reason for being more connected to nature is because they do not have a financial system that required them to grow for the last 700 years (Fuders, 2020).

13.15.4 Recapitulation

It is great that more and more people are interested in the common good and are thinking about how best to achieve it. Unfortunately, many fail to understand that the problems they correctly diagnose, especially regarding profiting at the expense of others and the natural environment, are mainly due to our current monetary system. I am convinced, with the implementation of natural money and a wholesale embrace of classic competition law we can live in harmony and peace with each other and the environment. Admittedly, "perfect competition" does not sound very attractive, and as explained earlier (Sect. 10.4.4) "perfect solidarity" would be a better term to accurately describe what the model of perfect competition is about. Perfect competition translates into perfect solidarity if it is implemented correctly, if governments (i) fight against monopolies (which are antithetical to perfect competition) and, in cases where monopolies cannot be avoided, the abuse of monopolistic power, (ii) fight cartels and unfair competition and, most importantly, (iii) implement natural money. Then, and only then can we achieve a market economy consisting of an indefinite number of small and medium-sized companies that strive to offer the best products at the best prices, and who in doing so, contribute to the fulfillment of fundamental human needs. On the contrary, in the long run without natural and neutral money we will neither truly live in harmony with nature nor with each other. We will never reach a *Gemeinwohl-Ökonomie*, nor a *buen vivir*.

13.16 Overview

Finally, I would like to end this chapter with an overview of the different proposals we discussed in Chaps. 12 and 13 (Table 13.2).

---

[41] As exception I would like to mention the concept of "Ecosocioeconomías" conceived by Carlos Sampaio, see Cioce Sampaio et al. (2018) and Cioce Sampaio et al. (2019).

Table 13.2 Overview

Proposal	Chapter	No solution at all + extra problems	No solution at all	No true solution but improvement	True solution
Governmentally imposed hoarding tax	12.1				X
Effective central bank negative interest rate policy	12.2				X
"100% money" and "sovereign money"	13.1			X	
Modern monetary theory (MMT)	13.2		X		
Marxism, communism and socialism	13.3	X			
Gold standard or other "backed money"	13.4		X		
Inflation adjusted accounting units like the Chilean UF	13.5		X		
Time banks and time vouchers	13.6			X	X (if subject to hoarding fee)
Social or local money	13.7			X	X (if subject to hoarding fee)
Credit cooperatives and trustee savings banks	13.8			X	
Cryptocurrencies	13.9		X		X (if subject to hoarding fee)
Prohibition of interest/ Islamic banking	13.10			X	
Ethical banking and microcredits	13.11			X	
Bailouts, "bad banks" & Co.	13.12	X			
Tobin tax and austerity	13.13		X		
Degrowth, decarbonize and circular economy	13.14			X	
Doughnut economy, Gemeinwohl-Ökonomie and Buen vivir	13.15			X	

Source: Own work

References

Adenauercampus. (2022). *Mythos: „In der DDR durfte jeder studieren"*. [Online] Available at: https://www.adenauercampus.de/ddrtutorium/mythos-und-wirklichkeit/-in-der-ddr-durfte-jeder-studieren. Accessed 14 Feb 2022.

Afhüppe, S. (2009). Einigung zu Bad Banks steht unmittelbar bevor – Steinbrück trifft Unionsministerpräsidenten – Länder zu politischer Absichtserklärung bereit. *Handelsblatt*, Issue 04.06.2009, p. 4.

Ahval. (2020). *Turkey budget deficit widens to record as spending surges on COVID-19*. [Online] Available at: https://ahvalnews.com/turkey-economy/turkey-budget-deficit-widens-record-spending-surges-covid-19. Accessed 25 May 2022.

Aizenman, J., & Pasricha, G. K. (2013). Net fiscal stimulus during the Great Recession. *Review of Development Economics, 17*(3), 397–413.

Andriotis, A., Rudegeair, P., & Hoffman, L. (2019). Facebook's new cryptocurrency, Libra, gets big backers. *The Wall Street Journal*. [Online] Available at: https://www.wsj.com/articles/facebooks-new-cryptocurrency-gets-big-backers-11560463312. Accessed 7 May 2022.

Aristotle. (1995a). Politik 1st Book. In *Aristoteles Philosophische Schriften in sechs Bänden. Band 4*. Felix Meiner Verlag.

Aristotle. (1995b). Nikomachische Ethik. In *Aristoteles Philosophische Schriften in sechs Bänden. Band 3*. Felix Meiner Verlag.

Azkarraga, J., Max-Neef, M., Fuders, F., & Altuna, L. (2011). *La evolución sostenible II*. Lanki (Mondragon Unibertsitatea).

Banco Central de Chile (BCCh). (2012). *Deuda externa de Chile 2011*. Banco Central de Chile.

Banco Palmas. (2022). *Instituto Banco Palmas*. [Online] Available at: https://www.institutobanco-palmas.org/. Accessed 20 Apr 2022.

Bank of England. (2021). *New forms of digital money*. [Online] Available at: https://www.bankofengland.co.uk/paper/2021/new-forms-of-digital-money. Accessed 10 May 2022.

Bank of England. (2022). *UK central bank digital currency*. [Online] Available at: https://www.bankofengland.co.uk/research/digital-currencies. Accessed 10 May 2022.

Bateman, M. (2017). *How the bank's push for microcredit failed the poor*. [Online] Available at: https://www.cadtm.org/How-the-Bank-s-push-for. Accessed 15 Oct 2021.

BDTI. (2020). *Economia R-evolucionaria*. [Online] Available at: https://www.economiarevolucionaria.cl/wp/2020/06/12/banco-del-tiempo-e-intercambio-valdivia/. Accessed 20 Apr 2022.

Beattie, A. (2010, November 7). Zoellick seeks gold standard debate. *Financial Times*.

Becker, J. (2020). Reasons to escape excess consumerism—And how to do it. *Forbes*. Online: https://www.forbes.com/sites/joshuabecker/2020/04/30/escape-excess-consumerism/?sh=7d1c5fd72fbb

Best, R. D. (2022). *Energy consumption of a Bitcoin (BTC, BTH) and VISA transaction as of April 2022*. [Online] Available at: https://www.statista.com/statistics/881541/bitcoin-energy-consumption-transaction-comparisonvisa/. Accessed 10 May 2022.

BitGesell. (2022). *Bitgesell*. [Online] Available at: https://bitgesell.ca/. Accessed 10 May 2022.

Blanc, J. (2011). Classifying "CCs": Community, complementary and local currencies' types and generations. *International Journal of Community Currency Research, 15*, 4–10.

Blume, G. (2010). Selbstmord einer großen Idee. *Zeit Online*. [Online] Available at: https://www.zeit.de/2010/47/Mikrokredite-Indien?utm_referrer=https%3A%2F%2Fwww.google.com%2F. Accessed 9 Dec 2022.

Bogdanoff, A. (1925). *A short course of economic science* (2nd ed.). Communist Party of Great Britain.

Böhm, F. (1961). *Kartelle und Monopole im modernen Recht*. J.-W.-Goethe-Universität.

Borchert, M. (2010). *Geld und Kredit: Einführung in die Geldtheorie und Geldpolitik* (8th ed.). Oldenbourg Wissenschaftsverlag.

Brundtland, G. H. (1987). *Our common future*. Oxford University Press.

Bundesbank. (2017). *Monthly report April 2017*. Bundesbank.

Bundesregierung. (2022). *Gefälschte Kommunalwahl in der DDR*. [Online] Available at: https://www.bundesregierung.de/breg-de/suche/gefaelschte-kommunalwahl-in-der-ddr-353694. Accessed 10 Feb 2022.

Catà, J. (2018). *Barcelona, ciudad de cooperativas*. [Online] Available at: https://elpais.com/ccaa/2018/06/01/catalunya/1527875472_060835.html. Accessed 2 May 2022.

Champ, B. (2008). Stamp scrip: Money people paid to use. In Federal Reserve Bank of Cleveland (Ed.), *Economic commentary* (pp. 1–4). Federal Reserve Bank of Cleveland.

Chiemgauer. (2022). *Chiemgauer Regiogeld*. [Online] Available at: https://www.chiemgauer.info. Accessed 20 Apr 2022.

CIA. (2022). *Factbook, country comparisons: Debt – external*. [Online] Available at: https://www.cia.gov/the-world-factbook/field/debt-external/country-comparison. Accessed 25 May 2022.

Cioce Sampaio, C., Henríquez, C., & Fuders, F. (2018). Ecosocioeconomías: Un concepto en construcción. *Educação e Contemporaneidade, 27*(52), 13–25.

Cioce Sampaio, C. A., Jurema Grimm, I., Schlemmer Alcântara, L. C., & Mantovaneli, O., Jr. (2019). *Ecossocioeconomias – Promovendo Territórios Sustentavéis*. Edifurb.

Cœuré, B. (2019). A tale of two money markets: Fragmentation or concentration. *ECB*. [Online] Available at: https://www.ecb.europa.eu/press/key/date/2019/html/ecb.sp191112~5808616051.en.html. Accessed 12 May 2020.

Cox, J. (2019). *The Fed is looking at a 'standing repo' operation to handle overnight funding issues*. [Online] Available at: https://www.cnbc.com/2019/11/20/the-fed-is-looking-at-a-standing-repo-operation-to-handle-overnight-funding-issues.html. Accessed 25 May 2022.

Daly, H., & Cobb, J. (1989). *For the common good: Redirecting the economy toward community, the environment, and a sustainable future*. Beacon Press.

Daly, H., & Farley, J. (2011). *Ecological economics – Prcinciples and applications* (2nd ed.). Island Press.

Danezis, G., & Meiklejohn, S. (2016). *Centrally banked cryptocurrencies*. Internet Society.

Dash, M. (2001). *Tulipomania: The story of the world's most coveted flower and the extraordinary passions it aroused*. G.K. Hall.

de Best, R. (2021). *Global bitcoin (BTC, BTH) energy consumption up until October 19, 2021*. [Online] Available at: https://www.statista.com/statistics/881472/worldwide-bitcoin-energy-consumption/. Accessed 10 May 2022.

de la Cuadra, F. (2015). Buen Vivir: ¿Una auténtica alternativa post-capitalista? *Polis, 14*(40). https://doi.org/10.4067/S0718-65682015000100001

DEAL (2020). *Doughnut economics*. [Online] Available at: https://doughnuteconomics.org/about-doughnut-economics. Accessed 25 May 2022.

Drukarczyk, J. (2003). *Finanzierung – Eine Einführung* (9th ed.). Stuttgart Lucius und Lucius.

ECB. (2015). *Virtual currency schemes – A further analysis*. ECB.

Ehnts, D. (2017). *Modern monetary theory and European macroeconomics*. Routledge.

Emmerich, V. (2001). *Kartellrecht*. Beck.

Engels, F. (2019) (reprint from 1882). *Die Entwicklung des Sozialismus von der Utopie zur Wissenschaft*. Henricus.

European Stability Mechanism (EMS). (2022). *Facts*. [Online] Available at: https://www.esm.europa.eu/. Accessed 25 May 2022.

Felber, C. (2010). *Die Gemeinwohl-Ökonomie: das Wirtschaftsmodell der Zukunft*. Deuticke.

Fikentscher, W. (1992). *Recht und wirtschaftliche Freiheit*. Mohr Siebeck.

Fisher, I. (1936a). *100% money* (2nd ed.). Adelphi.

Fisher, I. (1936b). 100% money and the public debt. *Economic Forum*, pp. 406–420.

Fisher, I. (1937). *Feste Währung – Zur Entwicklungsgeschichte der Idee*. Lautenbach.

Fisher, I., & Cohrssen, H. (1933). *Stamp scrip*. Adelphi.

Forstater, M., & Mosler, W. (2005). The natural rate of interest is zero. *Journal of Economic Issues, XXXIX*(2), 535–542.

Frank, R. (2015). *Micoreconomics and behaviour* (9th ed.). McGraw-Hill.

Freicoin. (2021). *Freicoin: A peer-to-peer digital currency delivering freedom from usury*. [Online] Available at: http://freico.in/. Accessed 10 May 2022.

Friedman, M. (1966). *Capitalismo y Libertad*. Ediciones RIALP.
Frühauf, M. (2009). Die Verstaatlichung des Anleihemarktes. *F.A.Z.*, Issue 02.04.2009, p. 9.
Fuders, F. (2008). Eine Wirtschaftsordnung ohne Zinsen. *Zeit-Fragen, 17*(23), 3–4.
Fuders, F. (2009a). *EG-Wettbewerbsrecht – Ein Kurzlehrbuch nach der Systematik der Prüfungsschemata aufgebaut und mit Fällen*. SVH-Verlag.
Fuders, F. (2009b). Zum so genannten freien Wettbewerb – Versuch einer Begriffsdefinition. *Aufklärung & Kritik, 16*(2), 146–150.
Fuders, F. (2010a). Alternative concepts for a global financial system – An answer to the present world financial crisis. *Estudios Internacionales, 166*, 45–56.
Fuders, F. (2010b). Warum der Zins auch moralisch nicht zu rechtfertigen ist. *Humane Wirtschaft*, 2, 26–29.
Fuders, F. (2011). Zum so genannten freien Wettbewerb und dem vermeintlichen Verbot vertikaler Kartelle im EU-Wettbewerbsrecht. *Zeitschrift für Wirtschaftspolitik, 60*(3), 271–284.
Fuders, F. (2014a). Indexierte Währungen und Recheneinheiten als Mittel gegen Inflation und Finanzkrisen? *Zeitschrift für Sozialökonomie, 51*(180/181), 15–24.
Fuders, F. (2014b). Die Mehrheit in Deutschland profitiert nicht vom Euro. *Humane Wirtschaft, 1/2014*, 21–23.
Fuders, F. (2016). Smarter money for smarter cities: How regional currencies can help to promote a decentralised and sustainable regional development. In *Decentralisation and regional development – Experiences and lessons from four continents over three decades* (pp. 155–185). Springer.
Fuders, F. (2017). Neues Geld für eine neue Ökonomie – Die Reform des Geldwesens als Voraussetzung für eine Marktwirtschaft, die den Menschen dient. In *Finanzwirtschaft in ethischer Verantwortung – Erfolgskonzepte für Social Banking und Social Finance* (pp. 121–183). Springer-Gabler.
Fuders, F. (2018). Karl Marx' historisches Missverständnis. *Fairconomy, 2*, 14–16.
Fuders, F. (2020). *The 'tragedy of the commons' and the role of the money interest rate*. Springer.
Fuders, F., & Louge, C. (2020). Effektive Negativzinsen statt Corona-Schuldenwachstum. *Fairconomy, 16*(2), 4–9.
Fuders, F., & Max-Neef, M. (2014). Dinero, deuda y crisis financieras. Propuestas teórico-prácticas en pos de la sostenibilidad del sistema financiero internacional. In *Economía Internacional. Claves teórica-prácticas sobre la inserción de Latinoamérica en el mundo*. LATIn.
Fuders, F., & Michaelis, N. (2016). Marktkonzentration: Ein großes Hindernis für eine Nachhaltige Ökonomie. In *Jahrbuch Nachhaltige Ökonomie* (pp. 271–294). Metropolis.
Fuders, F., & Nowak, V. (2019). The economics of love: How a meaningful and mindful life can promote allocative efficiency and happiness. In C. Steinebach & A. Langer (Eds.), *The economics of love: How a meaningful and mindful life can enhancing resilience in youth – Mindfulness-interventions in positive Environments*, ISBN 978 (pp. 259–277). Springer.
Fuders, F., & Schreiber-Martens, A. (2018). Bitcoin – eine Alternative für das Finanzsystem? *Fairconomy, 14*(3), 16–17.
Fuders, F., Mondaca, C., & Azungah Haruna, M. (2013). The central banks dilemma, the inflation-deflation paradox and a new interpretation of the Kondratieff waves. *Economía (U. de los Andes), 36*, 33–66.
Fuhrmann, H. (2003). *Überall ist Mittelalter – Von der Gegenwart einer vergangenen Zeit*. C.H. Beck.
Galbraith, J. K. (1983). *El Dinero – De dónde vino/Adónde fue* (2nd ed.). Hyspamerica.
Galbraith, J. K. (1993). *A short history of financial euphoria*. Whittle Books.
Garber, P. (1989). Tulipmania. *Journal of Political Economy, 97*(3), 535–560.
Gelleri, C. (2009). Chiemgauer regiomoney: Theory and practice of a local currency. *IJCCR, 13*, 61–75.
Gelleri, C. (2018). *Der Chiemgauer in der Praxis*. [Online] Available at: https://www.chiemgauer.info/fileadmin/user_upload/Basisinfo/Chiemgauer_praktisch.pdf. Accessed 20 Apr 2022.

Gelleri, C. (2020). *The phenomenon of complementary currencies.* [Online] Available at: https://justmoney.org/the-phenomenon-of-complementary-currencies/. Accessed 12 June 2022.

Gelleri, C. (2022). Reshaping the future of Europe with complementary currencies. In I. Staudinger et al. (Eds.), *Europarecht. Shaping the future of Europe: Contributions to the 3rd Young European Law* (pp. 199–213). Jan Sramek Verlag.

Gelleri, C., & Stodder, J. (2021). Chiemgauer complementary currency: Concept, effects, and econometric analysis. *International Journal of Community Currency Research, 25*(1), 75–95.

Georgescu-Roegen, N. (1975). Energy and economic myths. *The Southern Economic Journal, 41*(3), 347–381.

Gesell, S. (1892) (Reprint 2000). Die Verstaatlichung des Geldes. In: W. Onken (Ed.), *Gesammelte Werke Band 1* (p. 252 ff.). Verlag für Sozialökonomie.

Gesell, S. (1920) (Reprint 1988). *Die natürliche Wirtschaftsordnung durch Freiland und Freigeld* (4th ed.). Verlag für Spzialökonomie.

Ginsburg, H. J. (2011). Sturm auf islamische Geldanlagen. *Wirtschaftswoche.* Online: https://www.wiwo.de/politik/ausland/kapitalmarkt-sturm-auf-islamische-geldanlagen-/4648208.html

Gross, M., & Siebenbrunner, C. (2019). *Money creation in fiat and digital currency systems* (Working Paper No. 285). IMF.

GWÖ (2022). *Gemeinwohl-Ökonomie.* [Online] Available at: https://web.ecogood.org/. Accessed 28 May 2022.

Hauch-Fleck, M.-L. (2004). Wer profitiert von den Schulden? Die Gläubiger des Staates verstecken sich hinter dem Bankgeheimnis. *Zeit-Online*, Issue 13.05.2004 – No. 21.

Hein, C. (2011). *FAZ online: Grameen-Bank: Nobelpreisträger Yunus als Direktor abgesetzt.* [Online] Available at: https://www.faz.net/aktuell/wirtschaft/wirtschaftspolitik/grameen-bank-nobelpreistraeger-yunus-als-direktor-abgesetzt-1607720.html. Accessed 9 Dec 2022.

Herdzina, K. (1999). *Wettbewerbspolitik*. Lucius & Lucius.

Hern, A. (2013). This article is more than 8 years old Bitcoin hype worse than 'tulip mania', says Dutch central banker. *The Guardian.* Online: https://www.theguardian.com/technology/2013/dec/04/bitcoin-bubble-tulip-dutch-banker

Higgs, K. (2021). How the world embraced consumerism. *BBC.* Online: https://www.bbc.com/future/article/20210120-how-the-world-became-consumerist

Huber, J. (1998). *Vollgeld – Beschäftigung, Grundsicherung und weniger Staatsquote durch eine modernisierte Geldordnung*. Duncker & Humblot.

Huber, J. (2017). *Sovereign money – Beyond reserve banking*. Palgrave Macmillan.

Huber, J. (2022). *Vollgeld und 100%-reserve (Chicago Plan).* [Online] Available at: https://vollgeld.page/vollgeld-und-100-prozent-reserve-chicago-plan. Accessed 7 Feb 2022.

Huber, J., & Robertson, J. R. (2008). *Joseph Huber, James Robertson: Geldschöpfung in öffentlicher Hand – Weg zu einer gerechten Geldordnung im Informationszeitalter*. Gauke Verlag für Sozialökonomie.

Hucko, M. (2009). Allah, won't you buy me a Mercedes-Benz?. *Financial Times Deutschland,* Issue 8.10.2009, p. 1.

Ideo. (2018). *The 4 books Tim Brown can't put down.* [Online] Available at: https://www.ideo.com/blog/the-4-books-tim-brown-cant-put-down. Accessed 28 May 2022.

International Monetary Fund (IMF). (2020). *Policy responses to Covid-19.* [Online] Available at: https://www.imf.org/en/Topics/imf-and-covid19/Policy-Responses-to-COVID-19. Accessed 25 May 2022.

International Monetary Fund (IMF). (2021). *Policy responses to Covid-19.* [Online] Available at: https://www.imf.org/en/Topics/imf-and-covid19/Policy-Responses-to-COVID-19. Accessed 2 Feb 2023.

Irrera, A., & Wilson, T. (2020). *Criptomoneda respaldada por Facebook cambia de nombre a Diem en busca de aprobación reguladores.* [Online] Available at: https://www.reuters.com/article/facebook-criptomoneda-idLTAKBN28B5A6. Accessed 7 May 2022.

Jackson, A., & Dyson, B. (2013). *Modernising money: Why our monetary system is broken and how it can be fixed.* Positive Money.

JAK. (2022). [Online] Available at: https://www.jak.se/. Accessed 4 May 2022.

Jakab, Z., & Kumhof, M. (2015). *Banks are not intermediaries of loanable funds—And why this matters*. Bank of England.

Janisch, F., & Stapleton, L. (2021). Digital currencies and community empowerment in Austria: Gesell's concept of effective demand as a basis for local digital currencies. *IFAC-PapersOnLine, 54*(13), 698–703.

Kaiser, S. (2012, September 7). No limits – ECB president draghi reaches for the bazooka. Spiegel International.

Kant, I. (1868a). Metaphysik der Sitten – Erster Theil. Metaphysische Anfangsgründe der Rechtslehre. In *Immanuel Kant's sämmtliche Werke – In chronologischer Reihenfolge, Bd. VII* (pp. 1–173). Leopold Voss.

Kant, I. (1868b). Metaphysik der Sitten – Zweiter Theil. Metaphysische Anfangsgründe der Tugendlehre. In *Immanuel Kant's sämmtliche Werke – In chronologischer Reihenfolge, Bd. VII* (pp. 175–304). Leopold Voss.

Kantzenbach, E. (1967). *Die Funktionsfähigkeit des Wettbewerbs* (2nd ed.). Vandenhoeck u. Ruprecht.

Kayser, B. (2020). Kate Raworth, economista: "La 'economía del donut' satisface las necesidades de todas las personas, pero dentro de los límites del planeta". *El Diario*. Online: https://www.eldiario.es/ballenablanca/economia/economia-satisface-necesidades-personas-limites_128_3000932.html

Kelton, S. (2021). *Der Defizit-Mythos: Die Modern Monetary Theory und die Gestaltung einer besseren Wirtschaft*. Lola Books.

Kennedy, M. (2011). *Occupy Money – Damit wir zukünftig alle die Gewinner sind*. J. Kamphausen.

Kennedy, M., Lietaer, B., & Rogers, J. (2012). *People money – The promise of regional currencies*. Triarchy Press.

Keynes, J. M. (1936). *General theory of employment, interest and money*. Hartcourt.

Keynes, J. M. (1983). Economic articles and academic correspondence. In D. Moggrigde (Ed.), *The collected writings of John Maynard Keynes* (p. 402). Macmillan.

Kluge, F. (1989). *Etymologisches Wörterbuch der deutschen Sprache*. De Gruyter.

KPMG. (2020). *Indonesia Government and institution measures in response to COVID-19*. [Online] Available at: https://home.kpmg/xx/en/home/insights/2020/04/Indonesia-government-and-institution-measures-in-response-to-covid.html. Accessed 25 May 2022.

Krempl, S. (2017). Geldwäsche: EU-Gremien bekämpfen "Anonymität" von Bitcoin & Co. *Heise Online*. [Online] Available at: https://www.heise.de/newsticker/meldung/Geldwaesche-EU-Gremien-bekaempfen-Anonymitaet-von-Bitcoin-Co-3919854.html. Accessed 7 May 2022.

Kroll, J. A., Davey, I. C., & Felte, E. W. (2013). *The economics of bitcoin mining, or bitcoin in the presence of adversaries*. The Twelfth Workshop on the Economics of Information Security.

Latouche, S. (2003). Would the west actually be happier with less? The world downscaled. *Le Monde Diplomatique*, p. 1.

Leibenstein, H. (1966). Allocative efficiency vs. "X-efficiency". *American Economic Review, 56*(3), 392–415.

Liao, G. Y., & Caramichael, J. (2022). *Stablecoins: Growth potential and impact on banking* (International finance discussion papers 1334). Board of Governors of the Federal Reserve System. https://doi.org/10.17016/IFDP.2022.1334

Lietaer, B. (2002). *The future of money – Creating new wealth, work and a wiser world* (Vol. 13). Random House.

Lietaer, B., & Belgin, S. (2004). *Of human wealth: Beyond greed & scarcity*. Galley Edition.

Loncon Antileo, E. (2022). El Buen vivir: un paradigma para la vida, el equilibrio entre los pueblos y la Madre Tierra. *El Mostrador*. [Online] Available at: https://www.elmostrador.cl/destacado/2022/03/08/el-buen-vivir-un-paradigma-para-la-vida-el-equilibrio-entre-los-pueblos-y-la-madre-tierra/. Accessed 2 June 2022.

Louge, C. (2015). *Keynes & Gesell – ¿Nuevo Paradigma? – El Orden Económico Natural y breve historia monetaria Argentina*. Errepar.

Marcus, D. (2020). *Welcome to Novi.* [Online] Available at: https://about.fb.com/news/2020/05/welcome-to-novi/. Accessed 7 May 2022.

Martínez-Alier, J., Pascual, U., Vivien, F.-D., & Zaccai, E. (2010). Sustainable de-growth: Mapping the context, criticisms and future prospects of an emergent paradigm. *Ecological Economics, 69*(9), 1741–1747.

Marx, K. (1872). *Das Kapital – Kritik der politischen Oekonomie. Erster Band, Buch I: Der Produktionsprozess des Kapitals* (4th ed.). Meissner.

Max-Neef, M. (1986). *La Economía Descalza – Señales desde el mundo invisible*. Nordan.

Max-Neef, M., Elizalde, A., & Hopenhayn, M. (1991). *Human scale development – Conception, application and further reflections*. Apex.

Meadows, D. (1972). *The limits to growth*. New York: New American Library.

Mill, J. S. (1863). *Utilitarianism*. Parker.

Minuto-Regional. (2022). LebensZeit statt GeldSchein. *Minuto Regional.* [Online] Available at: https://www.minuto-regional.de/. Accessed 20 Apr 2022.

Mitchell, W. (2000). The job guarantee in a small open economy. In E. Carlson & W. Mitchell (Eds.), *The path to full employment, economic and labor relations review* (pp. 89–116). Kensington.

Monetative. (2022). *Monetative.* [Online] Available at: https://monetative.de/vollgeld/. Accessed 7 Feb 2022.

Mora, Z. (2020). El buen vivir, nuestro único destino. *El Mostrador.* [Online] Available at: https://www.elmostrador.cl/noticias/opinion/2020/10/23/el-buen-vivir-nuestro-unico-destino/. Accessed 2 Jan 2022.

Mosler, W. (1997). Full employment and price stability. *Journal of Post Keynesian Economics, 20*(2), 167–182.

Mühlbauer, P. (2018). *Bitcoin: Staatliche Alternativen, Trittbrettfahrer und der Stromverbrauch.* [Online] Available at: heise.de/tp/features/Bitcoin-Staatliche-Alternativen-Trittbrettfahrer-und-der-Stromverbrauch-3918719.html. Accessed 10 May 2022.

Namatama, N., Fuders, F., & Monti, G. (2018). *An empirical examination of the relationship between GDP growth rate and development in 100 countries and a city*. Universidad Austral de Chile.

New York Times. (1929, December 10). Asks house to vote tax on stock sales, representative Sabath introduces bill for 5% Levyon shorts. Penalty for non-payment Illinois man denounces 'Wall Street Gamblers' as 'Rapacious Parasites'". *New York Times*, p. 54.

Pérez, D. (2016). Decrecimiento: Necesitar menos para vivir mejor Fuente: Emol.com – https://www.emol.com/noticias/Tendencias/2016/03/18/793800/Decrecimiento-Necesitar-menos-para-vivir-mejor.html. *Emol*, online: https://www.emol.com/noticias/Tendencias/2016/03/18/793800/Decrecimiento-Necesitar-menos-para-vivir-mejor.html

Pittman, M., & Ivry, B. (2009). Financial rescue nears GDP as pledges top $12.8 trillion. *Bloomberg-online*, issue 03/31.

Positive Money. (2022). *Positive money.* [Online] Available at: https://positivemoney.org/. Accessed 7 Feb 2022.

Raworth, K. (2012). *A safe and just space for humanity – Can we live within the doughnut?* Oxfam Discussion Papers.

Raworth, K. (2017). *Doughnut economics: Seven ways to think like a 21st century economist.* Chelsea Green Publishing.

Reichmuth, K., & Reichmuth, R. (2001). *Der RealUnit – Zur Quelle der Geldwertstabilität*. Ott.

Reiermann, C. (2012, October 2). 'Dangerous territory' – Concerns mount that ECB Bond-Buying Program is illegal. *Spiegel International*.

Rösl, G. (2005). Regionalwährungen in Deutschland. *Wirtschaftsdienst, 85*(3), 182–190.

Rösl, G. (2006). *Regionalwährungen in Deutschland – Lokale Konkurrenz für den Euro?* (Diskussionspapier Reihe 1: Volkswirtschaftliche Studien (No. 43/2006)). Deutsche Bundesbank.

Rousseau, J.-J. (2012). *Du contrat social*. Flammarion.

Saeed, A. (1999). *Islamic banking and interest: A study of the prohibition of riba and its contemporary interpretation*. E.J. Brill.

Schachtschneider, K. A. (1994). *Res publica res populi – Grundlegung einer Allgemeinen Republiklehre. Ein Beitrag zur Freiheits-, Rechts- und Staatslehre*. Duncker & Humblot.

Schachtschneider, K. A. (2002). Grenzen der Kapitalverkehrsfreiheit. In K. A. Schachtschneider (Ed.), *Rechtsfragen der Weltwirtschaft* (pp. 253–327). Duncker & Humblot.

Schachtschneider, K. A. (2005). Demokratische und soziale Defizite der Globalisierung. In D. Siebold & A. Emmerich-Fritsche (Eds.), *Karl Albrecht Schachtschneider: Freiheit – Recht – Staat. Eine Aufsatzsammlung zum 65. Geburtstag* (pp. 617–630). Duncker & Humblot.

Schachtschneider, K. A. (2006). *Prinzipien des Rechtsstaates*. Berlin: Duncker & Humblot.

Schachtschneider, K. A. (2007). *Freiheit in der Republik*. Duncker & Humblot.

Schlemer Alcântara, L. C., et al. (2017). Bem Viver: discussões teórico conceituais. *Revista Pensamiento Actual, 17*(28), 66–77.

Schoon, N. (2016). *Modern islamic banking: products and processes in practice*. Wiley.

Schumacher, E. F. (1973). *Small is beautiful – A study of economics as if people mattered*. Blond & Briggs.

Seminar für freiheitliche Ordnung. (1984). Zins in der Geschichte. *Fragen der Freiheit, 167*(p), 30–36.

Sikora, J., & Hoffmann, G. (2001). *Vision einer Gemeinwohl-Ökonomie auf der Grundlage einer komplementären Zeit-Währung*. KSI.

Simon, G. (2018). Bitcoin und Co: Kryptische Perspektiven für die Finanzsysteme. *Telepolis*. [Online] Available at: https://www.heise.de/tp/features/Bitcoin-und-Co-Kryptische-Perspektiven-fuer-die-Finanzsysteme-3997553.html?seite=all. Accessed 7 May 2022.

Smith, P. B., & Max-Neef, M. (2012). *Economics unmasked: From power and greed to compassion and the common good*. Green Books.

Snyder, J. (2020). COVID-19: Federal deficit projected to reach $184B as economic response rolled out. *National Post*. Online: https://nationalpost.com/news/covid-19-federal-deficit-projected-to-reach-184b-as-economic-response-rolled-out

Soddy, F. (1933) (Reprint 1961). *Wealth, virtual wealth and debt – The solution of the economic paradox*. Briton.

Soddy, F. (1934). *The role of money – What it should be contrasted with what it has become*. Routledge.

Sovereign money. (2022). *Sovereign money*. [Online] Available at: https://sovereignmoney.site/. Accessed 7 Feb 2022.

Soza-Amigo, S., Fuders, F., & Aroca, P. (2021). La importancia del sector minero para el desarrollo de la economía chilena: la evolución de sus campos de influencia. *EL Trimestre Económico, LXXXVIII*(3), 831–872.

Statista. (2020). *Marktanteile an den Einlagen von Privatpersonen in Deutschland im Jahr 2020 nach Bankengruppen*. [Online] Available at: https://de.statista.com/statistik/daten/studie/161138/umfrage/marktanteile-der-banken-an-den-einlagen-von-privatpersonen/. Accessed 2 May 2022.

Steltzner, H. (2009, April 2). Auf dem Londoner Gipfel. *Frankfurter Allgemeine Zeitung (FAZ)*, p. 1.

Stiefl, J. (2005). *Finanzmanagement unter besonderer Berücksichtigung von kleinen und mittelständischen Unternehmen*. Oldenbourg.

Tirole, J. (2017). *Economics for the common good*. Princeton University Press.

Tobin, J. (1978). A proposal for international monetary reform. *Eastern Economic Journal, 4*(3–4), 153–159.

Tolksdorf, M. (1994). *Dynamischer Wettbewerb*. Springer Gabler.

Traverso, M. (2015). *Economía Consciente – La transformación espritual de la economía que comienza por uno mimso*. Kier.

Umara, M., et al. (2021). Bitcoin: A safe haven asset and a winner amid political and economic uncertainties in the US? *Technological Forecasting and Social Change, 167*, 120680.

US GAO (Government Accountability Office). (2004). *Federal debt – Answers to frequently asked questions*. US GAO.

Vollgeld-Initiative. (2022). *Vollgeld initiative.* [Online] Available at: https://www.vollgeld--initiative.ch/. Accessed 7 Feb 2022.

von Goethe, J. W. (1987). Natur und Kunst. In E. F. Friesenbiller (Ed.), *Natur und Kunst: Graphik und Gedichte*. Verlag Christian Brandstätte. p. Cover.

VR Bank eG. (2013). *100 Jahre VR Bank eG*. VR Bank eG Zentrale.

Wagemann, C. (2022). *What is a bad bank?*. [Online] Available at: https://www.thebalancemoney.com/what-is-a-bad-bank-5204263. Accessed 8 Feb 2023.

Walker, K. (1995). Die Überwindung des Kapitalismus - unter Beibehaltung des marktwirtschaftlichen Wettbewerbs. In W. Onken (Ed.), *Karl Walker Ausgwählte Werke* (pp. 232–246). Lütjenburg: Gauke.

Wash, R. (2022). *Behind the crypto hype is an ideology of social change.* [Online] Available at: https://theconversation.com/behind-the-crypto-hype-is-an-ideology-of-social-change-177981. Accessed 7 May 2022.

Weitkamp, H. (1993). *Das Hochmittelalter – ein Geschenk des Geldwesens*. HMZ-Verlag.

Wikipedia. (2022). *Entry "JAK members bank"*. [Online] Available at: https://en.wikipedia.org/wiki/JAK_Members_Bank. Accessed 4 May 2022.

WIR Bank. (2022). *Bank WIR.* [Online] Available at: https://www.wir.ch. Accessed 2 May 2022.

Wolf, M. (2017). Best books of 2017: Economics. *Financial Times.* [Online] Available at: https://www.ft.com/content/838ecc26-d62c-11e7-8c9a-d9c0a5c8d5c9. Accessed 28 May 2022.

Wolff, P. (2010). *Zeit-online: Die Wirkung von Mikrokrediten wird überschätzt.* [Online] Available at: https://www.zeit.de/politik/2010-03/mikrokredite-armut. Accessed 9 Dec 2022.

World Council of Credit Unions. (2021). *Statistical report 2020*. World Council of Credit Unions.

Wray, R. (1998a). Zero unemployment and stable prices. *Journal of Economic Issues, 32*(2), 539–545.

Wray, R. (1998b). *Understanding modern money: The key to full employment and price stability*. Edward Elgar.

Wray, R. (2012). *Theories of money and banking*. Edward Elgar.

Yunus, M., & Weber, K. (2010). *Building social business – The new kind of capitalism that serves humanity's most pressing needs*. PublicAffairs.

Chapter 14
Experiences with Gesell's Solution in the Past

14.1 Introduction

Gesell's natural money is not just a theoretical idea. In the 1930s it was actually implemented alongside the respective national currencies in cities in Germany, Austria, Liechtenstein, Switzerland, France, Spain, the Czech Republic and in a couple of cities in the U.S.A. In some cases, with extraordinary success. In several US States there were even bills proposed to officially introduce a natural (Gesell) money as legal tender, and later, Irving Fisher even advocated for nationwide introduction of such a non-hoardable dollar. Unfortunately, none of these bills were adopted (Fisher & Cohrssen, 1933; Ottacher, 2007; Champ, 2008). The most successful implementation took place in Germany in the village **Schwanenkirchen** and in Austria in **Wörgl** (Champ, 2008). In both villages unemployment rapidly fell to its pre-crisis levels, the municipalities recorded increased tax revenues, and many people were able to pay off their debts. The success of the Wörgl experiment in particular was so great that newspapers around the world reported about the "miracle of Wörgl" (Fisher & Cohrssen, 1933: IV; Onken, 1997). Nevertheless, both experiments were terminated by the respective national governments despite their great success. The official reasons given were the fear of inflation and violation of the note-issuing privilege of the central banks (Fisher & Cohrssen, 1933; Unger, 2007; Blanc, 2011).

© The Author(s), under exclusive license to Springer Nature Switzerland AG 2023
F. Fuders, *How to Fulfil the UN Sustainability Goals*,
https://doi.org/10.1007/978-3-031-37768-6_14

14.2 Natural Money Experiments

14.2.1 Schwanenkirchen/Wära Exchange Association

Schwanenkirchen is a village in Lower Bavaria. This Gesell-money experiment started as a nation-wide barter network, the "Wära exchange association" (Wära-Tauschgesellschaft). It began in October 1929 at about the very same time Black Friday hit the New York stock exchange. The association's purpose was to combat economic stagnation and unemployment by facilitating the exchange of goods and services among its members by issuing its own money, the "Wäras" (Onken, 1997). So Wära was a complementary currency that functioned alongside the legal tender national currency. The bill's name is a play on words that combines the German words "*währen*" (to last, continue, persevere) and "*Ära*" (era).

After 2 years, more than a 1000 companies from all parts and sectors of Germany had become members, among them grocery stores, bakeries, dairies, restaurants, health food stores, butcher shops, flower shops, hairdressing salons, handicraft stores, furniture stores, electrical goods stores, bicycle stores, printing houses, bookstores and coal shops. All of these businesses displayed signs reading "Wära accepted". Offices of the exchange association issued Wäras upon request in exchange for Reichsmarks or foreign currencies or against other securities (collateral). There were exchange offices in Berlin, Bielefeld, Bonn, Chemnitz, Dortmund, Düsseldorf, Eisenach, Erfurt, Freiburg, Halle, Hamburg, Cologne, Leipzig and Nuremberg (for further references see Onken, 1997).

The companies paid wages, in full or in part, in Wära, which could be used to buy the goods and services offered in associated shops and companies. Thus, a sort of complementary circulation of Wära gradually emerged alongside the circulation of the Reichsmark (Onken, 1997). The association even had its own newsletter entitled the "Wära-Handelsblatt" (ZDB, 2022).

In accordance with Gesell's ideas, Wära bills were subject to a hoarding fee. The bills were printed with 24 boxes on their reverse side (see Fig. 14.1), on each of which a stamp costing 1% of the nominal value had to be affixed every 2 weeks.

Fig. 14.1 Wära bill (front and reverse side) used in Schwanenkirchen (Germany) 1931. (Photo: Felix Fuders)

This first practical experiment of Gesell's natural money caused an international sensation at the end of 1930. The world economy was already in the midst of the Great Depression, and the 500-person village of Schwanenkirchen became a local hub. This is what happened:

Mining engineer Max Hebecker acquired the defunct Schwanenkirchen coal mine at auction. Initially, he lacked the necessary money to restart operations, as the banks would not give him a loan. He then turned to the Wära-Tauschgesellschaft, which co-founded the Wära financing consortium and raised the necessary 50,000 Reichsmark (RM) by issuing shares. A majority of this loan was issued in Wära and the rest in RM. With this money, Hebecker was able to restart the mine. He initially employed about 45 miners and soon hired more. 60–75% of wages were paid in Wära and the rest in Reichsmarks (Onken, 1997). The mine began to purchase goods and services from other companies which were members of the Wära exchange society. This helped build confidence among local businessmen who were initially skeptical of this unusual money.

Unfortunately, in 1939 a German Emergency Ordinance outlawed the use of Wära bills. The experiment came to an end. The coal mine went bankrupt again, the people lost their jobs. Schwanenkirchen fell back into deep economic depression. The silver lining is that this first experiment demonstrated the viability of Silvio Gesell's proposal. Gesell himself returned to Germany in 1927 and was able to see the success of his idea.

14.2.2 Wörgl

In the Austrian town of Wörgl, the Great Depression and the bankruptcy of the local ore mine caused the local economy to come to a standstill. This caused high unemployment and lower municipal tax revenues. The municipality had more than 1.3 million shillings in debt which it could not repay due to lack of tax revenue. New loans had to be taken out just to pay the interest. The mayor of the town, Michael Unterguggenberger, had read about Gesell's ideas in a publication (Der Physiokrat, 1913)[1] issued by the Gesellian movement. Unterguggenberger had also read about the Wära success story and convinced his colleagues to implement a Gesellian local currency.

The municipality printed 32,000 shillings of natural money bills, which they called "work confirmation certificates" (Arbeitsbestätigungsschein). Following Gesell's proposal these were subject to a hoarding fee. Every month the holders of the bills had to buy stamps costing 1% of the bill's face value and affix them to the bills. The revenue from the sale of the stamps was used to **help reduce poverty**. This feature probably helped build trust in this new unhoardable money. People

[1] The magazine was produced and supervised by Georg Blumenthal, probably the most prominent promoter of Gesell's Natural Economic Order in Germany as well as in the UK.

were more accepting of the fees since they knew that the revenue went to help the poor. The municipality paid public servants with this money and accepted it for payment of taxes (Onken, 1997).

The money was soon accepted in stores, enabling people to buy food or other goods. Even the local Raiffeisen credit cooperative (a cooperative bank, see Sect. 13.8) and the mutual savings banks participated, and bank accounts could be opened using the local currency. The citizen's desire to avoid the monthly fees caused the bills to circulate more consistently than the legal tender currency. Economic activity in Wörgl recovered quickly. The financial situation of the municipality also began to improve. People not only resumed paying their taxes, but some even paid in advance (Onken, 1997). The success of the Wörgl natural money experiment soon led other communities in the vicinity and later in the wider area to copy the idea. People spoke of the Wonder of Wörgl and "Mecca of the national economy" (Bourdet, 1933).

Politicians and scientists came to study the experiment. Irving Fisher's book "Stamp Scrip" was also influenced by the success in Wörgl. Due to Fisher's publications, and reports in newspapers Gesell's theory became more widely known in the US. Also in France, Spain, Liechtenstein and the Czech Republic Gesell's natural money was implemented. In June 1933 Michael Unterguggenberger gave a lecture in Vienna to 170 Austrian mayors. They all intended to follow Wörgl's "guide to the economic salvation of Austria" (Burgstaller, 1933: 55) and to introduce similar currencies in their cities and municipalities as well. On 9/15/1933 the government in Vienna ordered the municipality to withdraw the money from circulation; the mayor appealed, unfortunately without success (for further references see Schwarz, [1951] 2021; Onken, 1997). Based on the Wörgl experience a movie was produced (Das Wunder von Wörgl, 2018).

14.3 Natural Land Order Experiments

The successful experiments described above were implementations of Gesell's natural money proposal. But these experiments did not include Gesell's proposals regarding land ownership. The latter was successfully implemented in the "Eden land cooperative" (*Eden Genossenschaftssiedlung*) in Oranienburg, near Berlin. However, this initiative did not include Gesell's monetary proposals. The Eden cooperative was founded 1893, that is before Silvio Gesell's ideas were known in Germany, i.e., the Eden cooperative was not expressly based on Gesell's proposal.

The notion that land (naturally) belongs to everyone was a theme that many philosophers of the time had pondered. In this context, Silvio Gesell and Henry George can be seen as 'children' of the zeitgeist of that time. In the German-speaking countries, at the beginning of the twentieth century evolved a movement called "*Lebensreformbewegung*" what could be translated as "life-reform-movement". This included, next to the promotion of equal access to land, also a healthy life through organic nourishment. Maybe the Eden community could be seen as a forerunner to today's permaculture movement (Holmgren, 2002). The Eden land cooperative was founded not only with the aim to show that land can be commonly owned but also to produce heathy food, especially fruits and vegetables. The complete name of the cooperative was: "*Vegtarische Obstbau-Kolonie Eden*" (vegetarian fruit colony).

This is how it worked: people bought approx. 250 ha of land and donated the land to the cooperative. Every person admitted to the colony including the founders had to lease the land. Lease contracts were 30–50 years. If someone died their heirs were allowed to assume the contract. When contracts ended, the leaseholders had the right to renew the contracts before they were offered to others. This way each member of the colony had their own place, but the cooperative was the owner. Any buildings or improvements were **property of the people who built them**. In the event that someone left the cooperative, the improvements became property of the cooperative, but the original owner was compensated for his or her investment in the improvements. People thrived in the cooperative. In only 7 years they managed to plant 15,000 fruits trees, 50,000 berry bushes and 200,000 strawberry plants (Onken, 1997). The cooperative founded a company to process and sell the fruit. The cooperative functioned until after World War II when the Soviet controlled Eastern German government expropriated them. The cooperative's fruit processing and marketing company was taken over and became one of many State-owned conglomerates (for further references see Onken, 1997).

Eden was not only a cooperative to produce and market fruits and vegetables and associated products. The common property model was meant to **reunite man with nature** (Onken, 1997). In Eden, people could once again work privately and freely. This led to the physical, mental and spiritual regeneration of the people living there, even documented by a physician (Onken, 1997). A school was set up to educate the children according to the colony's principles. The example of the Eden cooperative shows that people can peacefully live in harmony together, work the land and even build beautiful houses without private land ownership. Indeed, Eden could be seen as a successful example of what Elinor Ostrom would later term a "common pool resource" (Ostrom, 1990). In Eden all eight principles Ostrom laid out for the successful governance of a common pool resource were fulfilled.[2] Finally, it is noteworthy that Silvio Gesell spent the last years of his life in Eden where he died 1930. His youngest and still living (as of February 2023) child Sonja Tomys was born in Eden.

[2]These eight principles are (Ostrom, 1990): (1) Commons need to have clearly defined boundaries, (2) Rules should fit local circumstances, (3) Participatory decision-making is vital, (4) Commons must be monitored, (5) Sanctions for those who abuse the commons should be graduated, (6) Conflict resolution should be easily accessible, (7) Commons need the right to organize, (8) Commons work best when embedded within larger networks.

14.4 Bracteates in the Middle Age

Probably the most successful form of the implementation of a Natural Economic Order happened by accident some 600 years before Silvio Gesell was born. Between the years 1000 and 1300, in most parts of Europe a currency known as "Bracteates" circulated and was probably an important contributing factor to the flourishing of that time (Walker, 1959; Weitkamp, 1993). A bracteate (lat. *Bractea* = a thin piece of metal) was a flat, thin, single-sided silver coin produced in Central Europe in the High Middle Ages. The thinness of the coins made cheap production of money possible for the first time in history. They also created a new way of levying taxes. The governor (king or duke) recalled all of the coins periodically. The coins were then melted down. The governor retained some of the silver as a tax and then issued new coins with a new design. Producing new series of coins with different designs was necessary to ensure that everyone turned in all of their money. The periodic recoining of the money (*renovatio monetae*) created pressure to keep the money circulating. Nobody wanted to keep the money for a prolonged period since the date of the recoining was not publicly known.[3]

In consequence, the money circulated rapidly, and interest did not exist. And life was more relaxed than today. Since money could not be hoarded (and investments bear risks) there was less incentive to produce more than is needed in the moment. It was a time of prosperity which today seems hard to believe (Walker, 1959; Weitkamp, 1993; Lietaer et al., 2013). People only worked 4 days a week. Monday was in many cases a regular holiday, some believe that this is where the expression "*Blue Monday*" finds its origin, people who went outside to relax, have a beer, with their blue working clothes on (Walker, 1959: 71 f.).[4] The average salary was 6-8 pence per day and people worked only 6 h per day. In comparison, a pair of shoes then cost 2 pence (Weitkamp, 1993). The inequality caused by interest did not exist.

People tended to invest in real goods like solidly built houses, tools, artwork or anything else that had intrinsic use value. It was during this period that most European cathedrals were built, which not seldom took a couple centuries to complete. Would anyone today undertake a construction project that would take centuries to finish? Probably not. The reason: time is money; and the reason for time being money is interest. One could say that this was truly development on a "human scale" to use the expression coined by my dear colleague Manfred Max-Neef. In those times land was owned by the king and his network of nobles. People could use the land for farming but had to pay tribute (lease) to the king. The more land someone used the more tribute they would have to pay. This is in some respects similar to Gesell's natural land property order (Sect. 10.5). That is to say, in those times

[3] See also: Karl Walker (1958, 1959); Weitkamp (1993).

[4] In German exists an expression frequently used until today for people wo do not go to work or for pupils who skip school: "make it a blue day" (*blau machen*). This locution could be connected to the "Blue Monday" (WDR, 2022).

next to natural money also a natural land property order was in place. No wonder that this period is usually referred to as the heyday of the Middle Ages.

This prosperity came to an abrupt end when the "*Denarius perpetuus*" (the eternal penny) was introduced. It initiated the descent into the dark Middle Ages (Walker, 1959; Weitkamp, 1993) with its characteristic pattern of poverty and diseases, which probably spread so widely because of poverty and associated malnutrition.[5] A reason for the introduction of the eternal penny and thus the poverty of the majority is believed to be the burgeoning banking system (Weitkamp, 1993).

References

Azkarraga, J., Max-Neef, M., Fuders, F., & Altuna, L. (2011). *La evolución sostenible II*. Eskoriatza.

Blanc, J. (2011). Classifying "CCs": Community, complementary and local currencies' types and generations. *International Journal of Community Currency Research, 15*, 4–10.

Bourdet, C. (1933). Une nouvelle Mecque économique – Wörgl ou l'argent fondant (zitiert nach Onken, 1997). *L'Illustration, 4723*(09.09.1933), 56.

Burgstaller, H. (1933). *Das Wörgler Beispiel (zitiert nach Onken, 1997)* (H. K. Sonderegger, Ed.) (p. 50 ff.). Hans Burgstaller.

Champ, B. (2008). Stamp scrip: Money people paid to use. In Federal Reserve Bank of Cleveland (Ed.), *Economic commentary* (pp. 1–4). Cleveland.

Das Wunder von Wörgl. (2018). [Film] Directed by Urs Egger. Epo-Film.

Der Physiokrat. (1913). *Erstrebt den vollen Arbeitsertrag für die Arbeiter aller Stände durch Grundbesitz- und Geldreform*. Physiokratischer Verlag.

Dick, H., Pringle, J., & Sloane, B. (2015). Detection and characterisation of Black Death burials by multi-proxy geophysical methods. *Journal of Archaeological Science, 59*, 132–141.

Fisher, I., & Cohrssen, H. (1933). *Stamp scrip*. Adelphi.

Holmgren, D. (2002). *Permaculture: Principles and pathways beyond sustainability*. Holmgren Design Services.

Lietaer, B., Arnsperger, C., Goerner, S., & Brunnhuber, S. (2013). *Geld und Nachhaltigkeit – Von einem überholten Finanzsystem zu einem monetären Ökosystem. Ein Bericht des Club of Rome/EU-Chapter*. Europa Verlag.

Morgan, J. (2014). *Black Death skeletons unearthed by Crossrail project*. [Online] Available at: https://www.bbc.com/news/science-environment-26770334. Accessed 3 June 2022.

Onken, W. (1997). *Modellversuche mit zozialpflichtigem Boden und Geld*. Kiel: Gauke.

Ostrom, E. (1990). *Governing the commons: The evolution of institutions for collective action*. Cambridge University Press.

[5]We previously theorized in an earlier publication (Azkarraga et al., 2011) that the millions of people who died in Europe from the pest (so-called "black death" or "bubonic plague") in the Middle Ages was not primarily due to the bacteria (*Yersinia pestis*) itself, but was actually caused by immunodeficiency due to malnutrition. Our theory was confirmed in 2013 when several mass graves from the Middle Ages were discovered during the expansion of the London underground railway (Dick et al., 2015). A DNA comparison revealed that the people had been infected with the very same pathogen that still exists today. But why do only relatively few people die of the plague today? It was found that the people suffered from severe malnutrition, because the bones show typical characteristics of malnutrition (porous bones). The scientists suspected, as we did 2 years earlier, that people's immune systems were severely compromised due to poor diet so that the otherwise relatively harmless pathogen was fatal to them (Morgan, 2014).

Ottacher, G. (2007). *Der Welt ein Zeichen geben – Das Freigeldexperiment von Wörgl 1932/1933*. Gauke.

Schwarz, F. (2021). *Das Experiment von Wörgl – ein Weg aus der Wirtschaftskrise*. Darmstadt: Synergia.

Unger, B. (2007). *Zeit online.* [Online] Available at: https://www.zeit.de/online/2007/37/besser-wirtschaften-brigitte-unger. Accessed 24 Apr 2020.

Walker, K. (1958) (Reprint 1995). Das Geld als Kulturfehler. In W. Onken (Ed.), *Karl Walker – Ausgewählte Werke* (pp. 251–255). Gauke.

Walker, K. (1959). *Das Geld in der Geschichte*. Rudolf Zitzmann Verlag.

WDR. (2022). *Warum heißt es „blaumachen"?* [Online] Available at: https://kinder.wdr.de/tv/wissen-macht-ah/bibliothek/kuriosah/blaumachen-100.html#:~:text=Wer%20blau%20macht%2C%20geht%20ohne,zur%20Schule%20oder%20zur%20Arbeit. Accessed 3 June 2022.

Weitkamp, H. (1993). *Das Hochmittelalter – ein Geschenk des Geldwesens*. HMZ-Verlag.

ZDB. (2022). *Wära-Handelsblatt.* [Online] Available at: https://zdb-katalog.de/title.xhtml?idn=019587570&view=full. Accessed 9 June 2022.

Part V
Outlook and Final Considerations

Chapter 15
Towards Development at Human Scale

15.1 The Human-Scale-Development Approach

In their *Human-Scale-Development* approach Max-Neef et al. (1991) proposed a matrix of nine *axiological* fundamental human needs and four *existential* dimensions of each need, in order to classify and identify potential means of fulfilling these needs in each cell of the matrix (see Table 15.1). According to this novel concept[1] fundamental human needs can be characterized as few, finite, classifiable and invariable through all human cultures and across all historical time periods. What might change are the means by which these needs are satisfied. The better people manage to satisfy their fundamental human needs, the more developed a country or region is considered.

15.2 The Necessary Market Structure

Manfred Max-Neef used to say that he was influenced by Schumacher's (1973) work "Small is beautiful". Indeed, development at "human scale" contains in its title the idea that people can satisfy fundamental human needs through a human scale (or small scale) market structure, i.e., small and medium-sized producers which, being close to the consumer, care about delivering good products at good prices. This is precisely the idea of the notion of perfect competition.

Perfect competition describes the utopian situation in which a large number of small and medium-sized producers do business with a large number of consumers. Producers compete with each other in trying to **best satisfy the needs of**

[1] Manfred Max-Neef was recognized as one of the key thinkers on development for his theory (Simon, 2016). On the impact of his teachings, see Barrera et al. (2023).

© The Author(s), under exclusive license to Springer Nature Switzerland AG 2023

F. Fuders, *How to Fulfil the UN Sustainability Goals*,
https://doi.org/10.1007/978-3-031-37768-6_15

Table 15.1 Example of the human-needs matrix with satisfiers in each cell

Need	Being (qualities)	Having (things)	Doing (actions)	Interacting (settings)
Subsistence	Physical and mental health	Food, shelter, work	Feed, clothe, rest, work	Living environment, social setting
Protection	Care, adaptability, autonomy	Social security, health systems, work	Cooperate, plan, take care of, help	Living environment, social setting
Affection	Respect, sense of humor, generosity, sensuality	Friendships, family, relationships with nature	Share, take care of, make love, express emotions	Privacy, intimate spaces of togetherness
Under-standing	Critical conscience, capacity, curiosity, intuition	Literature, teachers, policies, education	Analyze, study, meditate, investigate	Schools, families, universities, communities
Participation	Receptiveness, dedication, sense of humor	Responsibility, duties, work, rights	Cooperate, dissent, express opinions	Associations, parties, churches, neighborhoods
Leisure	Imagination, tranquility, spontaneity	Games, parties, peace of mind	Daydream, remember, relax, have fun	Landscapes, intimate spaces, places to be alone
Creation	Imagination, boldness, inventiveness, curiosity	Abilities, skills, work, techniques	Invent, build, design, work, compose, interpret	Spaces for expression, workshops, audiences
Identity	Sense of belonging, self-esteem, consistency	Language, religions, work, customs, values, norms	Get to know oneself, grow, commit oneself	Places one belongs to, everyday settings
Freedom	Autonomy, passion, self-esteem, open-mindedness	Equal rights	Dissent, choose, run risks, develop awareness	Possibility to communicate anywhere, social networks

Source: Own work, adapted from Max-Neef, et al. (1991: 32 f)

consumers, and no producer has the power to unilaterally increase prices without a corresponding improvement of the product in the eyes of the consumer. Since producers compete on quality and pricing, the quality of products and services tend to constantly improve while prices (and profits) tend to shrink. This leads to a situation where **everyone gains according to their effort** in satisfying consumer needs. In this model, when used in conjunction with neutral money, any inequality of income would result from different levels of effort (investment of production factors), as outlined in Sect. 10.4.1. Although companies and consumers do not intentionally strive to promote social welfare, the competitive equilibrium—through an "invisible hand" to use Adam Smith's ([1776] 1952: 194) famous expression— forces us to produce the best products and services (in the eyes of the consumer) at the best prices. Hence, it is arguable that perfect competition is the market structure that best assures development at human scale.

The reason why most markets usually do not even come close to the ideal of perfect competition is, in additional to a lack of enforcement of competition law, the non-neutrality of money. Remember, monetary neutrality is a concept that states that changes in the money supply only affect nominal variables and not real variables (see. e.g., Stiglitz, 1998: 187). Even in mainstream economic literature it is recognized that one reason for market failure is the non-existence of stable equilibrium. It is, however, not recognized why this is. A stable equilibrium can never be reached as long as the money supply constantly grows, and the real economy tries to keep up. Inflation, caused by a constantly growing money supply which is always a bit ahead of the growth of productive output, means that producers must constantly adjust prices. If some producers adjust their prices while others do not, prices will not accurately reflect supply and demand. As we discussed above (Sect. 10.4) implementing natural and neutral money would finally make it possible to achieve "workable" competition, i.e., a state as close as realistically possible to the theoretical ideal of "perfect competition". With neutral money a stable long-term equilibrium would finally be achievable.

Of course, this would also require that the State applies and enforces competition law: to fight against the abuse of dominant power in the market (monopoly), to fight against collusion and unfair competition. These are the three most recognized market failures; one could say the three pillars of perfect competition. If this was combined with natural and neutral money (Fuders, 2010, 2016, 2017, 2021; Fuders & Max-Neef, 2014a, b) we could finally achieve a result that approaches perfect competition. The economy would then serve people in their quest to satisfy fundamental needs and not vice versa. At the same time, it would be an economy in which everyone earns remuneration equivalent to the opportunity cost of factors invested as we say in microeconomics, i.e., everyone's earnings would correspond to their efforts.

15.3 Development at Human Scale and GDP Growth

Under the current monetary system, most economic growth causes increasing income inequality, while only a relatively small share (if any) promotes development in the sense of improving the satisfaction of fundamental human needs. However, I am convinced that with natural (and neutral) money nearly **100% of economic growth would translate into development**, as depicted in Fig. 15.1,—development understood as improvement in the fulfilment of fundamental human needs—provided that the government also vigilantly enforces competition law so that classical market imperfections and **monopolistic rents** are avoided thus allowing us to approach the ideal of perfect competition.

The sole driver of economic growth would then be the satisfaction of human needs. If further growth would not lead to improvements in the satisfaction of human needs, then we would simply not grow anymore. Unlike today such economic stagnation would not lead to more unemployment. Since natural (and

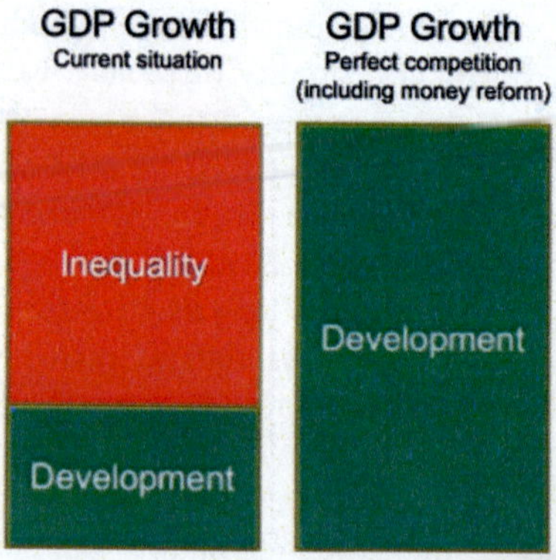

Fig. 15.1 Development, Inequality and GDP growth. Left side: current situation. Right side: reformed financial system plus vigilant enforcement of competition law enabling us to approach the ideal of perfect competition. (Source: Own work)

neutral) money cannot be hoarded for prolonged periods and since all investment entails risk, Gesell's solution provides a powerful incentive not to produce more than is actually needed. It is quite conceivable that without the obligation to grow, it would be possible to achieve a stable long-term equilibrium and to reach or at least come close to the "*steady-state-economy*" described by Herman Daly (1991, 1993), possibly even without the need for the government to regulate input and throughput of production processes.

The hypothesis advanced here is that for positive (desirable) human development, that is development where **every person can best satisfy their fundamental human needs** (development at human scale), we need to get as close as possible to the microeconomic notion of perfect competition. This requires monetary neutrality as explained in Sect. 10.4. On the other hand, the further away we are from the ideal of perfect competition (including neutral money) the more GDP growth leads to increasing income inequality rather than improved satisfaction of fundamental human needs.

To prove this hypothesis, we would need an index that measures the degree of development at human scale, i.e., the degree of fulfilment of fundamental human needs. The concept of such a "Human Scale Development Index" already exists (Fuders, 2011, 2015; Fuders et al., 2016; Fuders & Mora Motta, 2021), but has not yet been applied across different countries. Alternatively, we could use the UN Human Development Index (HDI), for which we have reliable data for all countries. In 2018, at Universidad Austral de Chile we conducted such a study using data for 100 countries. The outcome of the study provides empirical evidence for the hypothesis advanced above: We found a negative correlation between market concentration (monopolization of markets) measured via the Herfindahl-Hirschman Index and the HDI. Market concentration is the opposite of perfect competition. That is to say, the study showed that the more competitive markets are, the higher on average is the HDI. On the other hand, we did find a positive correlation between GDP *per capita* and HDI, but not between the GDP *growth rate* and HDI. The positive correlation between GDP *per capita* and HDI was to be expected since one of the variables that makes up the HDI is Gross National Income (GNI) per capita, a related concept to GDP per capita.

Admittedly, this positive correlation does not say anything about causality. Causality could go the other way around, i.e., the higher the degree of development, the higher the education level (years of schooling is a variable in the HDI), the more startups and therefore the more competition between producers. Causality is difficult to assess. In all likelihood, causality runs in both directions (for more references see Namatama et al., 2018). That is more competition leads to more diversity, higher quality and lower prices of products and services, all of which improve development in terms of satisfaction of fundamental human needs. On the other hand, a higher degree of satisfaction of human needs can lead to more competition.

15.4 Growth and Inequality

Some might doubt that GDP growth fosters inequality in the current system. For example, in the literature on Simon Kuznets' (1955) hypothesis Cerra et al. (2022) try to show that inequality and poverty can be reduced by economic growth. However, I only agree with the second half of the assertion. Poverty can indeed be reduced by growth, but as long as we maintain the current monetary system inequality will continue to grow. The Gini index, that measures income inequality, shows a consistently high degree of income inequality (before taxes and transfers) in many countries around the world (see, e.g., OECD, 2015; Bárcena et al., 2018). Some countries manage to mitigate this inequality via taxes and transfers, but such redistributive policies only address the symptoms.

If we succeed in implementing a system of neutral money and thereby create an economy that more closely approaches the ideal of perfect competition, consisting of a large number of small and medium-sized companies competing with each other to best satisfy consumer needs, there would be much less income inequality compared to today's situation. All differences in income would then be due to differences in effort (investment of production factors), i.e., income inequalities would then be just. Everyone would earn according to their effort. The other way around less income inequality would significantly decrease the possibility of the existence market power. For example, Elon Musk's empire is a good example of how financial power can lead to dominant market power in the real economy (Kane, 2022).

In order to foster competition, governments must provide support for start-ups and enforce competition law, i.e., prevent monopolistic power, collusion and unfair competition. Even though perfect competition is a utopian model that no economy in the real world comes close to, largely because of the lack of monetary neutrality (Suhr, 1989; Fuders, 2017, 2020a, b, 2021), we should still strive for the ideal of perfect competition in order to best satisfy fundamental human needs, i.e., to achieve development that serves people. Unfortunately, that we need more and not less competition and that a workable competition is crucial get closer to a *Development at Human Scale* (Max-Neef et al., 1991) seems to be understood only by few scholars and politicians. And increasing income inequality in many regions has led to a resurgence of socialism around the world. People demand to enforce ever more socialistic laws and regulations to mitigate inequality; symptoms curing instead of healing.

15.5 From Chrematistics Back to *Oikonomia*

The word "economics" stems from the Greek term "*oikonomia*". According to Aristotle, ([384 BC] 1995) *oikonomia* is the study of how to efficiently manage household resources. However, when we speak of economics today, we usually mean what Aristotle referred to as *chrematistics*, which is the pursuit of gaining money as an end in itself (Aristotle, [384 BC] 1995), a concept that could be seen as antithetical to *oikonomia*. Implementing natural and neutral money and thereby eliminating the necessity for interest in order to make money circulate would enable us to return to a focus on oikonomia. This would be truly a free-market economy: free of the interest burden and the resultant necessity to grow. In such a system, money would serve only as medium of exchange of goods and services. Let us call it a **market economy without capitalism**.

References

Aristotle. (1995). Politik 1st Book. In *Aristoteles Philosophische Schriften in sechs Bänden. Band 4*. Felix Meiner Verlag.

Bárcena, A., Cimoli, M., García-Buchaca, R., & Pérez, R. (2018). *La ineficiencia de la desigualdad*. Cepal.

Barrera, M., Belloy, P., Mougenot, B., & Doussoulin, J. P. (2023). The international impact of Manfred Max-Neef's scholarship: A bibliometric approach. *International Journal of Sustainable Development, 25*(1/2), 1–29.

Cerra, V., Lama, R., & Loayza, N. (2022). Links between growth, inequality, and poverty. In V. Cerra, B. Eichengreen, & A. El-Ganai (Eds.), *How to achieve inclusive growth* (pp. 32–73). Oxford University Press.

Daly, H. (1991). *Steady-state economics* (2nd ed.). Island Press.

Daly, H. (1993). Steady-state economics: A new paradigm. *New Literary History, 24*(4), 811–816.

Fuders, F. (2010). Alternative concepts for a global financial system – An answer to the present world financial crisis. *Estudios Internacionales, 166*, 45–56.

Fuders, F. (2011). *Política Económica – Class manuscript*. Universidad Austral de Chile.

Fuders, F. (2015). Do Calidad Humana and happiness depend on development? In *Calidad Humana – Sharing the Filipino Spirit* (pp. 94–110). Energy Development Cooperation.

Fuders, F. (2016). Smarter money for smarter cities: How regional currencies can help to promote a decentralised and sustainable regional development. In *Decentralisation and regional development – Experiences and lessons from four continents over three decades* (pp. 155–185). Springer.

Fuders, F. (2017). Neues Geld für eine neue Ökonomie – Die Reform des Geldwesens als Voraussetzung für eine Marktwirtschaft, die den Menschen dient. In *Finanzwirtschaft in ethischer Verantwortung – Erfolgskonzepte für Social Banking und Social Finance* (pp. 121–183). Springer-Gabler.

Fuders, F. (2020a). *The 'tragedy of the commons' and the role of the money interest rate*. Springer.

Fuders, F. (2020b). Freigeld und Freiland: für eine wirklich freie, nachhaltig funktionierende und gerechte Wirtschaftsordnung. *Humane Wirtschaft, 2*, 16–24.

Fuders, F. (2021). The role of money for a healthy economy. In *Urban health and wellbeing programme – Policy briefs: Volume 2* (pp. 17–22). Springer Nature.

Fuders, F., & Max-Neef, M. (2014a). Local money as solution to a capitalistic global financial crisis. In *From capitalistic to humanistic business*. Palgrave Macmillan.

Fuders, F., & Max-Neef, M. (2014b). Dinero, deuda y crisis financieras. Propuestas teórico-prácticas en pos de la sostenibilidad del sistema financiero internacional. In *Economía Internacional. Claves teórica-prácticas sobre la inserción de Latinoamérica en el mundo*. LATIn.

Fuders, F., & Mora Motta, A. (2021). El Índice de Desarrollo a Escala Humana: una primera evaluación. In C. Oyarzún, P. Donoso, & J. J. Nuñez (Eds.), *Historia natural, Servicios Ecosistémicos y Perspectivas de Desarrollo de la Cuenca de Llancahue* (pp. 173–194). Valdivia.

Fuders, F., Mengel, N., & Barrera, M. (2016). Índice de desarrollo a escala humana: propuesta para un indicador de desarrollo endógeno basado en la satisfacción de las necesidades humanas fundamentales. In R. Pérez Garcés, E. Espinosa Ayala, & O. Terán Varela (Eds.), *Seguridad Alimentaria, actores Terretoriales y Desarrollo Endógeno* (pp. 63–106). Iztapalapa: Laberinto.

Kane, M. (2022). *US: Tesla maintains domination in BEV and premium car segments in H1 2022.* [Online] Available at: https://insideevs.com/news/604580/us-bev-premium-car-sales-2022h1/#:~:text=Tesla%20accounts%20for%20more%20than,outpace%20Tesla%20later%20this%20year. Accessed 18 Aug 2022.

Kuznets, S. (1955). Economic growth and income inequality. *The American Economic Review, XLV*(1), 1–28.

Max-Neef, M., Elizalde, A., & Hopenhayn, M. (1991). *Human scale development – Conception, application and further reflections*. Apex.

Namatama, N., Fuders, F., & Monti, G. (2018). *An empirical examination of the relationship between GDP growth rate and development in 100 countries and a city*. Universidad Austral de Chile.

OECD. (2015). *Business and finance outlook*. OECD.

Schumacher, E. F. (1973). *Small is beautiful – A study of economics as if people mattered*. Blond & Briggs.

Simon, D. (2016). *Fifty key thinkers on development*. Roudledge.

Smith, A. (1952). *An inquiry into the nature and causes of the wealth of nations*. W. Benton.

Stiglitz, J. E. (1998). *Macroeconomía* (2nd ed.). Ariel.

Suhr, D. (1989). *The capitalistic cost-benefit structure of money – An analysis of money's structural nonneutrality and its effects on the economy*. Springer.

Chapter 16
Economic Resilience in the Face of External Shocks

16.1 Natural Money for a Sustainable Monetary System

16.1.1 Exponential Money Supply Growth, Debt and Interest

We analyzed above (Sect. 4.3) the fact that deposit balances grow exponentially because of interest. Constant exponential growth means that even the fastest computer in the world would not be able to continue to calculate the interest burden for lack of zeroes (Fuders & Max-Neef, 2014a). This is the logic of an exponential function. We can see why this inevitably leads to recurring financial crises (Kennedy, 1990, 2011; Creutz, 1993; Bichlmaier, 2010). History shows that such crises typically occur every 60–80 years depending on the average interest rate (Fuders, 2009, 2010). Large financial crises occurred in the industrialized world in 1857, 1929 and 2008. In the long run, no economic system can ever generate enough growth to satisfy the demands of interest.

The problem is that, as outlined previously in this book, demand and production cannot be increased indefinitely because we live on a planet with finite resources, a fact which the ancient Greek philosopher Aristotle ([384 BC] 1995) recognized. Extravagance, planned obsolescence, predominance of disposable products, governmental incentives to buy new products, the industrialization of agriculture, including the application of ever more pesticides and even the apparent invention of diseases[1] are symptomatic of the problem. We are forced to constantly pursue economic growth. For these reasons John Maynard Keynes remarked that the various "objectionable features of capitalism" are caused by interest, and that the solution is to reduce the marginal efficiency of capital to zero (Keynes, 1936: 221).[2]

[1] See e.g., Leitner, 2000; Blüchel, 2003; Coleman, 2003; McTaggart, 2005; Blech, 2005; Widmer et al., 2006; Engelbrecht & Köhnlein, 2006.

[2] On this concept see Footnote 3 in Chap. 7.

© The Author(s), under exclusive license to Springer Nature Switzerland AG 2023
F. Fuders, *How to Fulfil the UN Sustainability Goals*,
https://doi.org/10.1007/978-3-031-37768-6_16

In every economy there inevitably comes a point at which economic performance is no longer sufficient to keep up with the steady, exponential growth of interest. The inevitable collapse occurs when banks can no longer find enough creditworthy borrowers to continue to service deposits that constantly grow via interest and compound interest. Deposits continue to grow, even as banks lower the interest rates on deposits. But even with historically low interest rates on deposits (as of March 2022), banks still have trouble meeting their obligations because of the ever-growing size of deposits.

16.1.2 Subprime Lending Becomes Inevitable

Under this pressure banks begin to grant loans to borrowers with dubious (*subprime*) creditworthiness and ever-lower collateral ratios on loans. This dynamic describes not just what happened during the so-called subprime crisis of 2008, but it is a pattern that can be observed in practically all financial crises. The fact that the exponential growth of the money supply causes price bubbles in stock and real estate markets creates an incentive to grant loans even to 'ninjas' (no income, no job or assets) in order to buy assets on credit. The idea was as follows: after a while the prices of real estate (2008) or stocks (1929) would have increased, and even if loans defaulted this would not matter since the mortgaged real estate or stocks could be sold for a profit, and the obligations could be paid off with this profit. This business plan works as long as prices of real estate (2008) or stocks (1929), or whatever the collateral happens to be, are rising. But the moment the bubble bursts—which can happen at any time as a result of investors shifting resources into other asset classes—defaults are inevitable, putting at risk the solvency of individual banks and potentially of the whole system.

After the fact, bankers are accused of failing to control risk properly or gambling with their clients' money (Fuders, 2010), but the banks really have no choice if they want to stay in business. Even gambling on the stock market can be interpreted as a desperate attempt to maintain earnings in order to be able to continue paying interest. All of this is nothing new. After the 1929 crash bankers were found to be the culprits.[3] While it is true that individual bankers may behave incorrectly or even engage in illegal acts, e.g., delaying bankruptcy filings or getting rid of 'toxic' loans—loans they know are likely to default—by bundling them, securitizing them and selling them to unsuspecting investors. But such behaviors are not the primary cause of crises (either in 1929 or 2008). Instead, these behaviors are symptoms of a system that inevitably causes recurring crises by its nature.

The conclusion is that financial crises are not primarily caused by frivolous lending, speculation or lack of banking supervision. Nor are they caused by irresponsible borrowers or politicians who take on more debt than their countries can manage,

[3] See: New York Times, 1929a, b, c, d.

as many believe. Rather, these behaviors are symptoms of a monetary system in which banks must constantly expand lending in order to service the steadily growing interest requirements caused by deposits which grow according to the dynamics of compound interest.

16.1.3 Are the Subprime and Euro Crises Different?

We could ask whether the 2008 Subprime crisis and the 2010 European sovereign debt crisis are really two different crises? On the surface, one might say yes. The first crisis was ostensibly caused by irresponsible, greedy bankers who didn't properly analyze credit standards on mortgage loans. The second was caused by irresponsible governments which burdened their countries with unmanageable debts in order to finance generous pension and social welfare systems. The truth is that both crises are connected, and both are signs of a global financial system that is close to collapse (Fuders et al., 2013; Fuders & Max-Neef, 2014a, b).

We previously discussed the fact that speculation and price bubbles are not the primary causes but rather are symptoms of our current monetary system and precursors to the subsequent financial crises. The crises themselves are symptoms of the unending growth of money supply and debt. Banks must issue more and more loans if they want to be able to keep paying interest on deposits. In the same way, national governments are not the primary culprits of the European debt crisis. In both cases banks were forced to keep creating more debt since deposits are always growing. Banks did this not out of greed but because they must constantly increase the loan volume if they want to survive.

Remember, money is the counterpart of debt, or to quote Frederick Soddy (1934: 25), "Money is a credit-debt relation from which none can effectually escape". All financial crises start off with the weakest debtors, which in 2008 were mortgage borrowers in the US, and in Europe the weakest debtors were those countries with the lowest economic productivity (Fuders & Max-Neef, 2014a). The weakest ones are always the first to not be able to keep up with the exponentially growing interest burden.[4]

[4] In this context, note that the euro currency creates another problem, the same problem which faced Argentina from 1991 to 2001: the Euro is a basket of European currencies with fixed exchange rates. Fixed exchange rates between countries with huge differences in productivity generate disequilibrium in the trade balances and increasingly unmanageable debt burdens in those economies with lower productivity. Lower productivity cannot be compensated for through currency devaluations relative to economies with higher productivity. This, in turn, reduces their exports. On the other hand, the relatively strong common currency provides an extra incentive for those countries to import goods. The resulting deficit in the trade balance will be offset by increasing debt (Fuders, 2014). But this is just an additional problem specific to the Eurozone, that helps explain why southern European countries have higher debt burdens relative to GDP than the northern European countries. With flexible exchange rates financial crises would still happen. Economies with high productivity like Germany, Japan or the USA, are also struggling to keep up with their

Here, I would like to reiterate that the fact that money is the counterpart of debt is not necessarily a problem. The problem is that deposits and debt grow in tandem via interest and compound interest. At some point this causes the bankruptcy of the entire economic system.[5] Indeed, there is a clear pattern we can discern in all financial crises (Fuders et al., 2013).

16.1.4 All Financial Crises Follow the Same Pattern

We can now comprehend why **credit booms, speculation and financial crises go hand in hand** in virtually all financial crises ever recorded in economic history. But it is not that the crisis is caused by greedy bank clerks selling loans to people with subprime solvency who should not have been granted these loans. It is not that prices of stock and real estate markets increase because of banks encouraging clients to buy stocks or real estate on credit. But these interpretation are offered for both crises (regarding the Great Depression see Galbraith, 1983; Fisher, 1932, 1933; Dornbusch et al., 2009; regarding the subprime crisis see Krugman, 2009).

The causation works the other way around: banks are required to continually increase credit volume in order to keep up with the interest expense on ever-growing deposits. And because of the growth of deposits (and hence the money supply defined as cash plus deposits) price bubbles start forming. These price bubbles then represent an inducement for people to borrow more money. Banks then begin to grant loans even to people with questionable creditworthiness.

So, all financial crises seem to follow this same pattern: First interest causes a continuously increasing money supply, a money supply which seeks to be invested profitably. Since nowadays in industrialized nations, it is hardly possible to spend even more money on consumer goods, surpluses flow into real estate or stocks, and this is the reason why we find price bubbles in these markets. In their constant quest to find new borrowers, banks use the growing stock (crisis of 1929) and real estate prices (crisis of 2008) to encourage clients to take out loans to buy these assets. Therefore, credit boom, speculation and financial crisis go hand in hand (Fuders et al., 2013).

---

national debts. The USA is actually the most indebted nation in the world in terms of total debt and debt per capita (as of March 2022).

[5] Karl Marx (1872) was convinced that capitalism automatically self-destructs. What Marx did not properly understand is the fact that this is due to the unnatural design of our money. See Fuders (2018) and Sect. 13.3.

16.1.5 Do 'Kondratieff Waves' Correspond to the Cycles of Our Financial System?

One might ask if the long economic cycles described by Russian economist Nikolai Kondratieff (1926) may not be related to the invention of key technologies but, instead, correspond to the cycles of our financial system (Fuders, 2016; Fuders et al., 2013; Fuders & Max-Neef, 2014b). Kondratieff was one of the first economists to observe the existence of long economic cycles. Today, based largely on Schumpeter's (1939) interpretation, these long waves of economic activity are usually seen as a consequence of the invention of key technologies. According to this interpretation, investments in new technologies initially lead to self-sustaining economic growth. When those innovations are no longer new, they no longer attract investment, and this leads to an economic downturn (Fig. 16.1).

But if a key technology does not attract investment anymore, where does the money formerly invested in this technology go? It is likely that investors would either invest it in another productive technology or deposit the funds in a bank, which in turn would lend this money out in order to be able to pay the interest on deposits. Either way, the money keeps circulating in the economy (Fuders, 2016). So, why should this cause an economic downturn? In fact, the Kondratieff cycles could more accurately be interpreted as cycles of our financial system. This idea is supported by an interesting observation. Each period of economic growth, supposedly caused by a key innovation, has happened just after the financial system had experienced a severe crisis (in 1825: steam engine, 1875: railway, 1930: electronics and chemicals). After the recovery of the financial system, these inventions could be manufactured on a large scale. This explains why we observe this correlation between the introduction of key technologies and periods of economic growth.

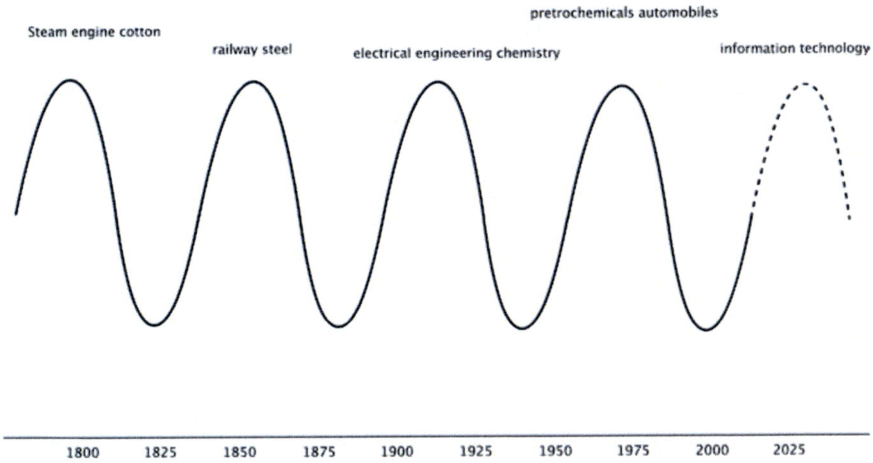

Fig. 16.1 Kondratieff waves. (Source: Own work)

It is also interesting to note that neither the light bulb nor the steam engine were actually invented by Edison or Watt respectively, as many believe. Rather, both inventions were introduced by less well-known engineers. The design of the first steam engine goes back to Thomas Newcomen in 1712 (Wikipedia, 2022a) and the first light bulb was invented by Heinrich Göbel in 1854 (Wikipedia, 2022b). At the time these technologies were invented the economy faced a financial crisis. Hence, these technologies could not be produced until later, when the financial system was stabilized and money began to circulate again, and investors were able to secure funding to finance the production of these inventions[6]. So, it was not the case that key innovations triggered the economic waves but the other way around, economic waves triggered the innovations. By the way, this is the interpretation that Kondratieff (1926) himself offered in his article.

16.1.6 The Classic Confusion of Causes with Symptoms

Unfortunately, few realize that the enormous national debts in most western economies is not just due to the irresponsibility of governments, but are mathematically inevitable, as shown above. Since money is the counterpart of debt, **if there were no debt, there would be no money**, with the exception of the relatively small monetary base created by the central bank. That is the reason, as already mentioned, why the wealthiest nations are also the most indebted ones (in absolute figures): U.S., UK, Japan, France and Germany (CIA, 2022). According to Irving Fisher's well-known "debt-deflation theory", deflation is supposedly caused by over-indebtedness. Fisher observed a number of effects that occur when debt bubbles burst (Fisher, 1933):

- Debt liquidation and distress selling
- Contraction of money supply as bank loans are paid off
- Fall in the level of asset prices
- Fall in net worth of businesses precipitating bankruptcies
- Fall in profits
- Reduction of production and employment
- Loss of confidence
- Hoarding of money

All these observations are true, but the causality goes the other way around: debt is not the direct cause of deflation but rather, an ever-growing total debt and the following financial crises leading to deflation, are both *symptoms* of our financial system, in which money is created by debt (Fuders et al., 2013). Irving Fisher's debt-deflation theory is a good example of misunderstanding the problem inherent

[6] In this context, it is interesting to note that in Edinburgh there exists a renowned university named after James Watt and his financier George Heriot: the Heriot-Watt University. Given the order in which their names appear, we can divine that the financier was deemed to be more important than the inventor.

in our financial system. Deposits grow via interest independently of the real performance of the economy, which is why total debt also grows independently of the real performance of the economy. The money supply therefore grows, causing inflation, until the system collapses. If the financial system collapses and banks go bankrupt people start hoarding money at home again. Money then fails to circulate, and deflation results. In this sense Dornbusch et al. (2009) are right in their conclusion that the Great Depression was caused by an increase in cash holdings relative to bank deposits, which caused the money multiplier to collapse.

We concluded above that in our current monetary system we see *either* inflation or deflation. **Stable money does not exist**. The fact that inflation is not always seen as a problem and appears relatively moderate is due to the fact that inflation is measured merely by prices of consumer goods, while households are likely to invest surpluses in stock or real estate markets, where we see steadily rising prices. This effect, which could be called the "inflation-deflation paradox" (see above Sect. 4.4.5), is perhaps a major reason why the underlying problems of our monetary system are not well understood by economists, (Fuders, 2011; Azkarraga et al., 2011; Fuders & Belloy, 2013).

16.2 Natural Disasters and Wars Benefit the Financial Sector

16.2.1 Natural Disasters: The Case of 2010 Chilean Earthquake

Natural disasters such as earthquakes which cause enormous destruction actually benefit the financial system. This dynamic could be observed after the devastating earthquake of February 27, 2010 in Chile, the fifth-strongest earthquake ever recorded, with a magnitude of 8.8 on the Richter scale. The most affected region was the area surrounding the coastal city of Concepción, where 80% of the Chilean population lives. Buildings that survived the quake were swept away by the subsequent tsunami. Although by some miracle only 400 people died, the catastrophe caused billions of dollars of damage. Yet there was also a beneficiary from the earthquake: the financial system (Fuders, 2011; Fuders & Max-Neef, 2014a).

Contrary to the predictions of Chilean central bank, whose experts predicted a decrease of economic growth (BCCh, 2010a, b), the earthquake actually led to additional economic growth: GDP growth in 2010 reached a level which had not been seen in years (Fierro, 2011: 58). While in 2009 Chile's economy had registered a negative GDP growth rate of −1.0%, it jumped to +6.1% in 2010 and +6% in 2011 and domestic demand rose 20.5% from 2009 to 2010 (BCCh, 2012). And because of the natural disaster banks in Chile issued more loans. While in the beginning of 2010 the issuance of loans was as low as it had been since 1999, the issuance of mortgage loans, commercial loans and consumer credit all increased (BCCh, 2010c, 2012, see also Fig. 16.2). In addition, the increased demand for loans has caused

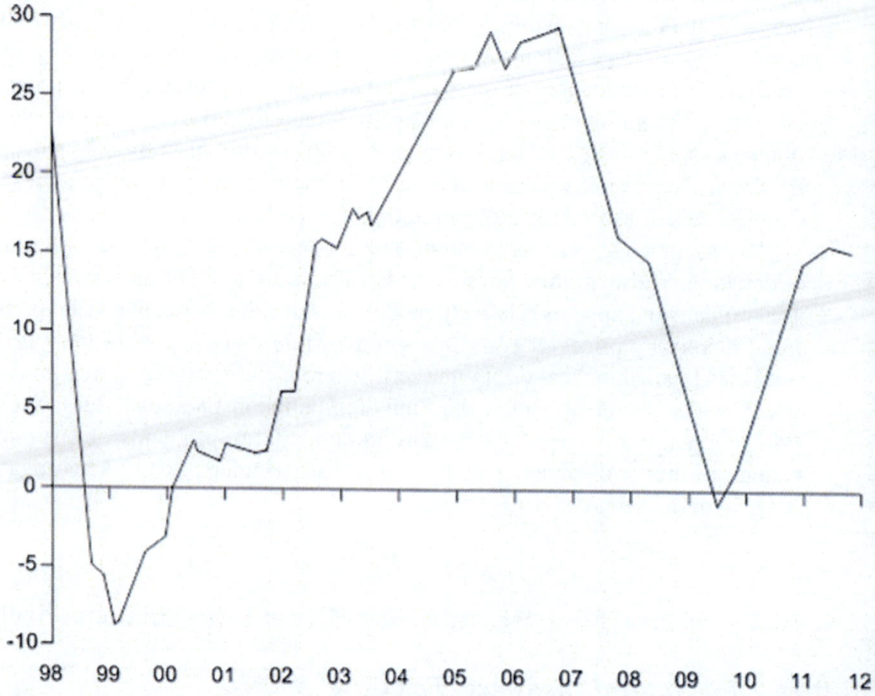

Fig. 16.2 Growth rate of credit issuance in Chile (average of all types of credits). (Source: Own work based on data provided by Banco Central de Chile)

interest rates in Chile to rise significantly. The Central Bank policy rate rose from 0.5% in February 2010 to 5.25% today (BCCh, 2011, 2012, see also Fig. 16.3).

Lenders quite obviously benefited from the earthquake (Fuders, 2011). In a saturated economy, however, in which there is hardly any possibility of additional economic growth, it is difficult for banks to find borrowers. But since they must pay interest on deposits, they are forced to grant loans anyway. If necessary, they lend to low quality borrowers, a situation which was on display in the US mortgage market leading up to the crisis of 2008. This dynamic can also be observed in the consumer credit market in Chile, where the least creditworthy households are encouraged to borrow money to buy consumer goods or even food. In Chile today rebates are offered to customers who buy on credit but not those who pay cash (Fuders & Max-Neef, 2014a).

How can this be? This form of marketing loans could be explained as follows. Chile (until 2020) had very low sovereign debt in relation to GDP. Unlike most governments around the world, the Chilean government has not been a good client for banks. A good client from the point of view of banks is one who demands loans and has the solvency to repay. This applies to public debt since the public sector in most countries demands huge sums and the public sector is very unlikely to default. But since the public sector in Chile is uncharacteristically reluctant to borrow,

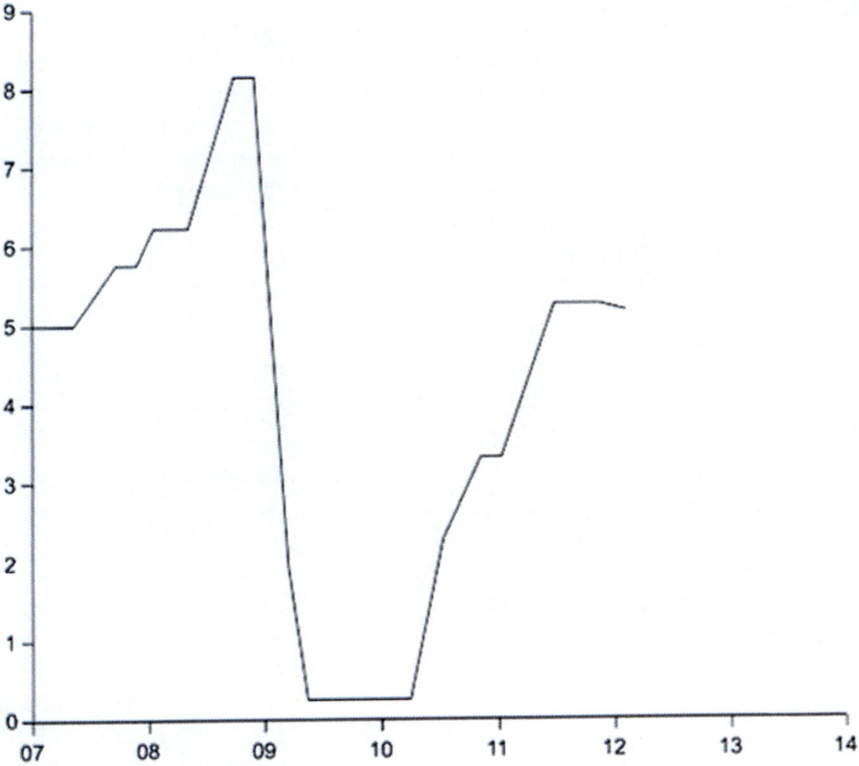

Fig. 16.3 Monetary policy rate in Chile. (Source: Own work based on data provided by Banco Central de Chile)

someone else has to pick up the slack: the private sector. In their desperate search for borrowers, banks pay rewards to malls, department stores and car dealers for inducing people to take credits. These rewards are then passed on to customers in the form of lower prices. Viewed in this light, an earthquake which causes households to borrow more would be viewed as a positive development by the financial sector. The reconstruction after the earthquake allowed new economic growth in Chile and banks were thus able to find borrowers more easily, delaying the collapse of the financial system.

16.2.2 Wars

Why wait for an earthquake? Of course, what applies to the destruction of real capital by natural disasters also applies to the destruction that wars wreak. Here we have a key clue to help us understand why wars seem to be linked to the world financial system (Gesell, 1949; Creutz, 1993, 2001). One reason: the interest burden of sovereign debt could be paid for with captured capital and resources (Creutz, 1993). The wars in Iraq, Afghanistan and Ukraine appear to support this supposition. Even more important: war is a most effective method to destroy wealth, making new economic growth and paying of interest possible (Creutz, 1993),[7] all the while increasing the demand for loans. Not only do households need to rebuild destroyed homes and replace destroyed goods, but it also causes the governments of the combatant nations to increase their national debts. So, wars, in theory, have the power to delay the collapse of the monetary system by enabling further economic growth on the one hand and providing an incentive to borrow on the other.

16.2.3 GDP as Indicator for Development?

In this context, it is interesting to explore how economic growth is possible after a devastating natural disaster or war, since the reconstruction only reestablishes the *status quo*; that is to say, no new wealth is created compared to the situation before the disaster. However, economic growth, as measured by GDP growth, is according to the prevailing view, an indicator of social welfare. The fact that destruction provides for GDP growth shows clearly that this indicator is not synonymous with wealth creation and even less with development of society (Max-Neef, 1986, 2007; Max-Neef et al., 1991; Daly & Farley, 2011; Lietaer et al., 2012). So, if the economic growth caused by an earthquake does not bring welfare gains, at least for the vast majority of households, but only a restoration of the pre-disaster *status quo*, who benefits from the growth of GDP? The answer to this question is obvious: once the reconstruction is complete, it is not the case that everyone is as well off as before. The new borrowers are burdened with the additional debt. Any natural disaster or war with huge destructive power therefore promotes the interest-induced redistribution of wealth (Fuders, 2010; Fuders & Max-Neef, 2014a).

16.3 A Negative Outlook

16.3.1 The Global Financial System Again on the Verge of a Meltdown in 2019

Those who follow the news from the financial world closely may have noticed that as early as September 2019, the world financial system was again on the verge of a 'meltdown'. Almost in a panic, the US Fed began in September 2019 to respond to liquidity shortages among commercial banks with a tool that the Fed later, from July

[7] Even disarmament treaties may support economic growth because the destroyed weapons in most cases are replaced by new ones (Creutz, 1993).

2021 on, would term "standing repo facility" (Fed, 2023). Basically, with this new instrument the Fed can provide virtually unlimited liquidity against US government bonds if required (Cox, 2019; Cœuré, 2019).

This could be seen as a new stage in an ultra-loose monetary policy that has been pursued since 2008 (and until spring 2022) not only by the Fed but also by other Central Banks, above all the ECB, in order to keep money flowing and to avoid deflation.[8] This new financial instrument represents a liquidity bomb. The volume of these repo transactions has been greatly expanded since they began in September 2019. What began with a volume of "only" 75 billion for overnight lending and 30 billion for longer maturities had been increased until May 2020 for one-month and three-month operations for USD 500 billion each.

16.3.2 *The Covid-19 Pandemic Is Playing into the Hands of the Banking Sector*

So, what happened? Why are interest rates rising again (as of spring 2022)? The answer is: Covid 19. Just like natural disasters and wars, the Covid-19 pandemic is playing into the hands of the banking sector because the massive State intervention that had become necessary promotes borrowing. Before the pandemic the banks had been struggling to find new creditworthy borrowers, similar to the situation prior to the subprime crisis of 2008. Most countries decided to stimulate their economies with expansionary fiscal policy, financed by debt. The public sector can issue bonds but also borrows from the private banking sector. For example, Germany, Japan, the US and Chile approved the largest economic stimulus packages ever (BR, 2020; Thomson & Fuentes, 2020; Welt, 2020; IMF, 2020). These enormous, unprecedented amounts of credit demanded by governments provided and continue to provide some breathing room for the beleaguered banking sector, which is desperate to expand credit.

From a banks' perspective, the public sector is an ideal borrower because it is always solvent (if the public sector goes bankrupt, everyone else is likely to go bankrupt first). Additionally, the demand for loans from private companies had also risen sharply in many countries. The latter is due to the fact that in many countries these loans are now backed by government guarantees. This is a safe business for banks. State intervention has never been so extensive. Even after the financial crisis in 2008, the economic stimulus packages and bank bailouts both in Europe and the USA were only about half the size of those that have been implemented in recent years. The increased demand for loans causes the price of borrowing money to increase.

[8] Even though the Fed had begun with this policy in the days of Ben Bernanke, who had then taken on the nickname "Helicopter Ben" in reference to a thought experiment once made by Milton Friedman in which a helicopter drops money over a village (Friedman, 1969).

16.3.3 Economic Stimulus Packages Instead of Bailouts

In the 2008 financial crisis the US government as well as many European governments implemented huge economic stimulus packages to mitigate the recession. The money was funneled to insolvent companies such as General Motors. A large part of the packages, however, went to the banks themselves (bank bailouts). In other words, the **banks were helped twice**. First because they received money from the governments. Second because the governments financed this financial aid by increasing the national debt. This way the State acted as a good and solvent customer demanding the loans banks so desperately needed to grant. From this second effect all banks benefited, not only those who received direct financial aid (Fuders, 2009). Possible that second effect was more important to save the monetary system than the bailout of some of the banks.

Two years later, the situation was similar in the so-called euro crisis, which in, in fact, was not a crisis of the euro as currency[9] but a sovereign debt crisis. The rescue fund known as the ESM (European Stability Mechanism), secures loans from private banks to Greece and other countries. Now, thanks to the ESM countries with questionable creditworthiness can still be a good client for the banking sector. The ESM is an instrument that helps the banking sector, not countries or the people living in countries. Probably no ordinary Greek citizen has ever seen a single cent of that money. It was used to restructure debts and to service pre-existing loans. The money flowed from one part of the monetary system to another (Fuders & Max-Neef, 2014b; Fuders, 2016).

14 years have passed since then, and because deposits and debts double approximately every 10–15 years, twice as much money is now needed to save the system. Banks again struggle to find solvent borrowers to be able to keep up with interest payments on deposits. The invention of new monetary-policy tools such as the "standing repo facility" (see Sect. 16.3.1) to provide overnight liquidity to banks is a clear symptom of this struggle. In this scenario Covid-19 and the war in the Ukraine is the best thing that could have happened to the financial sector. The pandemic made (predominantly) debt-based government intervention necessary. The war makes (predominantly) debt-based military aid necessary. For example, in March 2022 the German government announced a € 100 billion investment in national defense plus a € 200 billion aid package to help people and companies cope with higher energy prices. Both programs are financed via deficit spending. In February 2023 the US president announced a $ 500 billion military aid package for the Ukraine. This measure will add to the US national debt.

Raising such a large sum of money to help the banks again would probably not be accepted by most people. Even back in 2008, many were surprised at the large sums of money that were suddenly raised to save the banking sector, money that theoretically could have eliminated world hunger for 566 years (Max-Neef, 2010).

[9] Even though the common European currency creates other problems, especially large trade imbalances between States of the euro zone (Fuders, 2014, see also explication in footnote 4).

But few question increasing national debt for economic stimulus measures or for military aid.

16.4 A Positive Outlook

16.4.1 A Historic Opportunity for a New System

It may be a bit far-fetched to claim that the Covid-19 crisis or the war in Ukraine were deliberately brought about in order to postpone the collapse of the world financial system. But it is a fact that debt-financed stimulus packages help the financial sector. Even better than relying on regular (and growing) government intervention would be to establish a financial system that does not regularly fall into crisis. However, a financial system such as Silvio Gesell had in mind could not only prevent financial crises. As mentioned above, other major problems of our time, such as environmental degradation (caused by the economy's compulsion to steadily grow) and the exorbitant and ever-increasing income inequality in most parts of the world, are directly related to our monetary system, in which interest causes deposits and debt to grow in tandem.

As outlined above (Sect. 10.2), the role of banks in such a financial system would not differ much from their role today. Banks would be administrators of "natural money" deposits and intermediaries for interest-free loans just as they are today. Saving for the future would be possible in this system, but not through interest-bearing bank deposits. Savers could participate in productive investment, with banks playing the role of intermediators (Fuders, 2009, 2010; Fuders & Max-Neef, 2014a). Of course, and as discussed above, in a Gesellian economy it would still be possible to invest money and earn a profit. It would not be possible, however, to earn a return by lending money (to a bank or someone else).

16.4.2 Taking Advantage of the Crisis

The world is more than ever on a "collision course" (Max-Neef, 2010). So, this current crisis (pandemic, climate, war and upcoming famine) could and should be used as a chance to reform our financial system. The enforcement of effective negative interest rates, which are also applied to cash, i.e., the introduction of Gesell's natural money, is a means to avoid deflation as well as inflation that is now threatening many economies. As mentioned, John Maynard Keynes (1936) as well as Irving Fisher (1932; Fisher & Cohrssen, 1933) found that Gesell's proposal was suitable for pulling the economy out of the Great Depression. Today, we are standing at the precipice of new Great Depression, which could turn out to be even worse, since supply chains today are more globalized and interdependent. We now have a historic chance to do something different, so as not to repeat the mistakes that led to the current situation.

Such a reform would ideally be accompanied by a land ownership reform as Silvio Gesell (1949) and, before him, Henry George (1881) proposed (Fuders & Louge, 2020). The earth does not belong to man, but man to the earth.[10] A monetary reform as suggested by Gesell without an associated land reform would lead to people investing in land as a store of value. Eventually, most land would once again be in the hands of a few, just as today most financial assets are held by a small part of the population (see e.g., Piketty, 2014).

But it is not only about land in *stricto sensu*. Such a reform should also include the regulation of the exploitation of natural resources. These too should be regarded as the property of all mankind (Gesell, 1949), or at least as the property of the people in whose territory they are located. The practice in most countries where natural resources are exploited by private companies on behalf of the State but where these companies pay no (or almost no) royalty can be viewed as *de facto* privatization (Fuders, 2019). In Germany, coal mining by private corporations was even subsidized with taxpayer money for decades. The same is true in Argentina for the exploitation of silver, gold and oil.

16.4.3 The "Carrot-Method" Does Not Work in the Long Run

Achieving steady and reliable circulation of money via interest is not possible in the long run. A steady expansion of the money supply, accompanied by falling demand for credit (because households and businesses are ever more indebted) causes, as pointed out above, the price of lending money, the interest rate to fall. Therefore, in the long run, and irrespective of significant short-term fluctuations, we will see ever-lower interest rates all around the world. Unfortunately, due to the design flaw, money does not flow without interest, and is therefore increasingly hoarded under the mattress. It is precisely this dynamic that central banks throughout the industrialized world try to counteract by injecting more and more money into the economy.

It would be simpler, more natural and above all also fairer to introduce negative interest, which applies to cash as well, and to thereby break through the so-called "zero-lower bound". This would prevent the current situation which promotes a behavior that could be viewed as extortion, per the rationale: "either I hoard my money and thus do harm to everyone else, or else you have to pay me not to do so". One who harms others should be punished for it instead of being rewarded for not doing it.

[10] It is said that "Chief Seattle", one of the last chiefs of the North American Indians, said this in a speech to the then President of the United States.

16.4.4 Economic Resilience in the Face of Externals Shocks

Right now, the world economy is figuratively speaking dancing on the edge of a volcano. Any crisis anywhere in the world could trigger the next world financial crisis. When that happens, many will erroneously blame the virus, war or greedy bankers and fail to perceive that crises are inherent in our monetary system. However, with natural money as proposed by Silvio Gesell the decoupling of growth of the money supply (and debt) from the real economy could be avoided, and the financial system would be less susceptible to crisis. Indeed, the monetary system would be more resilient and less vulnerable to external shocks.

Money subject to a hoarding fee would circulate more reliably, and less economic stimulus packages would be needed in order to prevent the economy from plunging into deep depression. John Maynard Keynes (1936) and Irving Fisher (Fisher, 1932; Fisher & Cohrssen, 1933) both concluded that Gesell's proposal was suitable for pulling the economy out of the Great Depression. Yet, despite the endorsements of Keynes and Fisher, Gesell's proposal has not found its way into conventional economic teaching and hence his proposals are not being implemented. Right now, we are standing at the edge of a new Great Depression. To avoid this, 193 of 195 governments in the world have adopted debt-based economic stimulus packages of magnitudes never seen before (IMF, 2020).

However, we do not need more liquidity in the markets but rather liquidity that circulates and serves the economy, instead of more of the same 'easy-to-hoard' liquidity. We cannot solve our problems by repeating the same mistakes over and over again. We now have an historic chance to do something different, so as not to repeat the mistakes which led us to the current situation. Silvio Gesell, although almost completely forgotten today, can be seen as one of the greatest and most innovative thinkers in macroeconomics. Why not give his proposal a try?

References

Aristotle. (1995). Politik 1st Book. In *Aristoteles Philosophische Schriften in sechs Bänden. Band 4*. Felix Meiner Verlag.

Azkarraga, J., Max-Neef, M., Fuders, F., & Altuna, L. (2011). *La evolución sostenible II*. Eskoriatza.

Banco Central de Chile (BCCh). (2010a). *Boletín Mensual 2/2010*. Banco Central de Chile.

Banco Central de Chile (BCCh). (2010b). *Informe Política Monetaria (IPoM) 3/2010*. BCCh.

Banco Central de Chile (BCCh). (2010c). *Informe Política Monetaria (IPoM) 6/2010*. Banco Central de Chile.

Banco Central de Chile (BCCh). (2011). *Boletín Mensual 11/2011*. Banco Central de Chile.

Banco Central de Chile (BCCh). (2012). *Informe Política Monetaria (IPoM) 3/2012*. Banco Central de Chile.

Bayerischer Rundfunk (BR). (2020). *Konjunkturpakete gegen Coronakrise in den USA und Deutschland.* [Online] Available at: https://www.br.de/nachrichten/wirtschaft/do-konjunkturpakete-gegen-coronakrise-in-den-usa-und-deutschland,RuHOAgK. Accessed 21 July 2022.

Bichlmaier, S. (2010). Umlaufsicherung vs. Niedrig-Zins-Politik und Inflationsstrategie der Notenbanken. *Humane Wirtschaft, 2*, 6.

Blech, J. (2005). *Die Krankheitserfinder – Wie wir zu Patienten gemacht werden*. Fischer.

Blüchel, K. (2003). *Heilen verboten – töten erlaubt. Die organisierte Kriminalität im Gesundheitswesen*. Goldmann.

CIA. (2022). *Factbook, country comparisons: Debt – External*. [Online] Available at: https://www.cia.gov/the-world-factbook/field/debt-external/country-comparison. Accessed 25 May 2022.

Cœuré, B. (2019). *ECB – A tale of two money markets: Fragmentation or concentration*. [Online] Available at: https://www.ecb.europa.eu/press/key/date/2019/html/ecb.sp191112~5808616051.en.html. Accessed 12 May 2020.

Coleman, V. (2003). *How to stop your doctor killing you*. European Medical Journal.

Cox, J. (2019). *The Fed is looking at a 'standing repo' operation to handle overnight funding issues*. [Online] Available at: https://www.cnbc.com/2019/11/20/the-fed-is-looking-at-a-standing-repo-operation-to-handle-overnight-funding-issues.html. Accessed 25 May 2022.

Creutz, H. (1993). *Das Geldsyndrom. Wege zu einer krisenfreien Marktwirtschaft*. Wirtschaftsverlag Langen Müller/Herbig.

Creutz, H. (2001). Wirtschaftliche Triebkräfte von Rüstung und Krieg. *Zeitschrift für Sozialökonomie, 128*(Sonderdruck), 1–20.

Daly, H., & Farley, J. (2011). *Ecological economics – Prcinciples and applications* (2nd ed.). Island Press.

Dornbusch, R., Fischer, S., & Startz, R. (2009). *Macroeconomía* (10th ed.). McGraw-Hill.

Engelbrecht, T., & Köhnlein, C. (2006). *Virus-Wahn: Vogelgrippe (H5N1), SARS, BSE, Hepatitis C, AIDS – Wie die Medizin-Industrie ständig Seuchen erfindet und auf Kosten der Allgemeinheit Milliarden-Profite macht* (3rd ed.). Emu-Verlag.

Federal Reserve Bank (Fed). (2023). *FAQs: Standing repo facility*. [Online] Available at: https://www.newyorkfed.org/markets/repo-agreement-ops-faq. Accessed 21 Feb 2023.

Fierro, P. (2011, March 19). PIB de Chile llega a los US$ 200 mil millones en 2010 y supera en tamaño a Israel y Pakistán. *La tercera*, p. 58.

Fisher, I. (1932). *Booms and depressions – Some first principles*. Adelphi.

Fisher, I. (1933). The debt deflation theory of Great Depressions. *Econometrica, 1*(4), 339–357.

Fisher, I., & Cohrssen, H. (1933). *Stamp scrip*. Adelphi.

Friedman, M. (1969). *The optimum quantity of money and other essays*. Macmillan.

Fuders, F. (2009). Die natürliche Wirtschaftsordnung als Option nach dem Zusammenbruch. *Aufklärung & Kritik, 16*(2), 128–145.

Fuders, F. (2010). Alternative concepts for a global financial system – An answer to the present world financial crisis. *Estudios Internacionales, 166*, 45–56.

Fuders, F. (2011). Wie Zerstörung von Sachkapital dem Finanzsystem in die Hände spielt. *Humane Wirtschaft, 2*, 34–38.

Fuders, F. (2014). Die Mehrheit in Deutschland profitiert nicht vom Euro. *Humane Wirtschaft, 1*, 21–23.

Fuders, F. (2016). Smarter money for smarter cities: How regional currencies can help to promote a decentralised and sustainable regional development. In *Decentralisation and regional development – Experiences and lessons from four continents over three decades* (pp. 155–185). Springer.

Fuders, F. (2018). Karl Marx' historisches Missverständnis. *Fairconomy, 2*, 14–16.

Fuders, F. (2019). Boden, Wasser und Luft: Warum Privateigentum mit der Abschöpfung von Monopolrenten einhergehen sollte. In U. Kriese, D. Löhr, & H. Wilke (Eds.), *Grundsteuer: Zeitgemäß! Der Reader zum Aufruf* (pp. 229–246). Münster.

Fuders, F., & Belloy, P. (2013). Terremoto del 27.02.2010 en Chile – Cómo el sector financiero se beneficia de la destrucción de capital real. *Realidad Económica, 3*, 61–75.

Fuders, F., & Louge, C. (2020). Effektive Negativzinsen statt Corona-Schuldenwachstum. *Fairconomy, 16*(2), 4–9.

Fuders, F., & Max-Neef, M. (2014a). Local money as solution to a capitalistic global financial crisis. In *From capitalistic to humanistc business*. Palgrave Macmillan.

Fuders, F., & Max-Neef, M. (2014b). Dinero, deuda y crisis financieras. Propuestas teórico-prácticas en pos de la sostenibilidad del sistema financiero internacional. In *Economía Internacional. Claves teórica-prácticas sobre la inserción de Latinoamérica en el mundo*. LATIn.

Fuders, F., Mondaca, C., & Azungah Haruna, M. (2013). The central bank's dilemma, the inflation-deflation paradox and a new interpretation of the Kondratieff waves. *Economía (U. de los Andes), 36*, 33–66.

Galbraith, J. K. (1983). *El Dinero – De dónde vino/Adónde fue* (2nd ed.). Buenos Aires - Madrid.

George, H. (1881). *Progress and poverty – An inquiry into the cause of industrial depressions and of increase of want with increase of wealth – The remedy* (5th ed.). Appleton.

Gesell, S. (1949). *Die natürliche Wirtschaftsordnung durch Freiland und Freigeld* (9th ed.). Rudolf Zitzmann.

International Monetary Fund (IMF). (2020). *Policy responses to Covid-19.* [Online] Available at: https://www.imf.org/en/Topics/imf-and-covid19/Policy-Responses-to-COVID-19. Accessed 25 May 2022.

Kennedy, M. (1990). *Geld ohne Zinsen und Inflation – Ein Tauschmittel, das jedem dient.* Permakultur-Verlag.

Kennedy, M. (2011). *Occupy money – Damit wir zukünftig alle die Gewinner sind.* J. Kamphausen.

Keynes, J. M. (1936). *General theory of employment, interest and money*. Hartcourt.

Kondratieff, N. (1926). Die langen Wellen der Konjunktur. *Archiv für Sozialwissenschaft und Sozialpolitik, 56*, 573–609.

Krugman, P. (2009). *Die neue Weltwirtschaftskrise* (2nd ed.). Campus Verlag.

Leitner, L. (2000). *Mythos HIV – Eine Kritische Analyse der AIDS-Hysterie*. Videel.

Lietaer, B., Arnsperger, C., Goerner, S., & Brunnhuber, S. (2012). *Money and sustainability: The missing link – A report from the Club of Rome – EU Chapter to the Finance Watch and the World Business Academy*. Triarchy Press.

Marx, K. (1872). *Das Kapital – Kritik der politischen Oekonomie. Erster Band, Buch I: Der Produktionsprozess des Kapitals* (4th ed.). Meissner.

Max-Neef, M. (1986). *La Economía Descalza – Señales desde el mundo invisible*. Nordan.

Max-Neef, M. (2007). *La dimensión perdida – La deshumanización del gigantismo*. Icaria.

Max-Neef, M. (2010). The world on a collision course and the need for a new economy. *Ambio, 39*, 200–210.

Max-Neef, M., Elizalde, A., & Hopenhayn, M. (1991). *Human scale development – Conception, application and further reflections*. Apex.

McTaggart, L. (2005). *What doctors don't tell you: The truth about the dangers of modern medicine*. Thorsons.

New York Times. (1929a, November 14). Efforts to cut short selling and peg key stocks among day's development on the exchange. *New York Times*, p. 1.

New York Times. (1929b, November 17). Short sales stir conflicting views; some hold practice helpful, others harmful to the stock market. *New York Times*, p. N7.

New York Times. (1929c, November 25). Asks state inquiry on stock recession – Senator Hastings wants the governor to name Committee of Business Leaders. *New York Times*, p. 18.

New York Times. (1929d, December 10). Asks house to vote tax on stock sales, representative Sabath introduces Bill for 5% Levyon shorts. Penalty for non-payment Illinois man denounces 'Wall Street Gamblers' as 'Rapacious Parasites'". *New York Times*, p. 54.

Piketty, T. (2014). *Capital in the 21st century*. Harvard University Press.

Schumpeter, J. A. (1939). *Business cycles – A theoretical, historical and statistical analysis of the capitalist process*. McGraw-Hill.

Soddy, F. (1934). *The role of money – What it should be contrasted with what it has become*. Routledge.

Thomson, E., & Fuentes, V. (2020). *Chile's government announces 11.75 billion stimulus plan.* [Online] Available at: https://www.bloomberg.com/news/articles/2020-03-19/Chile-s-government-announces-11-75-billion-stimulus-plan. Accessed 21 July 2022.
Welt. (2020). *Welt.de: Japan beschliesst größtes Konjunkturpaket aller Zeiten.* [Online] Available at: https://www.welt.de/newsticker/dpa_nt/infoline_nt/wirtschaft_nt/article207087239/Japan-beschliesst-groesstes-Konjunkturpaket-aller-Zeiten.html. Accessed 21 July 2022.
Widmer, V., Lanka, S., & Brix, S. (2006). *Der Masern-Betrug*. Klein-Klein-Verlag.
Wikipedia. (2022a). *Entry "Thomas Newcomen"*. [Online] Available at: http://de.wikipedia.org/wiki/Thomas_Newcomen. Accessed 21 July 2022.
Wikipedia. (2022b). *Entry "Heinrich Göbel"*. [Online] Available at: http://de.wikipedia.org/wiki/Heinrich_G%C3%B6bel. Accessed 21 July 2022.

Chapter 17
Addendum: Some Final Considerations

17.1 Deviation from Nature: The Cause of Many Diseases

17.1.1 A Personal Experience

The reader might ask why am I now talking about diseases. Do I have expertise in this matter? The answer is, my understanding of the problems of interest are not originally based on my background in economics but are rather rooted in an illness I suffered, which helped me gain important insights regarding the causes of illness. Subsequently I applied this insight to economics.[1] In 2001 I was very sick, and no doctors were able to find the cause. After a total of four lengthy hospital stays and countless examinations, I had almost given up and felt I would soon leave this life, when it occurred to me that chronic-degenerative diseases—those strange diseases science does not have true answers for—are caused by deviation from nature.

I asked myself, what could be so unnatural in my life or environment that a young man could become so sick? First, I moved since the house I was living in at the time was next to a cellphone base station. Indeed, by eliminating the (what today I know are extremely unnatural) high frequency electromagnetic fields I was able to sleep much better, did not wake up frequently with cardiac arrhythmia as before. Ever since I have considered it important to have a radiofrequency-reduced place to sleep. This measure helped, but my illness was not cured. I still felt very weak and suffered from extreme nausea every day. What else could be going on? I checked my behavior, started to play more sports, changed to healthy (natural) organic food, but these measures—even though organic food is surely much healthier and tastier than unnatural industrially produced food—I was still unable to cure myself.

[1] These insights and the wish to help others encouraged me to study medicine at University Erlangen-Nürnberg where I was already a lecturer in the economics faculty. In 2009, when I moved with my family to Chile, I hoped to finish these studies. Unfortunately, this was not possible.

© The Author(s), under exclusive license to Springer Nature Switzerland AG 2023

F. Fuders, *How to Fulfil the UN Sustainability Goals*,
https://doi.org/10.1007/978-3-031-37768-6_17

Then I started considering my teeth. I used to have 20 gold fillings. Is it natural to have gold in my mouth, I asked myself? It is not. I started to search the internet and scientific publications on this topic but did not find much evidence against gold as a filling material. Gold was supposedly safe for filling cavities. However, gold fillings are not in fact pure gold, but include many other metals, some of them known to be toxic. And even if the fillings would be made of pure gold, when mixed with saliva colloidal gold and gold ions are released, which we then swallow. This, I said to myself, cannot be healthy in the long run. I decided to replace the gold inlays with ceramic ones. Even though ceramic is not 100% natural either (especially the glue used to adhere it to the teeth), the change helped and in a couple of months my health problems had ceased.

17.1.2 Intoxication Is Deviation from Nature

Based on this experience I developed the theory that many, if not most, chronic-degenerative diseases, especially cancer or so-called auto-immune diseases where our immune system attacks our own body, are probably caused by some sort of deviation from nature, from the **originally perfect equilibrium** that developed over millions of years of evolution. Can we really be so arrogant as to believe that we can improve upon evolution? Or for those who believe in God, can we improve upon divine creation? And if we cannot, then any deviation from nature **must have negative consequences**, at least in the long run. The larger the deviation, the more harmful the consequences.

Could it be that deviation from nature causes intoxication? The dictionary tells us that intoxication or poisoning is an interaction between a foreign chemical (toxin) and a biological system that results in damage to a living organism (e.g., Dear, 2014). But what is a foreign chemical, what is a toxin? Is there some sort of red line that tells us where nature ends, and chemistry begins? Does it depend on the quantity? An overdose is the "excessive use" of a drug (e.g., Chen & Chen, 2013). But according to what criteria do we define what the correct dose should be? Why do we die from just a tiny amount of mercury while it is almost impossible to ingest an overdose of apple juice? What is it that makes some things more toxic than others?

If we think about this question long enough, we might reach the conclusion that **intoxication is deviation from nature**. Mercury, for example, is not found in the natural environment of humans since, because of its heavy specific weight, it is naturally found only deep under ground. Only through human mining activity is it brought to the surface. Whether something is toxic or not depends on whether it is found naturally in our environment. Mercury, like other heavy metals, is not; and this is the reason why it is highly toxic for us. Apples are not toxic since they are found in our natural environment. In fact, it is natural (and therefore healthy) to eat apples. Our body needs exactly what nature provides for us. Not more and not less. In this sense, lacking vitamins or minerals, e.g., is also a deviation from nature and thus intoxication: it can be understood as a negative intoxication.

17.1.3 Deviation from Nature: The Cause of Many Diseases

So, concluding from the previous paragraph, if we suffer a chronic-degenerative disease, then some deviation from nature is likely to have taken place. Here, I am not referring to a cold or diarrhea, that lasts for a couple of days, and which can be normal and sometimes even be seen as a (natural) reaction precisely to get back to equilibrium. For example, a cold, as the name suggests, can be a reaction to having been exposed to low temperatures. The next day we develop a fever to counterbalance the cold. Likewise, diarrhea can be a natural and healthy reaction to purge the body. The flu can kill people, some will now object. But people do not in fact die of the flu alone but rather from its coincidence with old age or a compromised immune system. The latter can be due to lifestyle choices, including poor diet, that which are at odds with nature.

When the Spanish conquerors brought the flu to the native South American population, many of the indigenous people died from it. Conversely many of the Spanish conquerors died of dengue fever or malaria, while these illnesses were usually not fatal to the native population. What happened here? Did people of the old time deviate from nature? Maybe for the Spanish conquerors that could be the case. After a three-month journey on a vessel, they might have lacked a balanced nourishment. But the South American natives lived surely in harmony with nature. We could interpret it this way: For the native South American people malaria and dengue are part of their natural environment as much as the influenza virus was part of the natural environment of the Spanish conquerors. The reason why the conquerors died of malaria and the indigenous population died of flu is that these germs did not exist in their native (natural) environments.

Let me give one last example. Gastric ulcer as well as chronic gastritis can be caused by the bacterium *helicobacter pylori* (see, e.g., Berdel et al., 2004). Antibiotics can be used against bacterial infection. And yes, antibiotics will eliminate the ulcer, but most likely only temporarily since the bacteria is not the root cause. **More than half of the world's population** is infected with *helicobacter pylori* (Hooi et al., 2017; Khoder et al., 2019), in some regions this number exceeds 80% of the population (for further references see Khoder et al., 2019) but not all of these people develop gastric ulcer or gastritis. Some researchers therefore speak of *helicobacter pylori* as a risk factor instead of the cause of gastric ulcer (Khoder et al., 2019). However, is it accurate to refer to a germ that can be found in almost everyone as a risk factor? I prefer the following interpretation: In the case of gastric ulcer and chronic gastritis, prolonged stress has been shown to be a risk factor. Stress can change the composition or pH of the stomach fluids, and because of this, *helicobacter pylori* can then multiply and cause problems. This interpretation is supported by the fact that stress has been associated with gastritis even in the absence of *helicobacter pylori* (Megha et al., 2022). Of course, stressful situations also exist in our natural environment, e.g., getting attacked by a wild animal. But stress that persists over a long period of time is surely not natural and thus not healthy.

In this context it is necessary to understand that people have different genetic dispositions. For example, some people are hyper-sensitive to electropollution (electric fields, magnetic fields, high frequency electromagnetic fields[2]), which is a completely unnatural form of radiation and recognized as toxic by the WHO, while others are not (or they simply do not know that they are). That someone does not perceive this radiation does not mean that it is harmless.

17.1.4 Can We Improve Nature?

What is the difference between humans and animals? Man has pondered this question since time immemorial. Some believe that the use of the thumb is the defining difference between humans and animals. Some believe it is our intelligence, or the use of fire, or the use of language that permits us to communicate that distinguishes us from other species. Some say that the fact that we are self-conscious and are able to reflect about the difference between humans and animals makes us special. Some say that humor is the defining human characteristic. My dear colleague Manfred Max-Neef (who created the concept of human scale development) said in private conversations that in his interpretation "stupidity" is what differentiates humans from any other species. Humans are stupid since we do things knowing that they are wrong, while animals would never behave in this manner. I would say that we are the only species with the **arrogance to constantly try to improve upon nature**, and this can indeed be considered stupid. For those who believe in divine creation of nature, it is also a defiance of God.

Millions of years of evolution, generating an equilibrium state where everything was in balance with everything else, are certainly not improvable. **An equilibrium cannot be improved upon**, or else it would not be an equilibrium. For those who believe in divine creation of heaven and earth, we can say it this way: we were given a perfect world, a "paradise". **A perfect world cannot be improved upon**, otherwise it isn't perfect.

17.2 A Natural Economy Is a Healthy Economy

17.2.1 From a Healthy Life to a Healthy Economy

We discussed above the notion that we can hardly improve upon nature. Interest is unnatural, as Aristotle and Thomas Aquinas both understood. However, interest itself is a secondary phenomenon stemming from the fact that money is unnaturally

[2] Sometimes also referred to as "high frequency electromagnetic radiation", "microwaves", "millimeter waves", "millimeter radiation", "non-ionizing radiation". Diseases that could eventually evolve after a prolonged exposure, are brain cancer, leukemia strokes, psychiatric disease or infertility (see the following meta studies: Montenegro, 2008; Simkó & Mattsson, 2019; Verbeek et al., 2021; Pacchierotti et al., 2021).

storable. While everything in nature decays, our money seems to be immune to this universal tendency, as Gesell (1949) correctly observed. We do have inflation, but inflation too is, as outlined above, only a symptom of a steadily growing money supply, which originates from interest, and which makes deposits (and debt) grow constantly and usually faster than the output of the real economy. As we have discussed throughout this book, the constant growth of the money supply and debt are the real reasons why the real economy is required to grow.

Such an unnatural monetary system must be unhealthy for society. One might say that **our unnatural money intoxicates our society**. A healthy economy is a natural economy. i.e., an economy whose rules and monetary system do not deviate from nature (Fuders, 2021). Gesell called his work the "Natural Economic Order". Indeed, this is a fitting title.

17.2.2 A Healthy Economy Is Consistent with Holy Scripture

For those who believe in divine creation of nature, any deviation from nature is an offence against God, the creator of all that is. In this sense, it follows that an economy based on the laws of nature, would be **in line with the rules of Holy Scripture** (Bible, as well as Tanakh and Koran), as was previously discussed in Sect. 10.6. Let us recall:

- Interest is an unnatural, man-made construct as Aristotle and Thomas Aquinas both recognized. In indigenous cultures, as far as I know, interest is unknown. The Bible, Tanakh and Koran prohibit it.
- It is unnatural to prohibit private ownership of man-made products. Private ownership of artwork and any other man-made products is respected, as far as I know, in all indigenous cultures. The Bible, Tanakh and Koran endorse private ownership of man-made products. The commandment "thou shalt not steal" can be found in all three Scriptures.
- It is unnatural that naturally occurring products (land, air, sea, water, copper, minerals, etc.) should be treated as private property. As far as I know, indigenous cultures all over the world view nature as a sacred gift that naturally belongs to all. The Bible, Tanakh and Koran define land as God's property.

17.2.3 Without a Sustainable Financial System, a Sustainable Future Is Not Possible

Unfortunately, Gesell's proposals are still not well understood. Since in conventional economic teaching we believe erroneously that higher rates reduce inflation and economic growth while lower rates increase them, we cannot comprehend why through applying Gesell's solution (no interest on loans and negative rates on cash and demand deposits) we can get rid of inflation and the growth imperative. We will not grasp the fact that Gesell's proposals represents solutions to virtually ALL the big challenges we face today, especially ever-increasing income inequality, inflation (or deflation if the system crashes) and the ever-increasing unsustainability of our activities.

This, of course, does not mean that a financial system with characteristics like those proposed by Gesell would automatically stop any destruction of ecosystems or the production of goods and services with negative ecological balances. But in any event, what can be affirmed with complete assurance is that with the current, unsustainable financial system, a sustainable future is not possible.

If we aspire an economy that promotes the satisfaction of fundamental human needs, that is a Development at Human Scale (Max-Neef et al., 1991); if we want an economy that serves people and not *vice versa*, and in which gaining money is not an end in itself; if we want a true social-ecological transformation that **attacks the illness form its root** and doesn't just address the symptoms, then we have to reform our money first. This would then provide the basis to establish a truly free market economy, let us call it a market economy without capitalism. As we concluded above (Sect. 15.5), this is probably the *oikonomia* Aristotle used to refer to.

17.2.4 No More Economics of Accumulation

I remember that someone argued in a discussion that materialism would cause us to have a sort of "blind spot" when considering the wealth enhancing effects of free markets. And, according to him, this blind spot would consist of a concept of competition and freedom that allowed for accumulation and exploitation. Many would probably agree with him.

However, the "economy of accumulation" as we know it today is not due to competition but has its roots in the fact that money and land are (unnaturally) storable. This makes it possible to earn interest for lending money, the "right to the foreign labor yield" as Gesell (1920: 210) accurately called it. If we were to change the monetary system as Gesell proposed it, then we could come close to what in economics is described as "perfect competition". Perfect competition would then generate the positive effects economic theory describes to it. Perfect competition would generate "perfect solidarity", since it incentives us to make good (in the eyes of the consumer) products at good prices. In economics we term this *allocative efficiency*: prices fall to the lowest possible level, producers earn only the opportunity cost of production factors invested, i.e., each market participant earns according to their effort. The problem is not the concept of free markets, but the fact that markets are not free. We are indeed far away from markets that function as they should. Far away from the ideal of perfect competition. Far away from a mechanism that forces us to make good products at good prices.

It is understandable that many reject the concepts of the free market and perfect competition, since we are taught in microeconomics textbooks (e.g., Frank, 2015) that the positive results of competition are due to selfishness. However, the desirable results of competition do not come from selfishness, but on the contrary, the market, if it works, **limits our selfishness**. In the case of a monopoly, which is the opposite of perfect competition, we see what negative effects selfishness has: the monopoly owner usually takes shameless advantage. Competition, when it really works and when it is not perverted by the current unnatural monetary and territorial order, limits our selfishness as it forces us produce the best products at the best prices; it compels us to behave **as if we loved one another** (Fuders, 2017; Fuders & Nowak, 2019).

Now, better than a mechanism that causes us to behave as if we loved one another would be for us to really love one another. I refer to this as the **economy of neighborly love** (Fuders, 2017; Fuders & Nowak, 2019). Competition would then not be necessary. We would produce good products at good prices, not because of market pressure, but out of eagerness to contribute to the common good.

17.3 Towards an Economy of Neighborly Love

17.3.1 Love: The Transcendental Meaning of Life

Those who have reflected on the meaning of life may conclude that their true purpose is to promote the common good by practicing neighborly love (Fuders & Nowak, 2019). It is no coincidence that love for one's neighbor is the highest commandment in Christianity as well as Judaism and Islam.[3] Of course, altruistic behaviors can also be cultivated in a secular way (e.g., Condon, 2017). Love for one's neighbor translates into the 'golden rule', i.e., treating others as we would like to be treated (Bible, Matthew 7:12). We could say that loving our neighbor is the perfect realization of Immanuel Kant's (1868b) categorical imperative (Fuders & Nowak, 2019).

Neighborly love is not just about helping people in distress or observing the Ten Commandments. Instead, the commandment to love one another can be interpreted as a mandate to **make use of our talents** to promote the common good (e.g., Bible, Romans 12:1–8; Peter 4:10; 1 Corinthians 10:24). Part of the realization of neighborly love is therefore to find our vocation and make it our profession or life's work, so that we may earn our daily bread by employing our special skills. The following reflections are a summary of what has been more thoroughly outlined in Fuders (2017) and Fuders and Nowak (2019).

[3] For Christians, both the Old Testament and the New Testament state, as the two highest commandments and God's will, to love God and to love one another (Bible, Leviticus 19:18; Deuteronomy 6:5; Matthew 22:37–22:39; Galatians 5:14). According to Christ, both commandments are equally important (Bible, Mathew 22:40). For Islam see Koran, Surah 42:23; for Judaism see Torah, Leviticus 19:18.

17.3.2 The Great Puzzle of Love

Who is most important to the production of a good or service? Is it the worker on the assembly line or the head of the company or the engineer who designed the product or production process? The answer is, they are all important. The company owner, the engineer and the worker are *jointly* involved in the creation of value. **Everyone has their place in the world**. No one can do without the other. The engineer often is not a good entrepreneur. The entrepreneur cannot produce products without someone putting them together on the production line. The worker on the assembly line might not have the ability to run a business.

In this sense, every individual can be seen as part of what could be called the **great puzzle of 'love'**. Every person is a slice in this life-puzzle, and every person con contribute to its finalization by consistently making use of their talents. To recognize where we belong in this puzzle can be deemed a principal part of the realization of neighborly love, because we help others and promote the common good the most by utilizing the talents we have been endowed with.

So, when we make use of our talents and gifts to perform the tasks we are best suited to, we make our greatest possible contribution to the common good. On the contrary, one might argue that we harm our neighbor if we occupy a job for which we were not actually suited and thus prevent someone else from doing this work. Here, to use an expression made famous by Abraham Maslow (Maslow, 2005: 173 ff.), another person could reach "self-actualization" in this role. This could be likened to trying to force a piece of a puzzle into a place where it does not belong. In this way we not only destroy the puzzle piece, but the puzzle will never be whole again. The piece will be missing from the place where it was intended to fit by the creator of the puzzle. We can conclude that it is our very **duty** to use our gifts and talents for the benefit of all and not only as an activity through which we earn money.

17.3.3 Pareto Efficiency and True Happiness

In this sense so-called *Pareto efficiency* will never be reached as long as people occupy jobs in which they do not use their special talents. Remember, a society is Pareto efficient if it is impossible to reallocate resources so as to make any individual better off without making at least one individual worse off (Dascher, 2007; Frank, 2015). But can society be Pareto efficient while there are people occupying jobs in which they do not use their talents effectively? People who hold jobs they are not made for can improve the well-being of all (including their own) by looking for jobs in which they can actually make their special contribution to society.

And it is not only a question of efficiency. Self-actualization (in the sense of making one's maximum contribution to the common good) as Maslow (2005) called it also makes us happy. In fact, it may be the only way to be truly happy. There is no clear correlation between happiness and income. For example, those countries that achieve the highest scores on the so-called Happy Planet Index (Nef, 2022) are not those with the highest incomes. I am convinced that, if we truly want long-lasting happiness, we should aim to constantly help and serve others. Here, I am not referring to happiness as the superficial, materialistic joy we feel after having acquired new possessions or after eating a good meal, but to a **deep sense of inner fulfillment** that results from the intuition that we are contributing to the divine puzzle of love.

If we appreciate that the purpose of life is not to make as much money as possible, but to earn money through an activity that we would perform even in the absence of any monetary gain, then this will make us truly happy. It will make us happy because we feel that we are making a contribution to the common good. Unlike Aristotle ([384 BC] 1995b) in his Nicomachean Ethics, I do not consider felicitousness (*eudaimonia*) to be the highest good we can aspire to. The pursuit of happiness as an end in itself might even be considered as superficial as the pursuit of money or recognition. Rather, true happiness can be considered as the result of achieving the highest goal: the realization of neighborly love.[4]

17.3.4 Freedom, Dignity and Justice

Using our talents for the benefit of all, as the fulfillment of neighborly love and the transcendental meaning of life might even contribute to the realization of freedom. Freedom, as we learn from Kant, is the state of being unconstrained by other people's arbitrary choices (Kant, 1868a). This also means that the freedom of others should not be limited by our own choices. That, in turn, means we should not occupy jobs that other people are better suited to. For example, someone sitting in their office all day may be less exposed to the open air than someone in prison (who might be allowed to do sports in the prison's yard). However, this person does not necessarily feel imprisoned, at least not if he or she enjoys and loves their job. Instead, they feel that their work is making a valuable contribution to the well-being of all. On the other hand, someone who works outside all day, e.g., on a construction site, may feel less free than someone sitting in an office all day if they feel compelled to work in order to survive. The difference between slaves and free people is that free people do their work out of *free will*, i.e., voluntarily, and not only in order to survive.

That we should make use of our talents is also connected to justice. Plato found justice to be realized when all people perform the work to which their natures are best suited. According to Plato, "to do one's own business and not to be a busybody is justice" (Plato, [427 BC] 2011: 332 ff.—author's translation). In the Bible we learn that God is just (Psalm 116: 5) and also that God is love (1 John 4:8). Thus, infringing against the commandment to love one another is a violation of justice.

[4] This can also be derived from the German word for felicitousness (Glückseligkeit), in which the words "happiness" (Glück) and "soul" (Seele) are combined. Felicitousness is the state in which the soul is happy.

Furthermore, Holy Scripture teaches us that there is righteousness when one lives in love and keeps God's commandments (Bible, Ezekiel 18:9; Tanakh, Ezekiel 18:9). As the use of one's talents for the benefit of all can be seen as part of the realization of neighborly love, performing any activity solely for the sake of monetary gain while not making use of our special abilities and talents is not only a violation of the commandment to love one another, but also of justice. In this understanding of justice, and in accordance with Plato ([427 BC] 2011), the government of a State should ensure, or at least incentivize, that everyone carries out their tasks according to their abilities and without interfering with the responsibilities of others.

17.3.5 Recapitulation

If we understand that it is essential to use our talents for the benefit of society and that the purpose of working is not just to make money, then we can build an economy built on the principle of neighborly love: an economy in which the wish to best serve one another, not market competition, is the principal driver behind the production of good products at good prices. Under such a system, people would be happy because they would feel that they are making a special contribution to the well-being of all, and in so doing, filling their lives with greater meaning. Then we serve our neighbor, the common good, and ultimately ourselves. Since voluntarily people would offer the best possible products at the best possible prices, we could achieve what in microeconomic is termed allocative efficiency without the need of competition.

And surely, we would not charge interest for lending money nor hoard huge amounts of money under the mattress, as this would hinder money to adequately circulate and thus restrict the possibility to exchange our talents this way. Since money would then flow without interest being necessary there would be no obligation to grow. The prerequisite for the latter is that economists succeed in educating people about the economic and moral-ethical problems of interest as well as the hoarding of money.

However, if we were all angels we would already be in heaven. Such an economy of neighborly love is, and probably will remain for a very long time, only a utopia. And as long as a majority does not understand that a purpose of life is to contribute to the common good by making use of our individual talents, we need competition to force us to make good products at good prices. Therefore, we need a government that courageously enforces competition law and above all implements monetary reform. This book tries to contribute to this understanding.

17.4 Concluding Remarks

Today, many people are once again doubting the wealth-enhancing effects attributed in economic theory to a competition based free-market economy and are even considering communism again as an alternative economic model. Karl Marx observed the inequality of income distribution, which was in those times similarly extreme as it is today, and looked for a solution. Unfortunately, he did not recognize the problems inherent in the unnatural design of money and therefore came up with solutions that did not solve the actual problem, but instead added others. Neither communism nor the competition-based market economy will work in the long run unless we reform money.

In this book, I have explained that our monetary system is the **main cause of market failure**, which unfortunately few economists recognize today. The fact that money in our current system is NOT neutral is the reason why we cannot approach the ideal of perfect competition. The problem is not competition as many believe, but the unnatural design of our money. With money as proposed by Silvio Gesell, micro and macroeconomic theory would finally bring the desired results. If we reform our money as Silvio Gesell proposed, we could establish a sustainable and just economic order: Let us call it a market economy that serves people.

With the introduction of Gesell's natural money, the self -destructive character of the monetary system as well as the economic growth imperative and the associated environmental problems would probably disappear or at least be significantly mitigated. Such natural money would produce a **natural and healthy economy**. And most importantly, implementing Gesell's proposals would enable us to get closer to the ideal where everyone gains according to their own efforts. This would be an economy that can finally reach a long-term equilibrium, an equilibrium that today due to a constantly growing money supply and debt is unachievable.

Nevertheless, the Gesellian reforms would not automatically eradicate all undesirable features of our economic system, such as the urge to monopolize (Suhr, 1988). One of the essential lessons of Goethe's Faust is that we strive to control nature in order to increase prosperity and recognition but overlook the fact that every new technology, every deviation from the divine equilibrium always bears risks (Binswanger, 1985). As long as we do not recognize that nature as an original perfect equilibrium cannot be improved upon and that the meaning of life does not lie in the accumulation of material wealth and power, many of our vices, which the current system promotes (Suhr, 1988) might still exist in less pronounced form. Therefore, we need to recognize that making money is not an end in itself and that the purpose of our existence is to use our talents to contribute to the common good. We could then transition from a market economy that serves people to a true economy of neighborly love, in which people achieve happiness by serving one another through the use of their talents.

References

Aristotle. (1995a). Politik 1st Book. In *Aristoteles Philosophische Schriften in sechs Bänden. Band 4*. Felix Meiner Verlag.

Aristotle. (1995b). Nikomachische Ethik. In *Aristoteles Philosophische Schriften in sechs Bänden. Band 3*. Felix Meiner Verlag.

Berdel, W., et al. (2004). *Repetitorium Innere Medizin*. Urban & Fischer/Elsevier.

Binswanger, H. C. (1985). *Geld und Magie – Deutung und Kritik der modernen Wirtschaft anhand Goethes Faust*. Edition Weitbrecht.

Chen, C.-Y., & Chen, W. J. (2013). Overdose. In P. M. Miller (Ed.), *Principles of addiction comprehensive addictive behaviors and disorders, Volume 1* (pp. 187–192). Academic Press/Elsevier.

Condon, P. (2017). Mindfulness, compassion, and prosocial behaviour. In J. C. Karremans & E. K. Papies (Eds.), *Mindfulness in social psychology* (pp. 124–138). Routledge.

Dascher, K. (2007). *Kristof Dascher, Ökonomie in Bausteinen: Einführung in die Volkswirtschaftslehre*. Oldenbourg.

Dear, J. W. (2014). Poisoning. In W. Marshall, M. Lapsley, A. Day, & R. Ayling (Eds.), *Clinical biochemistry: Metabolic and clinical aspects* (pp. 787–807). Edinburgh.

Frank, R. (2015). *Microeconomics and behaviour*. McGraw-Hill/Irwin.

Fuders, F. (2017). Neues Geld für eine neue Ökonomie – Die Reform des Geldwesens als Voraussetzung für eine Marktwirtschaft, die den Menschen dient. In *Finanzwirtschaft in ethischer Verantwortung – Erfolgskonzepte für Social Banking und Social Finance* (pp. 121–183). Springer-Gabler.

Fuders, F. (2021). The role of money for a healthy economy. In *Urban health and wellbeing programme – Policy briefs: Volume 2* (pp. 17–22). Springer Nature.

Fuders, F., & Nowak, V. (2019). The economics of love: How a meaningful and mindful life can promote allocative efficiency and happiness. In C. Steinebach & A. Langer (Eds.), *The economics of love: How a meaningful and mindful life can enhancing resilience in youth – Mindfulness-interventions in positive environments* (pp. 259–277). Springer. ISBN 978.

Gesell, S. (1920) (Reprint 1988). *Die natürliche Wirtschaftsordnung durch Freiland und Freigeld* (4th ed.). Verlag für Spzialökonomie.

Gesell, S. (1949). *Die natürliche Wirtschaftsordnung durch Freiland und Freigeld* (9th ed.). Rudolf Zitzmann.

Hooi, J., et al. (2017). Global prevalence of helicobacter pylori infection: Systematic review and meta-analysis. *Gastroenterology, 153*(2), 420–429.

Kant, I. (1868a). Metaphysik der Sitten – Erster Theil. Metaphysische Anfangsgründe der Rechtslehre. In *Immanuel Kant's sämmtliche Werke – In chronologischer Reihenfolge, Bd. VII* (pp. 1–173). Leopold Voss.

Kant, I. (1868b). Metaphysik der Sitten – Zweiter Theil. Metaphysische Anfangsgründe der Tugendlehre. In *Immanuel Kant's sämmtliche Werke – In chronologischer Reihenfolge, Bd. VII* (pp. 175–304). Leopold Voss.

Khoder, G., et al. (2019). Prevalence of helicobacter pylori and its associated factors among healthy asymptomatic residents in The United Arab Emirates. *Pathogens, 8*(2), 44.

Maslow, A. (2005). *El hombre autorealizado* (16. ed.). Kairós.

Max-Neef, M., Elizalde, A., & Hopenhayn, M. (1991). *Human scale development – Conception, application and further reflections*. Apex.

Megha, R., Farooq, U., & Lopez, P. (2022). *Stress-induced gastritis*. Pearls Publishing.

Montenegro, R. (2008). Efectos de la radiación de microondas y radiofrecuencias sobre la salud y el ambiente. In N. Camparelli, N. Mata, R. Montenegro, & B. Aliciardi (Eds.), *Contaminación por antenas de telefonía celular* (pp. 85–182). Buenos Aires.

Nef. (2022). *Happy Planet Index.* [Online] Available at: http://www.happyplanetindex.org

Pacchierotti, F., et al. (2021). Effects of radiofrequency electromagnetic field (RF-EMF) exposure on male fertility and pregnancy and birth outcomes: Protocols for a systematic review of experimental studies in non-human mammals and in human sperm exposed in vitro. *Environment International, 157*(106806). https://doi.org/10.1016/j.envint.2021.106806

Plato. (2011). Politeia. In U. Wolf (Ed.), *Platon – Sämtliche Werke Band 2* (pp. 211–537). Rowohlt.

Simkó, M., & Mattsson, M.-O. (2019). 5G Wireless Communication and Health Effects—A Pragmatic Review Based on Available Studies Regarding 6 to 100 GHz. *International Journal of Environmental Research and Public Health, 16*(18), 3406. https://doi.org/10.3390/ijerph16183406

Soddy, F. (1933a) (Reprint 1961). *Wealth, virtual wealth and debt – The solution of the economic paradox*. Briton.

Soddy, F. (1933b). *Money versus man*. Dutton.

Soddy, F. (1934). *The role of money – What it should be contrasted with what it has become*. Routledge.

Suhr, D. (1988). *Alterndes Geld – Das Konzept Rudolf Steiners aus geldtheoretischer Sicht*. Novalis Verlag.

Verbeek, J., et al. (2021). Prioritizing health outcomes when assessing the effects of exposure to radiofrequency electromagnetic fields: A survey among experts. *Environment International, 146*(106300). https://doi.org/10.1016/j.envint.2020.106300